Symbolic Projection for Image Information Retrieval and Spatial Reasoning

Signal Processing and its Applications

Symbolic Projection for Image Information Retrieval and Spatial Reasoning

SHI-KUO CHANG
Professor of Computer Science
University of Pittsburgh
Pittsburgh
USA

ERLAND JUNGERT
Swedish Defence Research Establishment
Linkoping
Sweden

ACADEMIC PRESS
Harcourt Brace & Company, Publishers
London • San Diego • New York • Boston • Sydney • Tokyo • Toronto

ACADEMIC PRESS LIMITED
24–28 Oval Road
LONDON NW1 7DX

U.S. Edition Published by
ACADEMIC PRESS INC.
San Diego, CA 92101

This book is printed on acid free paper

A catalogue record for this book is available from the British Library

ISBN 0-12-168030-4

Typeset by Mackreth Media Services (MMS), Herts, UK.
Printed in Great Britain by Hartnolls Ltd., Bodmin, Cornwall.

Series Preface

Signal processing applications are now widespread. Relatively cheap consumer products through to the more expensive military and industrial systems extensively exploit this technology. This spread was initiated in the 1960s by introduction of cheap digital technology to implement signal processing algorithms in real-time for some applications. Since that time semiconductor technology has developed rapidly to support the spread. In parallel, an ever increasing body of mathematical theory is being used to develop signal processing algorithms. The basic mathematical foundations, however, have been known and well understood for some time.

Signal Processing and its Applications addresses the entire breadth and depth of the subject with texts that cover the theory, technology and applications of signal processing in its widest sense. This is reflected in the composition of the Editorial Board, who have interests in:

(i)   Theory – The physics of the application and the mathematics to model the system;

(ii) Implementation – VLSI/ASIC design, computer architecture, numerical methods, systems design methodology, and CAE;

(iii) Applications – Speech, sonar, radar, seismic, medical, communications (both audio and video), guidance, navigation, remote sensing, imaging, survey, archiving, non-destructive and non-intrusive testing, and personal entertainment.

Signal Processing and its Applications will typically be of most interest to postgraduate students, academics, and practising engineers who work in the field and develop signal processing applications. Some texts may also be of interest to final year undergraduates.

Richard C. Green
The Engineering Practice,
Farnborough, UK

Contents

Preface

In the last few years, there has been an explosion in the number of publications, special issues, workshops, conferences, etc. devoted to image information systems. The increase in interests is largely due to technological advances, which make it possible to bring images into nearly every computer large or small, with a corresponding increase in applications. The evolution of high-volume low-cost image storage devices, the development of multimedia communications across networks and the availability of fast parallel-processing image processors, etc. have led to the proliferation of information systems with a preponderance of image and graphics data.

However, until now there was no uniform terminology, and the terms image information systems, pictorial information systems, visual information systems, and their counterparts image databases, pictorial databases, visual databases, are often used more or less interchangeably. A possible distinct is that images refer to what are captured by imaging devices, pictures are images together with their data structures which facilitate computer processing, and visual data emphasize the presentation of images as well as non-image data to the human users. Depending on the emphasis of the design or the researcher, the same system can be called an image information system, a pictorial information system or a visual information system.

From a historical viewpoint, the terms used by researchers were pictorial information systems and pictorial databases, reflecting the emphasis on image data structures. An early series of workshops devoted to this field was the IEEE Workshop on Picture Data Description and Management. The first of the IEEE PDDM workshops, organized by K. S. Fu and S. K. Chang, was held in Chicago in 1977. As the title of the workshop suggests, the two themes are the description and the management of pictorial data. The two themes subsequently led to two separate fields – visual languages (emphasizing picture description among other things) and image databases (emphasizing picture management). After nearly two decades of research, these two fields have grown and developed their respective research agenda, but interestingly are getting interwoven again, because in the emerging image information systems it is equally important to manage, describe and visualize images, and to actively employ images in the reasoning process for human-computer interactive problem solving.

This book addresses image information retrieval and spatial reasoning, focusing on an approach called Symbolic Projection, which supports descriptions of the image content on the basis of the spatial relationships between the objects in the image. Symbolic Projection forms the basis of a growing number of image information retrieval algorithms, and also supports query by picture. As often is the case in the advancement of science, the original idea first came as an intuition to one of the authors. In early 1984, while working on image information retrieval, Chang recalled his earlier research on the reconstruction of binary and gray-level images from projections, and thought it would be nice if symbolic images can be treated in the same way. In other words, if the symbolic image can be projected onto the two orthogonal axes so that the spatial relationships among the objects in the symbolic can be preserved in the projections, then the symbolic image perhaps can also be reconstructed from such projections. This intuitive idea led to the 2D strings for iconic indexing, and later the theory of Symbolic Projection.

The book represents many years of work on the part of both of the authors and is also a testimony of our friendship. When we first started our collaboration the Internet was in its infancy and long-distance cooperation across two continents was still relatively uncommon. We somehow have managed to work together for nearly fifteen years. The book documents the contributions by research groups from all over the world on the theory of Symbolic Projection, as well as our work on spatial reasoning which was among the first in developing this new research area. Also we decided to include many examples, practical applications and system design considerations so that it will serve as a useful overview for designers, practitioners and researchers of image information systems and multimedia systems. Image information systems have a wide variety of applications, including robotics, medical pictorial archiving, computer-aided design, and geographical informations systems, and this book is comprehensively illustrated with examples from these areas. It can also be used as a reference or textbook in a graduate seminar on advanced information systems, image information systems or multimedia information systems.

Many people contributed to this book, but above all we wish to thank our wives, Judy and Britt, without whose understanding and support this book would not have been possible. We also wish to thank Peter Holmes for providing the image produced by his research work as the background of cover page. Thanks are due to our colleagues Timothy Arndt, Brandon Bennet, Alberto Del Bimbo, C. C. Chang, C. Chronakis, Anthony Cohn, Louise Comfort, Francesca Costabile, Gennaro Costagliola, Max Egenhofer, Andrew Frank, Christian Freksa, Bill Grosky, John Hildebrand, Masahito Hirakawa, F. J. Hsu, Tadao Ichikawa, Ramesh Jain, Gene Joe, S. Kostomanolakis, C. M. Lee, S. Y. Lee, Y. Li, S. Y. Lin, M. Lourakis, S. D. Ma, Amitabna Mukerjee, Jonas Persson, G. Petraglia, David Randell, Ralf Röerig, Hanan Samet, M. Sebillo, Terry Smith, Guo Sun, Kim Tang, M. Tang, Genoveffa Tortora, M.

Tucci, E. Vicario, W. P. Yang and Cui Zhan for providing figures, allowing us to quote their work or contributing references. However, we are solely responsible for any mistakes in quoting or interpreting their results.

S. K. Chang
E. Jungert

1

Introduction

This book presents the *theory of Symbolic Projection*, which is a theory of spatial relations. This theory is the basis of a conceptual framework for image representation, image structuring and spatial reasoning. Within this framework, we will explore the applications of this theory to image information retrieval and spatial reasoning, illustrated by many examples in geographic information systems, medical pictorial archiving and communications, computer aided design, office automation, etc. The intended audience of this book are the designers, practitioners and researchers of image information systems, visual database systems and multimedia systems. It can also be used as a reference book or textbook in a graduate seminar on advanced information systems, image databases or multimedia computing.

What are symbolic projections? Why is the theory of Symbolic Projection of practical significance to the applications in image information retrieval and spatial reasoning? A simple example will first be presented to illustrate the concept.

Figure 1.1(a) shows a picture with a house, a car and a tree. This is called a *symbolic picture*, as opposed to an actual image, because it contains objects that have symbolic names: house, tree, car, etc. How to obtain the symbolic picture from the actual image will be discussed later. Suppose the objective is to find out whether there is a tree to the southeast of the house. The x-projection of the above symbolic picture can be constructed by projecting the names of objects in each column of the symbolic picture onto the x-axis. The "$<$" symbol is inserted to distinguish the objects belonging to different columns. Thus the x-projection is:

$$x\text{-projection:} \quad \text{house car} < \text{tree}$$

Similarly, the y-projection is:

$$y\text{-projection:} \quad \text{car} < \text{tree} < \text{house}$$

Unlike the projections of a mathematical function, the projections of a symbolic picture are strings. A pair of two symbolic projections is called a *2D string*.

The statement "there is a tree to the southeast of a house" corresponds to the symbolic picture shown in Figure 1.1(b). This picture has the following

(a)   **(b)**

Figure 1.1 A symbolic picture (a) and its subpicture (b).

symbolic projections:

$$x\text{-projection:}\quad \text{house} < \text{tree}$$

$$y\text{-projection:}\quad \text{tree} < \text{house}$$

We immediately notice "house < tree" is a subsequence of "house car < tree" and "tree < house" is a subsequence of "car < tree < house". In this case, the two symbolic pictures can be perfectly reconstructed from the two corresponding pairs of symbolic projections. Therefore, the above statement can be verified to be true, just by checking the subsequence property of the 2D strings involved.

The theory of Symbolic Projection was first developed by Chang and co-workers (1987) based on intuitive concept described above. It forms the basis of a wide range of image information retrieval algorithms. It also supports pictorial-query-by-picture, so that the user of an image information system can simply draw a picture and use the picture as a query.

Many researchers have since extended this original concept, so that there is now a rich body of theory as well as empirical results. The extended theory of Symbolic Projection can deal not only with point-like objects, but also objects of any shape and size. Moreover, the theory can deal with not only one symbolic picture, but also multiple symbolic pictures, three-dimensional (3D) pictures, a time sequence of pictures, etc. Both the theoretical findings and the empirical results will be presented in this book.

A word about terminology. A symbolic picture can have at least two symbolic projections, and in general more than two symbolic projections can be defined for a symbolic picture. We will use the plural form *symbolic projections* to refer to them. However, the capitalized singular form *Symbolic Projection* is reserved to refer to this theory of spatial relations.

In this chapter, section 1.1 describes what an image information system is. In section 1.2, we present a conceptual framework for image information system design. This conceptual framework also forms the basis for this book. Examples are described in section 1.3 to motivate the theoretical considerations. The organization of the book is explained in section 1.4.

1.1. IMAGE INFORMATION SYSTEMS

Before we introduce our conceptual framework for image representation, image structuring and spatial reasoning, let us first describe what an image information system is.

An image information system is an information system for the input, storage, processing, output and communication of a large quantity of images. An image information system typically has the following five components:

- image input subsystem
- image processing subsystem
- image output subsystem
- image database subsystem
- image communications subsystem.

This is illustrated schematically in Figure 1.2.

Recently, advances in image storage technologies have made the creation of very large image databases feasible. Multimedia communications also greatly facilitate the distribution of images across communication networks. Parallel computers lead to faster image processing systems. High resolution graphics and dedicated co-processors enable the design of image output subsystems with superior image quality. State-of-the-art image information systems have found their way into many application areas, including geographical information

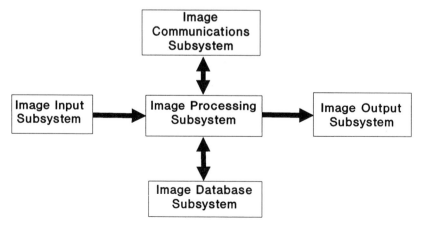

Figure 1.2 Component subsystems of an image information system.

systems (GIS), office automation (OA), medical pictorial archiving and communications systems (PACS), computer-aided design (CAD), computer-aided manufacturing (CAM), computer-aided engineering (CAE), robotics, and scientific databases (SD) applications.

Wider applications also lead to more sophisticated end users. Image information systems, like other types of information systems, have increasingly become knowledge-based systems, with capabilities to perform many sophisticated tasks by accessing and manipulating domain knowledge.

The emphasis on the user has also led to changes in terminology. Increasingly people are using the term "visual information systems" instead of "image information systems" to emphasize the visual presentation and visualization of stored data, which often include numerical and textual data. For example, the scientific data from a chemistry experiment are not images, but can be visualized by the user as such. The distinction between image information systems and visual information systems tends to be blurred, because image information systems can also handle artificial images created from non-image data. We will stick with the former, although many techniques described in this book often can be applied to both.

So far, image information systems have been designed on an *ad hoc* basis. The technological advances mentioned above dictate a better methodology for the design of knowledge-based, user-oriented image information systems. The design methodology, taking into consideration the diversified application requirements and users' needs, should provide a unified framework for image representation, structuring, as well as spatial reasoning. Such a framework will be presented below.

1.2. A CONCEPTUAL FRAMEWORK

As discussed in section 1.1, an image information system typically consists of image input subsystem, image output subsystem, image processing system, image database system, and image communications subsystem. We will now concentrate on the image processing system and image database system, which constitute the heart of the image information system.

A traditional image processing system performs the tasks of image analysis, image enhancement and pattern recognition. Within the context of an image information system, the image processing system and image database system must perform the following three functions, which can also be regarded as three stages in knowledge-based image processing. The three stages are illustrated in Figure 1.3.

1. *Image analysis and pattern recognition*. The raw image is analyzed, and the image objects recognized. This stage is almost always present in any image processing system. Extensive techniques are available for image enhancement,

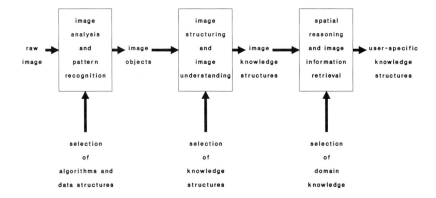

Figure 1.3 Stages in knowledge-based image processing.

normalization, segmentation and pattern recognition. The end result is a collection of recognized image objects. These image objects are usually encoded in some data structures for ease of access and further manipulation. For example, the image objects may be encoded in run-length codes, polygonal contour codes, quad-trees, oct-trees, etc. The set of image objects constitutes a symbolic picture.

In general, the image objects are objects with attributes including coordinates, geometric properties, etc. The unanalyzed or unprocessed parts of the image can be regarded as image entities which will be analyzed later if needed. Therefore, the system input to this stage includes the selection of various image processing algorithms, and the selection of data structures.

2. *Image structuring and understanding.* For some applications, it may be sufficient to access and manipulate the image objects, and there is no need for further structuring. However, for many applications, the image objects must be converted into image knowledge structures, so that spatial reasoning and image information retrieval can be supported. The particular image knowledge structure emphasized in this book is based upon 2D strings of symbolic projections. For some applications, 2D strings will be sufficient. For other applications, we must use generalized versions of 2D strings. The 2D strings can also be embedded into hierarchical image knowledge structure called the σ-tree (to be described in Chapters 2 and 14). Therefore, in this book, the 2D string-based image knowledge structure will be the basic knowledge representation for symbolic pictures. However, it may be desirable to use other image knowledge structures, such as directed graphs of spatial relations, semantic networks, etc. Therefore, the system input to this stage includes the selection of knowledge structures. When 2D strings are used as the image knowledge structure, the system input to this stage includes the selection of cutting mechanism and operator set (to be described in Chapters 4 and 5).

3. *Spatial reasoning and image information retrieval.* An image information system supports the information gathering needs and problem solving activities

of the end users. Some applications require spatial reasoning. Other applications are primarily concerned with image information retrieval. In medical information systems, for instance, the clinician may want to retrieve computer assisted tomography (CAT) images of all patients having a tumor similar to the shape present in a specific CAT image. Generally speaking, both spatial reasoning and image information retrieval are needed and may complement each other. To solve specific problems in a domain, a domain knowledge base is needed. It is also necessary to perform various transformations upon the image knowledge structure, so that the desired image knowledge can be easily accessed and/or manipulated. The transformations include rotation, translation, change of point-of-view, projection from 3D to 2D views, addition/deletion of image objects, etc. The final result of this stage is a user-specific knowledge structure, such as a navigation plan, a path, a set of retrieved images, an image index or indices, etc.

In the conceptual framework presented above, the three stages can be regarded as three generalized transformations, to transform a raw image first into an image data structure, then into an image knowledge structure, and finally into a user-specific knowledge structure.

1.3. TWO EXAMPLES OF IMAGE INFORMATION RETRIEVAL AND SPATIAL REASONING

We now present two examples of image information retrieval and spatial reasoning to explicate the conceptual framework presented in Section 1.2. Both examples involve path finding in an image. However, the first example is primarily a path finding problem, whereas the second example involves more complex image information processing and spatial reasoning.

Figure 1.4 illustrates a raw image of a coastal region. The specific problem for this application is to find a path for navigating a ship from point A to point B (Holmes and Jungert, 1992).

At the image analysis and pattern recognition (IAPR) stage, minimal enclosing rectangles for each island object are constructed, as shown in Figure 1.5. The image objects are run-length encoded and stored in the image database. This is the basic image data structure created by the IAPR stage.

At the image structuring and image understanding (ISIU) stage, the run-length encoded data structure is processed, so that a connectivity graph for the "tiles" (the "runs" of white pixels) can be constructed. This is part of the resultant image knowledge structure created by the ISIU stage. A small symbolic picture extracted from the image of Figure 1.4 is shown in Figure 1.6, where the "tiles" are the "runs" of white pixels. The corresponding tile graph is illustrated in Figure 1.7. The spatial knowledge structure can be

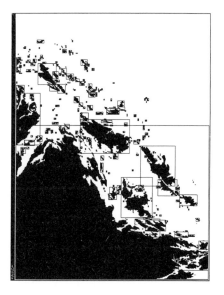

Figure 1.4 An image illustrating the coastal region of the Swedish Archipelago.

Figure 1.5 Identification of island objects in Figure 1.4, where each object is enclosed in a minimal enclosing rectangle.

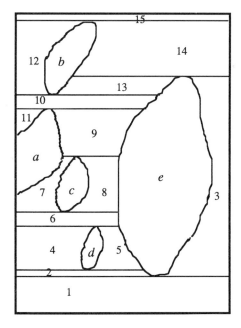

$\langle a < b < c < d < e, d < ce < a < b \rangle$

Figure 1.6 A small symbolic picture containing five island objects. The 2D string is given below the figure.

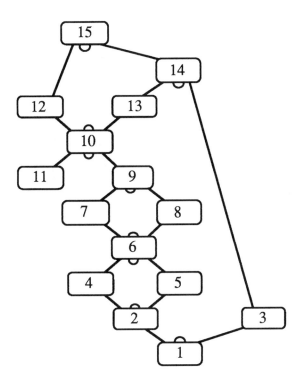

Figure 1.7 The tile graph constructed from Figure 1.6.

expressed by the 2D string as follows:

$$(a < b < c < d < e, d < ce < a < b)$$

The 2D strings will be further explained in Chapter 3. Intuitively, the first string says that "*a* is to the left of *b*, which is to the left of *c*, which is to the left of *d*, which is to the left of *e*", and the second string says that "*d* is below *c* and *e*, which are below *a*, which is below *b*". Therefore, the 2D strings express the approximate spatial relations among image objects.

The spatial relations can be expressed even more precisely, if the image objects are segmented using horizontal and vertical cutting lines, as illustrated in Figure 1.8. After segmentation, the generalized 2D strings are as follows:

$$(a \mid ab \mid cab \mid cb \mid dcb \mid db \mid d < e,$$
$$e \mid de \mid e \mid ace \mid ae \mid e \mid be \mid b)$$

The generalized 2D strings will be further explained in Chapters 4 and 5. Suffice it to say that the tile graph and the generalized 2D strings are equivalent image knowledge structures, and there are transformations to convert from one representation into another and vice versa.

Suppose the user wants to find a navigation path from the region of tile 7, to

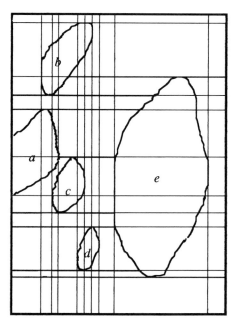

Figure 1.8 Segmentation of the symbolic picture of Figure 1.6 using cutting lines.

the region of tile 3. At the spatial reasoning and image information retrieval (SRIIR) stage, the tile graph is examined by a path-finding algorithm so that different navigation plans can be generated. The resultant user-specific knowledge structure is a plan, or a number of alternate plans, as illustrated in Figure 1.9. Figure 1.9(a) shows the tiles which lie in the navigation path. Figure 1.9(b) is the navigation plan. Details of the spatial reasoning techniques will be explained in Chapter 10.

We can summarize the knowledge-based image processing activities for this path-finding example as follows:

Activities	Results
Image analysis and pattern recognition (IAPR)	Identify image objects (islands) and produce run-length encoded image data structure
Image structuring and image understanding (ISIU)	Produce tile graph as the image knowledge structure
Spatial reasoning and image information retrieval (SRIIR)	Produce navigation plans as the final output

The above example illustrates the transformations from raw images to image data structures, image knowledge structures, and finally navigation plans (user-specific knowledge structures). For a more complicated planning problem, such as the Forest Fire Crisis Management system, the transformations are more complex, as will be explained below (Cohen *et al.*, 1989).

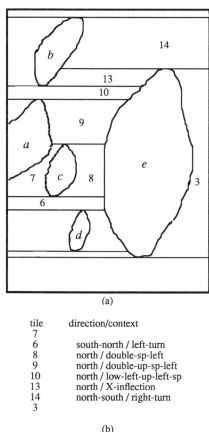

(a)

tile	direction/context
7	
6	south-north / left-turn
8	north / double-sp-left
9	north / double-up-sp-left
10	north / low-left-up-left-sp
13	north / X-inflection
14	north-south / right-turn
3	

(b)

Figure 1.9 The resultant navigation plan from Figure 1.7.

Figure 1.10 illustrates a map which can be displayed on the screen of the Forest Fire Crisis Management system. The raw images which are the maps are either manually digitized or automatically digitized. The map objects such as towns, cities, forest, rivers, roads, etc. are stored in the image database. The image data structure could again be the run-length code.

This image database for maps, with its run-length encoded data structure, can be used to answer many map related queries, such as "find the roads within the city boundary of city A". However, to support the planning and problem solving activities for a system such as the Forest Fire Crisis Management system, it will be too time-consuming to process the image database to answer some queries. Therefore, a symbolic spatial data structure called the σ-tree, which is a hierarchical structure with embedded 2D strings, can be created by the ISIU stage. From this spatial knowledge structure, which will be presented in detail in Chapters 2 and 14, an efficient 3D model can be generated and displayed. An example of such an approximate 3D model is illustrated in Figure 1.11. (When there are multiple approximate 3D models, a pyramid is

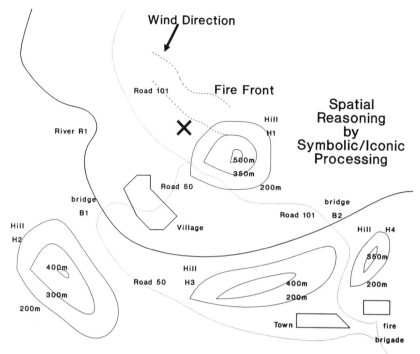

Figure 1.10 A map showing the forest fire.

formed and each slice of this pyramid corresponds to an image with a different level of detail.)

Therefore, the user of the system can observe both the original map, and the approximate 3D model, on separate screens or separate windows of the same screen. We now have two image structures: a run-length encoded image data structure for the maps, which will be regarded as the low-level image structure; and a spatial knowledge structure, the σ-tree, which will be regarded as the high-level image structure. In what follows, we illustrate a scenario where the user's problem solving activities are supported by the image information system, which performs both high-level and low-level image processing activities.

Suppose the user poses the following query: "From the current fire front, compute fire fronts 1 hour, 2 hours, 3 hours, etc. from the present time". This query can be answered first by high-level processing of the symbolic spatial data structure as follows. From the symbolic structure, compute the fire fronts in areas other than the hill H1. Since all objects and their locations are stored in the symbolic structure, this computation can be done quickly. However, insufficient information is carried by the symbolic structure as far as H1 is concerned. Therefore, low-level processing of the image data structure is necessary, to find out the areas in H1 covered by forest. After such processing, we can also compute the fire front in the H1 area.

It is worth noting that the high-level and the low-level processing tasks can

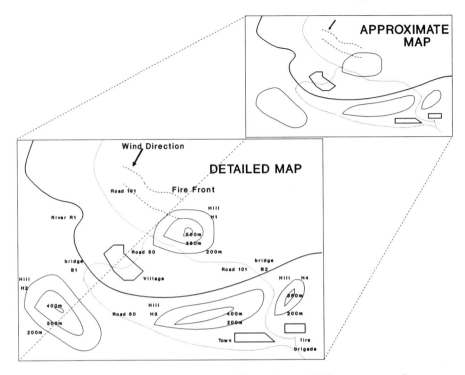

Figure 1.11 The symbolic picture constructed from the spatial data structure, the σ-tree.

be carried out simultaneously by parallel processes, so that the user will first receive a quick but incomplete (and less accurate) reply, while more detailed information will be provided later.

In an active information system, the detection of the fire front may lead to automatic recalculation of the predicted fire front positions. The detected object is a *hot spot* in an active index system (to be explained in Chapter 12) which enable an image to automatically react to certain events.

The second query is as follows. The user places a cross on the symbolic picture, indicating the approximate location of the fire corridor to be constructed by the firemen, and ask, "What are the estimated times needed to deploy firemen".

Again, both high-level and low-level processing are necessary. In the high-level processing, we can estimate the approximate times needed to travel from town T to bridge B1 and bridge B2. These are the two alternate paths that the firemen may take. Again, since object locations are known, the travel times can be estimated from the symbolic structure. In the low-level processing, the approximate times of travel between T and B1 or B2 are replaced by more accurate results, by taking into consideration road types, road conditions, terrain conditions, etc.

Similarly, we can first do high-level processing to estimate the travel times

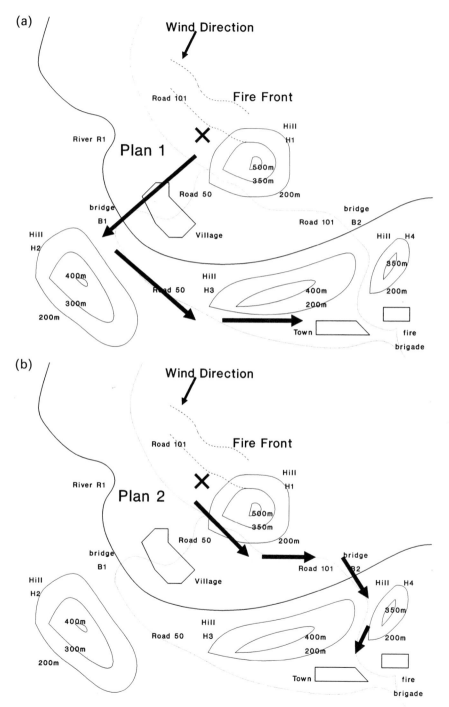

Figure 1.12 Multiple plans on the symbolic picture.

from B1 or B2 to the cross ×, and at the same time do low-level processing to recompute the travel times. The refined results are supplied to the user when available. Such information can also be displayed on the symbolic picture which the user can manipulate at will.

The third and final query from the user, is "Determine how to construct a feasible fire corridor." As shown in Figure 1.12, two plans are available. The first plan is displayed on the symbolic picture. The user may decide bridge B1 could be blocked by evacuating traffic. Certain additional costs (time delays) are added by the user to the symbolic picture, i.e. the symbolic knowledge structure, and the deployment time is recalculated by high-level processing of the symbolic picture. Low-level processing is then initiated, to compute the cost of constructing the fire corridor. The user may move the fire corridor in the symbolic picture, causing recomputation of the cost by the low-level processing. Additional queries may be posed by the user to determine the best location for the fire corridor. For example, he may ask "what is the nearest road to the north of the town", because it will be much less costly if the fire corridor can be constructed alongside the road. Such queries can be answered by high-level processing of the symbolic structure.

When the second plan is displayed, the user can again decide whether the terrain in hill H1 will cause a problem or not. Since a crude terrain model is part of the 3D model of the symbolic structure, high-level processing will give the user initial estimates of the terrain passability. Low-level processing will lead to fine tuning of the travel times through H1. Finally, cost in constructing fire corridor can again be estimated.

This second example illustrates the desirability of having two levels of knowledge structures, the high-level image knowledge structure for quick computation and spatial reasoning, and the low-level image data structure for more elaborate accurate computation. However, we must be able to switch efficiently between the two levels of processing, by quickly relating image knowledge structure and image data structure.

As illustrated by these two examples, image information retrieval and spatial reasoning are complex tasks. They include the sub-tasks of image retrieval (to find images satisfying certain query conditions), image measurement, reasoning and interactive decision making. The conceptual framework presented in Section 1.2 can be used to analyze the image information retrieval and spatial reasoning tasks in a logical manner. Within this framework, the theory of Symbolic Projection and its applications can be studied.

1.4. ORGANIZATION OF THE BOOK

The remainder of the book is divided into three parts. In Part I, we cover the basic theory of Symbolic Projection, and illustrate the applications to image information retrieval. Part I consists of five chapters – Chapter 2 to Chapter 6.

Chapter 2 covers the most important image data structures. It also defines the concept of image objects as used in this book.

Chapter 3 discusses the representation of point objects as 2D strings, which become the basic knowledge structure for symbolic image information representation. The fundamental spatial relations are introduced. The basic operator set consists of just two operators, "=" and "<".

Chapter 4 introduces the variable-sized grid, which motivates the cutting mechanisms to obtain more efficient and more compact symbolic image knowledge representations called generalized 2D strings. The operator set is now extended to include the "edge-to-edge" operator, "|".

Since spatial relations are the most important for spatial reasoning, we investigate the theory of spatial relations in Chapter 5. Different classes of local spatial operators are defined, which enable us to develop more sophisticated cutting mechanisms and construct different symbolic image knowledge structures, such as generalized 2D strings, 2D C-strings, etc.

In Chapter 6 we discuss applications of the theory of Symbolic Projection to image information retrieval. Different approaches to image database queries are surveyed. An intelligent image database system is presented, which supports spatial reasoning, flexible image information retrieval, visualization, and traditional image database operations. The pictorial data structure based upon 2D strings provides an efficient means for iconic indexing in image database systems and spatial reasoning. Several application examples are described to illustrate the diversity of potential applications.

Part II presents the advanced theory of Symbolic Projection, with applications to spatial reasoning. It consists of four chapters – Chapter 7 to Chapter 10.

In Chapter 7, the image algebra is introduced as a unified notation for image manipulation. This is followed by the discussion of various transformations in Chapter 8. These two chapters provide the theoretical foundation to define different image knowledge structures and show their equivalence.

Chapter 9 deals with the generalization of symbolic projections to slope projections for 2D spatial object representation, which can be used for determining distance and directions among objects.

Chapter 10 discusses spatial reasoning concepts, and presents various techniques for spatial reasoning. Knowledge representation and reasoning mechanisms are presented, followed by examples in path finding and similarity retrieval.

Part III extends the theory and deals with various issues in active image information systems design. It consists of five chapters – Chapter 11 to Chapter 15.

In Chapter 11, the differences between visual reasoning, visual-aided reasoning and spatial reasoning are discussed. Multiparadigmatic visual queries are then presented.

Chapter 12 describes active index system, and the application of active index and symbolic projections for spatio-temporal reasoning.

In Chapter 13, we discuss the various extensions to higher dimensions; 3D strings are introduced. The encoding of a sequence of symbolic pictures into 2D strings, and the retrieval of time-sequenced images by spatial/temporal logic, are then presented.

Chapter 14 presents a unified data model for 2D and 3D symbolic projections, called the σ-tree data model, which is shown to be useful in 3D and higher-dimensional spatial knowledge representation. The recursive evaluation of queries on the σ-tree is presented.

Finally, Chapter 15 concludes this book by presenting a taxonomy of image indexing techniques and a survey of various approaches to image information system design. Future directions in active image information systems are then discussed.

REFERENCES

Chang, S. K., Shi, Q. Y. and Yan, C. W. (1987) Iconic indexing by 2D strings. *IEEE Transactions on Pattern Analysis and Machine Intelligence* **9**, 413–28.

Cohen, P. R., Greenberg, M. L., Hart, D. M. and Howe, A. E. (1989) Trial by fire: Understanding the design requirements for agents in complex environments. *AI Magazine*, Fall issue, 32–48.

Holmes, P. D. and Jungert, E. (1992) Symbolic and geometric connectivity graph methods for route planning in digitized maps. *IEEE Transactions on Pattern Analysis and Machine Intelligence* **14**, 549–65.

PART I

BASIC THEORY WITH APPLICATIONS TO IMAGE INFORMATION RETRIEVAL

2

Image Data Structures

2.1. INTRODUCTION

We require basic spatial data structures that can be used to represent spatial objects and, more specifically, for symbolic projection string generation. This is important, because the generation of projection strings must be efficient. The list of structures to be described here is by no means complete. However, the structures are representative although some of them have characteristics that do not entirely fit with the use of Symbolic Projection. Another important aspect is concerned with object attributes required for generation of the symbolic projections, known as *characteristic attributes*. Generation of characteristic attributes generally depends on the spatial data structure that is used: consequently, this is an aspect which affects not only the way the projections are generated but also the efficiency of the method. A hierarchical spatial data structure which can be used to logically represent more complex relationships, for instance in 3D, is also introduced from a basic point of view. This structure is called the σ-*tree* and is part of the attempt to generalize Symbolic Projection and to improve its efficiency.

Section 2.1 gives a brief introduction to some image data structures. In Section 2.2 the object oriented data model is discussed generally, while in Sections 2.3 and 2.4 spatial data structures that can be used to represent abstract objects are introduced. Section 2.5 is concerned with the characteristic attributes that are used in Symbolic Projection. Finally, in Section 2.6 the σ-tree is briefly introduced.

Image data structures obviously play an important role in all systems dealing with manipulation of and reasoning in images and pictures. For that reason, efficient image data structures are required since, sooner or later, all kinds of low-level data manipulation will be necessary in such systems. For instance, spatial reasoning or image retrieval cannot always take place on a high abstraction level since at some stage the processes must proceed on lower levels of data manipulation. The low-level image data structures must fit with the high-level structures, otherwise the methods simply will not work as intended. Consequently, the problem of low-level image data representation must be dealt with before we proceed to higher abstraction levels.

Subsequently, various types of image data structures will be discussed. The structures considered here are mainly, but not entirely, of raster data type. The motivation for this is that a large number of such structures is available, but also because such data structures are very well suited both for high-level data manipulation and for spatial reasoning and image retrieval.

2.2. THE OBJECT DATA STRUCTURE

The general approach or paradigm used throughout this book is that of object orientation; see for example work on object-oriented database systems by Ullman (1988) or the anthology by Nahouti and Petry (1991). However, the definition of object orientation need be clarified. Traditionally, the commonly used definition is presented in the following way:

- *Object identity*, which means that the system can distinguish two objects that seem to be the same, i.e. the objects are differentiated through *unique identifiers*.
- *Encapsulation*, that is, a way of isolating objects such that they only can be exempted by particular types of procedures of which each one only can be applied to a specified class of objects.
- *Complex objects*, which normally means that hierarchical or nested structures are allowed to be defined in the system. Normally, this includes the ability to inherit various types of properties from objects on higher levels in the hierarchies. Furthermore, this kind of structure also permit grouping of related objects into classes.

These aspects are generally included in commercially available object-oriented systems. Among products that are object-oriented in this sense are Smalltalk and C++. All such systems are categorized as belonging to a specific type of programming environment based on the same programming paradigm. However, here the discussion is generally concerned with objects in a way that does not exclude the object orientation paradigm. That is, an object normally has an identity which can be used in all the various techniques that will subsequently be discussed, but in image retrieval systems and in reasoning *characteristic patterns* can also be used to identify objects uniquely. The main difference between the approach given here and the general object-oriented approach is that the latter is part of an object-oriented programming system while Symbolic Projections does not require this, since it can be implemented using a conventional programming paradigm: i.e. in this context the type of systems that are dealt with are object-based systems. Consequently, the methods discussed are generally independent of the programming paradigm that may be used to implement them.

2.3. RUN-LENGTH CODE

Run-length code (RLC) is normally not object-oriented. Originally it was developed for compacting image data, i.e. as a simple storage structure. Within an object-oriented paradigm, an object in an image can be represented by a set of RLC lines. The structure, which is well adapted to the relational data model, is discussed by Jungert (1986). Jungert also demonstrates the use of RLC in a query language called Graqula (Jungert 1993) developed for a geographical information system (GIS). RLC can also be applied to various other types of image information systems. Generally, a line in its simplest case, is described in terms of its start x- and y-coordinates and its length corresponding to the triplet:

$$y, x, \text{length}$$

The line records are sorted with respect to their coordinates and stored in sequential order. The structure is homogeneous since the representation of all types of entities including points, lines and entities of extended type are permitted. For point entities the length of a RLC line is by definition set to 1. The line length of a linear object varies depending on its orientation horizontally and vertically. Logically, their representation is not much different from extended objects of general type. Given a RLC database, i.e. a geometric database (GDB), a large number of operations, such as set-theoretical operations, can efficiently be applied.

A geometric database can simply be organized as a flat file with one single access method. The normal access method used ought to be a B-tree, see for instance Wirth (1977). There are two main reasons for the choice of the B-tree. It is self-organizing, and it is also a convenient indexing technique for direct access of records from flat files. Normally, sequential sets of line-records are accessed from the geometric database. A complete access is therefore composed of a single search through the B-tree, directly followed by a sequential access of the flat file corresponding to a predefined interval. The set of records that are read may correspond either to a particular image or to a specific object. Hence, a full access is fast since the tree-access is performed just once. When accessing the geometric database this will normally cause some unnecessary records to be read. Such records are simply filtered out without slowing down the process.

Even if the RLC records permit the storage of logically homogeneous information in the geometric database, further information than just the triplet above is normally necessary in an object-oriented environment in order to differentiate the objects from each other. For this case an *object-id* is necessary as well. Other information that may be useful in some applications is, for instance, the *entity type*, which gives the following record structure:

$$y, x, \text{length}, \text{type}, \text{object-id}$$

In a GIS, for instance, spatial information which does not correspond to objects will appear. An example of this is terrain elevation data. In cases where that type of information occurs the object-id must be dropped and substituted with other information – in a GIS context it could be any kind of characteristic attribute, such as the slope. For terrain elevation data, elevation and slope directions would be suitable attributes. Consequently, there is no change in the way the information is stored, only in the way it is interpreted. An illustration of the RLC structure is given in Figure 2.1. In the right-hand figure both the white and the black areas correspond to run-length lines required to complete the object.

The process of accessing and displaying RLC data is the part of the data model that deals with how image data are accessed from the geometric database and displayed on the screen. This process must be fast, since the volume of data is often large. In this approach it is attained as a result of the sequential file-access method.

The access and display process is demonstrated in Figure 2.2. The first step corresponds to the identification of an interval of the geometric database, defined by the user and normally corresponding to either a complete image or to an arbitrary object in an image. Those records that belong to the interval are simply accessed in sequence and transformed into lines that are drawn on the

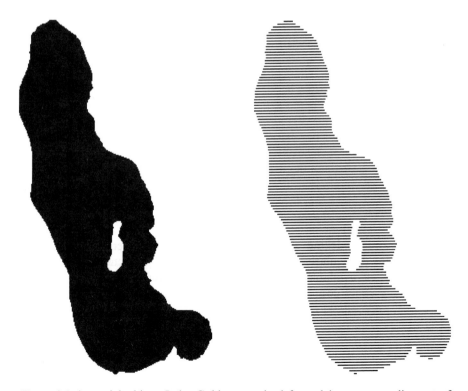

Figure 2.1 A spatial object, Lake Gubben, to the left, and its corresponding set of run-length lines to the right.

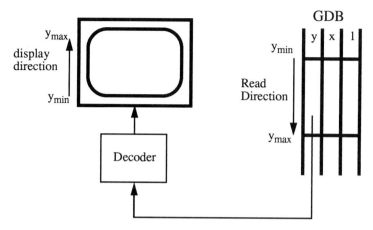

Figure 2.2 Illustration of the access and display process.

screen; records outside the image are simply filtered out. Lines that are on the edge of the image, that is, lines that cross the edge of the image on either the left or the right, are simply cut off. This procedure does not require any complicated computations, unlike, for instance, situations when a polygon has to be cut. The display process is fully sequential. In other words, the image on the screen is built up in the same order as the line records are accessed and read from the database. The advantage of this approach is apparent compared to polygon-based systems, for example. Another advantage with the RLC structure is that no fill operations are required; all objects are automatically filled at the same pace as they are displayed. Furthermore, in polygon-based systems, holes must be displayed separately, while here there is no need to bother with holes at all since they are outside the objects.

2.4. OTHER SPATIAL DATA STRUCTURES

Most spatial data structures other than RLC can be used as a basis for generation of symbolic projections, and consequently they can be used in, for instance, iconic indexing and spatial reasoning. However, not all spatial structures are particularly well adapted to symbolic projections. A complete overview of all available structures cannot be given here: we will restrict discussion to just a couple of the most important and widely used structures, quad-trees and polygons.

2.4.1. The Quad-Tree

The quad-tree structure is not one single spatial data structure but rather a family of related structures that can be used for description of extended objects

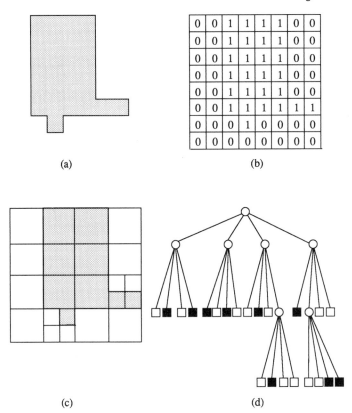

0	0	1	1	1	1	0	0
0	0	1	1	1	1	0	0
0	0	1	1	1	1	0	0
0	0	1	1	1	1	0	0
0	0	1	1	1	1	0	0
0	0	1	1	1	1	1	1
0	0	0	1	0	0	0	0
0	0	0	0	0	0	0	0

(a)   (b)

(c)   (d)

Figure 2.3 (a) Region, (b) the corresponding binary image, (c) the block decomposition of the region and (d) the quad-tree representation.

and in some cases also for the identification of simple relations between various point objects. The structure is well documented in the literature, e.g. by Samet (1990).

The *region quad-tree* is the most commonly used of the family of quad-trees. Similar to RLC, it describes the extension of an object of regional type using rasters. However, the structure is hierarchical, unlike RLC. The basic idea of the quad-tree is to recursively split the regional object into four equal rectangles such that a rectangle is a terminal node in the tree when it covers a part of the region completely. For those subregions that only partly cover a region of the object, the process of splitting the region into four equal rectangles proceeds until a terminal rectangle is reached, which may be a single pixel in the worst case. A consequence of this method is that the size of the top rectangle in which the regional object resides must be $2^k \times 2^k$ pixels. Figure 2.3 illustrates the principles of the regional quad-tree. Figure 2.3(a) shows the region while in Figure 2.3(b) the corresponding $2^3 \times 2^3$ binary array can be seen. The resulting set of rectangles corresponding to the region can be seen in Figure 2.3(c).

Finally, the complete quad-tree is presented in Figure 2.3(d). The rectangular nodes in the tree are terminal, while the circles correspond to nonterminal nodes. The black terminal nodes illustrates the regional nodes, while the white are nonterminal. In this particular example the tree is of degree 4. The sons of a node in the tree are labelled in order NW, NE, SW and SE.

2.4.2. The Polygon Structure

The polygon representation is in many ways the simplest of all spatial data structures and is also commonly used. It is a compact structure from a storage point of view, with a high degree of accuracy, and has therefore become very popular. However, it has also some disadvantages, such as finding an arbitrary vertex. An extended object represented with a closed polygon of n coordinate points can be described as a set of coordinate points $((x_0, y_0), (x_1, y_1), \ldots, (x_{n-1}, y_{n-1}))$ where the points (x_0, y_0) and (x_{n-1}, y_{n-1}) are identical. This representation can obviously be used as a contour description of closed objects, and it is also used for description of linear objects. A disadvantage of this structure is that holes cannot be handled in a simple way because a hole must somehow have a representation of its own. A simple example of a polygon representation is shown in Figure 2.4. Polygons may also be represented with a variation of the quad-tree, e.g. the Hunter and Steiglitz quad-tree.

A variation of the polygon structure is the vector structure. This requires the storage of each vertex twice since both the start and end of each vector must be stored, consequently giving rise to redundancy in the spatial database (unlike the polygon structure).

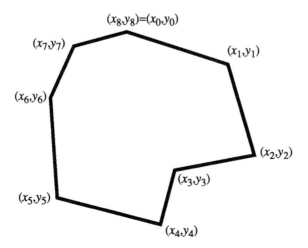

Figure 2.4 A simple closed polygon.

2.5. CHARACTERISTIC ATTRIBUTES

The characteristic attributes of concern in Symbolic Projection are generally spatial attributes that are somehow essential for the generation of the syntactic descriptions that are produced as a result of the projections that are made in Symbolic Projection. Which of these attributes will be of concern in a particular application depends on the variation of Symbolic Projection that is used. The total number of characteristic attributes required, considering all variations, is small and can therefore easily be stored in the image database, if not calculated directly when needed.

The most important of the characteristic attributes are those that describe the corresponding projection intervals, in other words x_{min}, x_{max}, y_{min} and y_{max}. These attributes correspond also to the extreme points that describe the minimal bounding rectangle (MBR) of the object and can therefore serve other purposes as well. A fifth necessary characteristic attribute is one corresponding to the centroid of the MBR of the object. A drawback with the centroid attribute is that it may, in some applications, be outside the object itself. However, the centroid attribute was used mainly in the original approach of Symbolic Projection while the four former attributes are used in more recent versions of the method.

The complexity of the generation of the characteristic attributes varies depending on which spatial data structure is used. In RLC all the attributes can be generated on the fly without too much effort, since the lines are stored in sequence and y_{min} and y_{max} correspond to the y-coordinates of the first and last lines respectively among the set of lines that describes the object. Furthermore, the two remaining extreme points of the object can be determined almost as simply, since this can be done with just a single pass of the line set. The fifth and final attribute can be generated just as simply once the four first are known. For this reason, none of these five attributes needs to be preprocessed, unless they are needed frequently, since they do not require much extra computational time.

For the other structures discussed in section 2.4 the situation is somewhat different. For quad-trees the use of MBRs is almost mandatory. The rectangles must therefore be preprocessed since they are used in a large variety of search methods, for instance plane-sweep methods, although other search techniques are used as well. The generation of an MBR from a quad-tree requires the full quad-tree to be traversed, since the bounding rectangle is in general smaller than the $2^k \times 2^k$ rectangle that is used for generation of the quad-tree for the object. Of course, an alternative is to determine the bounding rectangle at the same time as the quad-tree is created. Here, as well as for the RLC, the centroid can simply be calculated from the bounding rectangle.

The generation of the bounding rectangle for a polygon requires all vertices in the polygon to be traversed, which can be time consuming if the object has a large number of points. It is therefore better to preprocess the bounding rectangles for all objects. The algorithm for determination of the bounding rectangle of an arbitrary polygon is, however, trivial.

There may be other attributes as well as the characteristic attributes of the MBR, but they are mainly application dependent. They include the characteristic patterns that were discussed in section 2.2. Such patterns are of particular interest since they can be used to identify an object or a subobject as well.

2.6. THE σ-TREE STRUCTURE

The σ-tree structure is a 3D generalization (Jungert and Chang, 1992) of the original approach to symbolic projection. The basic principles of the σ-tree will be introduced in this section, while later, in Chapter 14, there will be a deeper discussion of the method including also some basic examples showing how the method can be applied to a simple model.

Basically, the σ-tree is a symbolic hierarchical representation of a 3D space. The structure of the hierarchy is designed to support spatial reasoning and image retrieval. We first define some basic concepts: these are the universe, the spatial model and the cluster.

Definition 2.1. A *universe* is an arbitrary three-dimensional space containing any collection of objects.

Definition 2.2. A *spatial model* is a logical description of a universe.

Definition 2.3. A *cluster* is an arbitrary cubic subspace of a universe that either is empty or contains an arbitrary number of objects or subclusters.

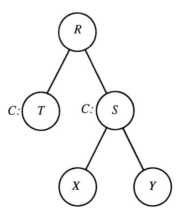

Figure 2.5 A σ-tree with two clusters, S and T, and two objects, X and Y, which are members of S.

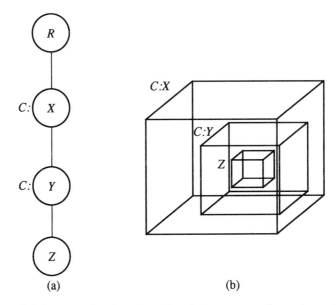

Figure 2.6 An example of a σ-tree (a) and its corresponding universe (b).

Clusters may be of arbitrary size, allowed to coincide completely or partly. A cluster may fill up a universe completely, although logically it resides on a lower level of the tree. A cluster may have parameters or attributes, to allow transformations such as translation, rotation, scaling, etc. of the entire cluster.

Let R correspond to a universe and let σ be an operator that transforms R into a logical description (the spatial model) of the universe. Consequently, σR is a logical description of a universe containing an arbitrary number of objects and clusters. The syntactic description of σR expressed in terms of symbolic projection is described in depth in Chapter 14, while here the hierarchical structure of the tree structure is illustrated in Figure 2.5. A cluster is labelled C:. For example, the cluster S is denoted $C : S$. The submodel S of R, i.e. the logical description of S, is denoted by σS. An object in the hierarchy corresponds to a terminal node of the tree. An object may have parameters or attributes such as coordinates of its location, size, colour, etc. Figure 2.5 is an illustration of a σ-tree made up by two clusters, S and T, and the two objects, X and Y, which are both located inside S while T is empty. An example of a cluster that fills the whole universe is given in Figure 2.6.

REFERENCES

Jungert, E. (1986) Run length code as an object-oriented spatial data structure. *Proceedings of the IEEE Workshop on Languages for Automation, Singapore*, pp. 66–70.
Jungert, E. (1993) Graqula – a visual information-flow query language for a geographical information system. *Journal of Visual Languages and Computing* **4**, 383–410.

Jungert, E. and Chang, S. K. (1992) The σ-tree – a symbolic spatial data model. *Proceedings of the 11th IAPR International Conference on Pattern Recognition, The Hague*, 461–465.

Nahouti, E. and Petry, F. (eds) (1991) *Object-Oriented Databases*. IEEE Computer Society Press, Los Alamitos, CA.

Samet, H. (1990) *The Design and Analysis of Spatial Data Structures*. Addison-Wesley, Reading, MA.

Ullman, J. D. (1988) *Principles of Database and Knowledge-based Systems*. Computer Science Press, Rockville, Maryland.

Wirth, N. (1977) *Algorithms + Data Structures = Programs*. Prentice-Hall, Englewood Cliffs, NJ.

3

2D Strings

This chapter formalizes the basic concepts of symbolic projections, based upon the work by Chang and co-workers (1987). In section 3.1, we describe the 2D string representation as the symbolic projection of a picture, and define the absolute 2D string, normal 2D string and reduced 2D string. Section 3.2 presents algorithms for picture reconstruction from the 2D string, and defines the augmented 2D string. Picture matching by 2D subsequence matching is discussed in Section 3.3. Section 3.4 outlines the methodology of iconic indexing. The problem of ambiguous pictures and their characterization is analyzed in Section 3.5. In Section 3.6, we summarize the algorithms and ambiguity results for the various 2D string representations.

3.1. 2D STRING REPRESENTATION OF SYMBOLIC PICTURES

Let Σ be a set of symbols, or the vocabulary. Each symbol might represent a pictorial object, a pixel, etc.

Let A be the set $\{=, <, :\}$, where "$=$", "$<$" and "$:$" are three special symbols not in Σ. These symbols will be used to specify spatial relationships between pictorial objects.

A *1D string* over Σ is any string $x_1 x_2 \ldots x_n$, $n \geq 0$, where the x_i's are in Σ.

A *2D string* over Σ, written as (u, v), is defined to be

$$\left(x_1 y_1 x_2 y_2 \ldots y_{n-1} x_n, x_{p(1)} z_1 x_{p(2)} z_2 \ldots z_{n-1} x_{p(n)} \right)$$

where $x_1 \ldots x_n$ is a 1D string over Σ, $p: \{1, \ldots, n\} \rightarrow \{1, \ldots, n\}$ is a permutation over $\{1, \ldots, n\}$, $y_1, \ldots, y_{n-1}$ is a 1D string over A and $z_1, \ldots, z_{n-1}$ is a 1D string over A.

We can use 2D strings to represent pictures in a natural way. As an example, consider the picture shown in Figure 3.1.

Figure 3.1 A picture f.

The vocabulary is $\Sigma = \{a, b, c, d\}$. The 2D string representing the above picture f is

$$(a = d < a = b < c, \; a = a < b = c < d)$$

$$= (x_1 y_1 x_2 y_2 x_3 y_3 x_4 y_4 x_5, \; x_1 z_1 x_3 z_2 x_4 z_3 x_5 z_4 x_2)$$

where $x_1 x_2 x_3 x_4 x_5$ is *adabc*, $x_1 x_3 x_4 x_5 x_2$ is *aabcd*, p is 13452, $y_1 y_2 y_3 y_4$ is $=<=<$ and $z_1 z_2 z_3 z_4$ is $=<=<$.

In the above, the symbol "$<$" denotes the left–right spatial relation in string u, and the below–above spatial relation in string v. The symbol "$=$" denotes the spatial relation "at the same spatial location as". The symbol "$:$" denotes the relation "in the same set as". Therefore, the 2D string representation can be seen to be the *symbolic projection* of picture f along the x- and y-axes.

A *symbolic picture* f is a mapping $M \times M \rightarrow W$, where $M = \{1, 2, \ldots, m\}$, and W is the power set of Σ (the set of all subsets of V). The empty set $\{\ \}$ then denotes a null object. In Figure 3.1, the "blank slots" can be filled by empty set symbols, or null objects. The above picture is

$$
\begin{array}{lll}
f(1, 1) = \{a\} & f(1, 2) = \{\ \} & f(1, 3) = \{d\} \\
f(2, 1) = \{a\} & f(2, 2) = \{b\} & f(2, 3) = \{\ \} \\
f(3, 1) = \{\ \} & f(3, 2) = \{c\} & f(3, 3) = \{\ \}
\end{array}
$$

We now show that, given f, we can construct the corresponding 2D string representation (u, v), and vice versa, such that all left–right and below–above spatial relations among the pictorial objects in Σ are preserved. In other words, let R_1 be the set of left–right and below–above spatial relations induced by f. Let R_2 be the set of left–right and below–above spatial relations induced by (u, v). Then R_1 is identical to R_2, for the corresponding f and (u, v).

It is easy to see that from f, we can construct the 2D string (u, v). The above example already illustrates the algorithm. In the formal algorithm, we will first construct the 2D string (u', v') where u' and v' contain symbols in W. Then, u' and v' can be rewritten as follows: if we see a set $\{a, b, c\}$, we write $a : b : c$. If we see a null set $\{\ \}$, we simply remove it (i.e. we write a null string). We can rewrite "$=<$", "$<=$", and "$<<$" as "$<$", and "$==$" as "$=$", so that redundant spatial operators are removed. The operators "$=$" and "$:$" can also be omitted. The algorithm now follows.

procedure **2Dstring**(f, m, u, v, n)
begin

```
/*we assume f is m × m square picture*/
n = m × m;
for j from 1 until m
    for k from 1 until m
    /*we are looking at f(j,k) */
    /*construct string x₁...xₙ */
        begin
        i = k + (j − 1)m;
        xᵢ = f(j,k);
        /*construct string x_{p(1)}...x_{p(n)} */
        p(i) = j + (k − 1)m;
        end
/*construct y string and z string */
for i = 1 until n − 1
    if i is multiple of m
        then yᵢ = zᵢ = "<"
        else yᵢ = zᵢ = "=";
/*rewrite strings u and v*/
while strings u and v contain rewritable substrings
    apply the rewriting rules (see following explanation):
        r₁.{a₁, a₂, ..., aₖ} is rewritten as a₁ : a₂ : ... : aₖ
            { } is rewritten as (null string)
        r₂."==" is rewritten as "="
            "=<" is rewritten as "<"
            "<=" is rewritten as "<"
            "<<" is rewritten as "<"
        r₃."=" is rewritten as (null string)
        r₄." : " is rewritten as (null string)/*reduced string only*/;
end
```

As an example, for the picture f shown in Figure 3.1, if we apply the procedure **2Dstring** without the rewriting rules, we obtain:

$$(\{a\} = \{\ \} = \{d\} < \{a\} = \{b\} = \{\ \} < \{\ \} = \{c\} = \{\ \},$$

$$\{a\} = \{a\} = \{\ \} < \{\ \} = \{b\} = \{c\} < \{d\} = \{\ \} = \{\ \}) \quad (3.1)$$

Now we apply rewriting rule r_1 to obtain:

$$(a == d < a = b =<= c =,\ a = a =<= b = c < d ==) \quad \text{(absolute 2D string)}$$
$$(3.2)$$

Since the 2D string of (3.2) contains all the spatial operators, it is a precise encoding of the picture f. The 2D string of (3.2) is called an *absolute 2D string*. This coding is obviously inefficient. In fact, one string suffices to represent f precisely. Therefore, we can apply rewriting rule r_2 to obtain:

$$(a = d < a = b < c,\ a = a < b = c < d) \quad (3.3)$$

In the above 2D string representation, we only keep the *relative* positioning information, and the *absolute* positioning information is lost. If we also omit the "=" symbols by applying rewriting rule r_3, we obtain:

$$(ad < ab < c, \ aa < bc < d) \qquad \text{(normal 2D string)} \qquad (3.4)$$

The same procedure can be applied to pictures whose "slots" may contain multiple objects (i.e. object sets). For example, if in Figure 3.1, $f(1,3)$ is $\{d, e\}$ instead of $\{d\}$, then the 2D string representation is

$$(ad : e < ab < c, \ aa < bc < d : e) \qquad \text{(normal 2D string with sets)} \qquad (3.5)$$

The 2D strings of (3.4) and (3.5) are called *normal 2D strings.*

If the ":" symbols are also omitted by applying rewriting rule r_4, we obtain the following *reduced 2D string*:

$$(ade < ab < c, \ aa < bc < de) \qquad \text{(reduced 2D string)} \qquad (3.6)$$

We note that in a reduced 2D string representation, there is no apparent difference between symbols in the same set and symbols not in the same set. In the above example, the local substring (i.e. the substring between two "<"s, or one "<" and an end-marker) *ade* might also be encoded as *aed, dae, dea, ead*, or *eda*. In other words, for reduced 2D strings, a local substring is considered to be equivalent to its permutation string.

3.2. PICTURE RECONSTRUCTION FROM A 2D STRING

From the 2D string (u, v), we can reconstruct f. As an example, suppose the 2D string is $(x_1x_2 < x_3x_4 < x_5, x_2x_3x_4 < x_1x_5)$. We first construct the picture shown in Figure 3.2(a), based upon the 1D string u, by placing objects having the same spatial location (i.e. objects related by the "=" operator) in the same "slot". Next, we utilize the 1D string v to construct the final picture. The algorithm now follows.

Procedure **2Dpicture**(f, m_1, m_2, u, v, n)
begin
   /*find out the size of the picture*/
   $m_1 = 1 +$ number of "<" in string u;

Figure 3.2 Objects with the same spatial location placed in the same slot.

$m_2 = 1 +$ number of "<" in string v;
/*find out the x-rank and y-rank of each object*/
for i from 1 until n
    begin
      x-rank of $x_i = 1 +$ number of "<" preceding x_i in u;
      y-rank of $x_{p(i)} = 1 +$ number of "<" preceding $x_{p(i)}$ in v;
    end
/*we construct an $m_1 \times m_2$ picture f */
for j from 1 until m_1
    for k from 1 until m_2
      $f(j,k) =$ the set of all objects
        with x-rank j and y-rank k;
end

In the procedure **2Dpicture**, we assume the permutation function p is given. If all the objects in a 2D string are distinct, the permutation function p is unique. If, however, there are identical symbols in the 2D string, then in general there are many permutation functions. For example, the picture f of Figure 3.1 has 2D string representation $(u, v) = (ad < ab < c, aa < bc < d)$. Let the u string be $x_1x_2 < x_3x_4 < x_5$. Then the v string is either

$$x_1x_3 < x_4x_5 < x_2 \quad \text{or} \quad x_3x_1 < x_4x_5 < x_2$$

We can use either permutation function in the above procedure **2Dpicture**, which will always reconstruct a picture. However, the reconstructed picture could be different from the original picture. In the above example, the reconstructed picture is actually unique and unambiguous. However, this may not always be the case. The problem of ambiguity will be treated in section 3.5.

Another consideration is the uniqueness of the 2D string representation. As stated above, given any 2D string (u, v) and its permutation function p, the procedure **2Dpicture** will always reconstruct a picture. However, only for the reduced 2D string representation is the 2D string generated from the reconstructed symbolic picture always identical to the original 2D string, because the local substrings and their permutation strings are considered equivalent.

The above considerations suggest that we can include the permutation function p in the 2D string representation. Then we will have no ambiguity in picture reconstruction, and the 2D string representation is also unique.

Therefore, we define an *augmented 2D string* (u, v, p), where (u, v) is the reduced 2D string as previously defined, and p is the permutation function. With the augmented 2D string, picture f of Figure 3.1 is represented by

$$(ad < ab < c, aa < bc < d, 13452) \quad \text{(augmented 2D string)} \quad (3.7)$$

In actual coding, the second string v can be replaced by $w = p(1)z_1 p(2)z_2 \ldots z_{n-1}p(n)$. In other words, the symbols in v are replaced by the permutation indexes:

$$(ad < ab < c, 13 < 45 < 2) \qquad \text{(augmented 2D string)} \qquad (3.8)$$

We will use the notation (u, v, p) to denote an augmented 2D string, although in actual coding we will use (u, w). For augmented 2D strings, the procedure **2Dpicture** can be applied directly, by including p as an additional input parameter. Procedure **2Dstring** should be replaced by the following procedure:

Procedure **2DstringA**$(f, m; u, w)$
begin
   /* we assume f is $m \times m$ square picture */
$n = 0$
for i from 1 until m do
   begin
      for j from 1 until m do
         begin
           if $f(i, j)$ is not empty
             for s in $f(i, j)$
                begin
                   $n = n + 1$
                   $q(i, j) = q(i, j) \cup \{n\}$
                end
         end
   end
$l = 0; u = $ "null" /* "null" denotes null string */
for i from 1 until m do
   begin
      for j from 1 until m do
         begin
           if $f(i, j)$ is not empty
           $l = l + \text{size-of}(f(i, j))$
           $u = u \cdot f(i, j)$ /* "$\cdot$" denotes concatenation operator */
         end
      if $i < m$ and $l < n$
         $u = u \cdot <$
   end
$k = 0; w = $ "null"
for j from 1 until m do
   begin
      for i from 1 until m do
         begin
           if $f(i, j)$ is not empty
            $k = k + \text{size-of}(f(i, j))$
            $w = w \cdot q(i, j)$
         end

    if $j < m$ and $k < n$
        $w = w \cdot \;<$
  end
end

The above procedure **2DstringA** generates an augmented 2D string representation, where the function size-of(S) returns the number of elements in a set S.

3.3. PICTURE MATCHING BY 2D STRING MATCHING

2D string representation provides a simple approach to perform subpicture matching on 2D strings. The *rank* of each symbol in a string u, which is defined to be 1 plus the number of "$<$" preceding this symbol in u, plays an important role in 2D string matching. We denote the rank of symbol b by $r(b)$. The strings $ad < b < c$ and $a < c$ have ranks as shown in Table 3.1.

Table 3.1 Ranks of strings.

String v	String u
$a_1 d_1 < b_2 < c_3$	$a_1 < c_2$

A substring where all symbols have the same rank is called a *local substring*.

A string u is *contained* in a string v, if u is a subsequence of a permutation string of v.

A string u is a *type-i 1D subsequence* of string v, if (a) u is contained in v, and (b) if $a_1 w_1 b_1$ is a substring of u, a_1 matches a_2 in v and b_1 matches b_2 in v, then

    (type-0) $r(b_2) - r(a_2) \geqslant r(b_1) - r(a_1)$ or
$$r(b_1) - r(a_1) = 0$$
    (type-1) $r(b_2) - r(a_2) \geqslant r(b_1) - r(a_1) > 0$ or
$$r(b_2) - r(a_2) = r(b_1) - r(a_1) = 0$$
    (type-2) $r(b_2) - r(a_2) = r(b_1) - r(a_1)$

Now we can define the notion of type-i ($i = 0, 1, 2$) 2D subsequence as follows. Let (u, v) and (u', v') be the 2D string representation of f and f', respectively. (u', v') is a *type-i 2D subsequence* of (u, v) if (a) u' is type-i 1D subsequence of u, and (b) v' is a type-i 1D subsequence of v. We say f' is a *type-i subpicture* of f.

In Figure 3.3, f_1, f_2 and f_3 are all type-0 sub-pictures of f; f_1 and f_2 are type-1 subpictures of f; only f_1 is a type-2 subpicture of f. The 2D string representations are:

$$\begin{array}{ll}
f & (ad < b < c,\ a < bc < d) \\
f_1 & (a < b,\ a < b) \\
f_2 & (a < c,\ a < c) \\
f_3 & (ab < c,\ a < bc)
\end{array}$$

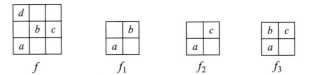

Figure 3.3 Picture matching example.

Therefore, to determine whether a picture f' is a type-i subpicture of f, we need only determine whether (u', v') is a type-i 2D subsequence of (u, v). The picture matching problem thus becomes a 2D string matching problem.

For augmented 2D strings, we can define the notion of 2D subsequence as follows: (u', v', p') is a type-i 2D subsequence of (u, v, p), if (a) (u', v') is a type-i 2D subsequence of (u, v), and (b) if x'_i of u' matches x_j of u, then $x'_{p'(i)}$ of v matches $x_{p(i)}$ of v.

In type-1 subsequence matching, each local substring in u should be matched against a local substring in v. In Table 3.1, substring a in u is a subsequence of ad in v, and substring c in u is a subsequence of c in v. Notice the skipping of a rank is allowed in type-1 subsequence matching. Therefore, the type-1 subsequence matching problem can be considered as a two-level subsequence matching problem, with level-1 subsequence matching for the local substrings, and level-2 subsequence matching for the "superstring" where each local substring is considered as a supersymbol, and supersymbol u_1 matches supersymbol v_1 if u_1 is a subsequence of v_1.

Type-2 subsequence matching is actually simpler, because the rank cannot be skipped. That is to say, if local substring u_1 of u matches local substring v_1 of v, then substring u_i of u must match substring v_i of v for any i greater than 1. In the example shown in Table 3.1, the v string $a < c$ is not a type-2 subsequence of $ad < b < c$.

The following is a procedure for augmented 2D string matching. Setting the parameter i to 2 and applying procedure **2DmatchA** once, we can check if (u', w') is a type-2 2D subsequence of (u, w). Therefore, **2DmatchA** can be used to list all type-2 subsequences of a 2D string. A similar procedure **2Dmatch** can be written to handle absolute, normal, or reduced 2D string matching.

Procedure **2DmatchA**$(u', w'; u, w; i)$
begin
1. convert(u', w') to $(x', r', s', p') = (x'(1) \dots x'(N), r'(1) \dots r'(N), s'(1) \dots s'(N),$
  $p'(1) \dots p'(N))$ using procedure $CA(u', w'; x', r', s', p')$
  convert (u, w) to $(x, r, s, p) = (x(1) \dots x(M), r(1) \dots r(M), s(1) \dots s(M),$
    $p(1) \dots p(M))$ using procedure $CA(u, w; x, r, s, p)$
  while $N \leqslant M$ execute the following steps:
  /* check if (u', w') is type-i 2D subsequence of (u, w) for $i = 0, 1, 2$ */

2. for j from 1 until M
   begin
      if $x(p(j)) = a$, let j belong to match(a)
   end
   for n from 1 until N
   begin
      $MI(n) = \text{match}(x'(p'(n)))$
   end
3. for n from 1 until $N - 1$
   begin
      $MC = \{ \}$ /* $\{ \}$ denotes empty set */
      while k belongs to $MI(n)$ and j belongs to $MI(n + 1)$
      begin
         call subroutine $d(n + 1, j; n, k)$
         if return "yes"
            let k belong to $a(j, n + 1, 1)$
         while $n > 1$
         begin
            $AP = a(k, n, 1)$
            for m from 2 until n
            begin
               $a(j, n + 1, m) = \{ \}$
               while la belongs to AP
               begin
                  call subroutine $d(n + 1, j; n - m + 1, la)$
                  if return "yes"
                     let la belong to $a(j, n + 1, m)$
               end
               $AP = \{lp \,|\, \text{whenever } la \text{ belongs to } a(j, n + 1, m) \text{ such that}$
                     $lp \text{ belongs to both } a(la, n - m + 1, 1) \text{ and } a(k, n, m)\}$
            end
         end
         if all $a(j, n + 1, m)$, $m = 1, \ldots, n$ are not empty
            let j belong to MC
      end
      if MC is empty
         emit "no" and stop
      else $MI(n + 1) = MC$
   end
   emit "yes"
end
subroutine $d(m, j; n, k)$
begin
   if $j = k$

```
        return "no"
        if [r(p(j)) − r(p(k))][r(p'(m)) − r(p'(n))] ⩾ 0 and
           (when i = 0) [s(j) − s(k)][s'(m) − s'(n)] ⩾ 0
                |s(j) − s(k)| ⩾ |s'(m) − s'(n)|
                |r(p(j)) − r(p(k))| ⩾ |r'(p'(m)) − r'(p'(n))|
           (when i = 1) s(j) − s(k) ⩾ s'(m) − s'(n) > 0 or s(j) − s(k) = s'(m) − s'(n) = 0
                |r(p(j)) − r(p(k))| ⩾ |r'(p'(m)) − r'(p'(n))| > 0 or
                r(p(j)) − r(p(k)) = r'(p'(m)) − r'(p'(n)) = 0
           (when i = 2) s(j) − s(k) = s'(m) − s'(n)
                r(p(j)) − r(p(k)) = r'(p'(m)) − r'(p'(n))
        return "yes"
     else
        return "no"
  end
  procedure CA(u, w; x, r, s, p)
  begin
     /* convert (u, w) = (u(1) … u(L), w(1) … w(K)) to (x, r, s, p) = (x(1) … x(N),
        r(1) … r(N), s(1) … s(N), p(1) … p(N)) */
     m = 0
     for l from 1 until L
     begin
        if u(l) ≠ "<"
           x(l − m) = u(l)
           r(l − m) = m + 1
        else m = m + 1
     end
     n = 0
     for k from 1 until K
     begin
        if w(k) ≠ "<"
           s(k − n) = n + 1
           p(k − n) = w(k)
        else n = n + 1
     end
  end
```

For type-0 and type-1 matching, the algorithm **2DmatchA** is inexact, and there may exist instances in the answer set which do not actually match the query. The conditions of occurrence of such "false drops" are discussed in Drakopoulos and Constantopoulos (1989) and an algorithm which avoids false drops has been proposed to list all matched subsequences (Petrakis, 1993), and the original algorithms have been extended to take into account any number of object properties and the inclusion relationships between objects.

3.4. ICONIC INDEXING

The 2D string representation is ideally suited to formulating picture queries. In fact, we can easily imagine that the query can be specified graphically, by drawing an iconic figure on the screen of a personal computer. The graphical representation can then be translated into the 2D string representation using the procedure **2Dstring** described in section 3.2. This approach combines the advantage of the query-by-example approach, where a query is formulated by constructing an example, and the concept of icon-oriented visual programming systems (Chang and Clarisse, 1984).

Pictorial information retrieval is then transformed into the problem of 2D subsequence matching. The query, represented by a 2D string, is matched against the *iconic index*, which is the 2D string representation of a picture. Those pictures whose iconic indexes match the query 2D string are retrieved.

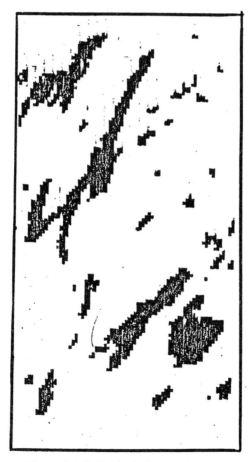

Figure 3.4 A digitized picture of lakes.

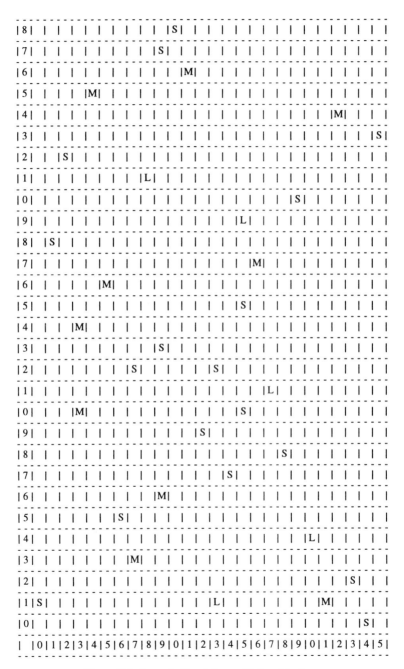

Figure 3.5 A symbolic picture of lake objects.

The iconic index not only can be used in pictorial information retrieval, but also provides an efficient means for *picture browsing*. Since we can reconstruct a picture from its 2D string representation, we can apply procedure **2Dpicture** to construct an *icon sketch* from the iconic index. In this picture browsing mode, we need only access the iconic indexes, instead of retrieving the actual pictures from the pictorial database. For some applications, the picture browsing technique can be very useful.

To construct the iconic index, we proceed as follows. First, we apply edge detection and boundary detection techniques, to construct icon sketches. The icon sketches are then classified into objects, and translated into 2D string representation. The location of each object is its centroid for planar objects, or the central point of its media axis for linear objects. The icon sketches can be displayed in a browsing mode for user visualization. However, the retrieval is actually by 2D subsequence matching.

Some experimental results will now be described. Figure 3.4 shows a map overlay of lakes. This picture was digitized, and the lake objects were classified into small, medium-sized and large lakes. The resultant symbolic picture is shown in Figure 3.5. The user can specify a query by a query pattern, as shown in Figure 3.6. Figure 3.6(a)–(c) are queries to retrieve type-0, type-1, and type-2 subpictures respectively. By applying the 2D string matching algorithms, the matched subpictures are found. When there are several matched subpictures, the results (the coordinates of retrieved objects) are all displayed.

Figures 3.6(d)–(f) illustrate queries containing variables. For example, Figure 3.6(d) shows a query pattern, "find the object with a large lake at its northwest and another large lake at its southeast by type-0 matching". The corresponding 2D string for this query contains the symbol Z, which is a variable that can match any symbol other than the operator symbols. The 2D string matching algorithms was modified to handle 2D strings with variables. These examples illustrates the potential applications of our approach to pictorial information retrieval.

The above approach is applicable to pictures containing objects whose convex hulls are mutually disjoint, so that the objects have only left–right and below–above spatial relations. More complex objects need to be segmented first, so that they can be described in terms of the constituent objects. The segmentation problem, and the introduction of cutting lines, will be discussed in Chapter 4.

3.5. CHARACTERIZATION OF AMBIGUOUS PICTURES

We define a picture f to be *ambiguous* if there exists a different reconstructed picture g from its 2D string representation (u, v).

The ambiguity problem for binary pictures was first treated in Chang (1971). The same technique can be applied to the general case, by regarding the binary

QUERY PATTERN (TYPE-0)
,L
L,

RESULT:
*(13, 1) (20, 4)
*(13, 1) (17, 11)
*(13, 1) (15, 19)

(a)

QUERY PATTERN (TYPE-1)
L, ,
,L,
, ,L

RESULT:
*(20, 4) (17, 11) (15, 19)
*(20, 4) (17, 11) (8, 21)
*(20, 4) (15, 19) (8, 21)
*(17, 11) (15, 19) (8, 21)

(b)

QUERY PATTERN (TYPE-2)
S, ,
, ,
, ,M

RESULT:
*(7, 3) (6, 5)
*(11, 26) (10, 28)

(c)

QUERY PATTERN (TYPE-0)
, ,L
,Z,
L, ,

RESULT:
*(13, 1) (20, 4)
*(13, 1) (14, 7) (15, 10) (17, 11)
*(13, 1) (14, 7) (15, 19)

(d)

QUERY PATTERN (TYPE-1)
,L, ,
, ,Z,
, , ,L
L, , ,

RESULT:
*(13, 1) (20, 4) (18, 8) (17, 11)
*(13, 1) (20, 4) (18, 8) (17, 11) (16, 17) (15, 19)
*(13, 1) (17, 11) (16, 17) (15, 19)

(e)

QUERY PATTERN (TYPE-2)
, , , , L, , ,
, , , , , , ,
, , , , , , ,
, , , , , Z, ,
, , , , , , ,
, , , , , , ,
, , , , , , , ,L
, , , , , , ,
, , , , , , ,
L, , , , , , ,

RESULT:
*(13, 1) (20, 4) (18, 8) (17, 11)

(f)

Figure 3.6 Picture queries and results.

picture as a special case, where the vocabulary set V contains a single symbol, say a. For example, as shown in Figure 3.7(a), the 2D string $(a < a, a < a)$ has ambiguous reconstruction. However, if the symbols are distinct, as shown in Figure 3.7(b), there is no ambiguity. Since an absolute 2D string represents a picture precisely, it cannot be ambiguous. An augmented 2D string is also unambiguous. Therefore, we should investigate normal and reduced 2D string representation.

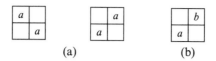

(a)   (b)

Figure 3.7 Ambiguous picture (a) and unambiguous picture (b).

For reduced 2D strings, we can prove the following theorem.

Theorem 3.1. Let F be the class of all symbolic pictures without subpatterns of the form:

$$\begin{matrix} a & a \\ a & a \end{matrix}$$

A symbolic picture f in F is ambiguous under the reduced 2D string, if and only if the reduced 2D string contains the subsequence $(a < a, a < a)$, where a is some symbol in V.

Proof. Suppose $(a < a, a < a)$ is a subsequence of (u, v), the 2D string representation of f. If we apply the procedure **2Dpicture** to reconstruct the picture, we will find x- and y-indices i_1, i_2, j_1, j_2, for the subsequence $(a < a, a < a)$.

$$\text{The subsequence:} \quad (a < a, \ a < a)$$
$$x\text{- and } y\text{-indices:} \quad i_1 \quad i_2 \ j_1 \quad j_2$$

We will have $f(i_1, j_1) = \{a\} \cup K_{11}$, and $f(i_2, j_2) = \{a\} \cup K_{22}$. If we have $f(i_1, j_2) = K_{12}$ and $f(i_2, j_1) = K_{21}$, then we can change the assignments as follows:

$$g(i_1, j_1) = K_{11}, \qquad g(i_2, j_1) = K_{21} \cup \{a\}$$
$$g(i_1, j_2) = K_{12} \cup \{a\}, \qquad g(i_2, j_2) = K_{22}$$

This would construct a different picture g with the same 2D string representation, unless the four K_{ij}'s all contain a – which is impossible, because K_{11} and K_{22} do not contain a.

Now suppose the picture f is ambiguous. Let its 2D string be (u, v). If the symbols in u are all distinct, then f cannot be ambiguous. Therefore, there must be at least two identical symbols in string u (or string v). Let this symbol be a. We construct a new picture g, where

$$g(i, j) = f(i, j) \cap \{a\}$$

In other words, we only preserve the symbol a in $f(i, j)$. For some a, g must be ambiguous. By the results proven in Chang (1971), g must contain a "switching component", as illustrated in Figure 3.7(a). Therefore, the 2D string of f contains a subsequence $(a < a, a < a)$. ☐

The above condition is for reduced 2D strings. For normal 2D strings, we have the following result.

Suppose f and g are symbolic pictures, Q and Q_l $(l = 1, 2, \ldots, k)$ are sets over V. Suppose f and g satisfy the following conditions:

    (i) $Q = Q_1 \cup Q_2 \cup \cdots \cup Q_k$ and $Q_i \cap Q_j = \{\ \}$ for $i \neq j$

    (ii) $g(i, j) = f(i, j) - Q$

    (iii) $g(i, j_l) = f(i, j_l) \cup Q_l$ for $l = 1, 2, \ldots, k$

We define the moving operation $O[r(i): j(Q) \to j_l(Q_l), \ldots, j_k(Q_k)](f) = g$.

If we interchange the roles of i and j, we can see that the moving operation $O[c(j): i(Q) \to i_l(Q_l), \ldots, i_k(Q_k)](f) = g$.

Let $O_1, O_2, \ldots, O_n$ be moving operations. We define

$$O_1, O_2, O_3, \ldots, O_n(f) = O_1(O_2(O_3(\ldots (O_n(f) \ldots)$$

We define $fr(i)$ as the string obtained by concatenating all the symbols in $f(i, 1), f(i, 2), \ldots, f(i, m)$ in sequence, $fc(j)$ as that by concatenating all the symbols in $f(1, j), f(2, j), \ldots, f(m, j)$, where m is the size of f. $fr(i)$ and $fc(j)$ are called a *row string* and a *column string*, respectively, and both are called *segment strings*. Now we give the definition of the *ambiguous loop* for the normal 2D string as follows. Suppose

$$(i_1, j_1), (i_2, j_1), (i_2, j_2), \ldots, (i_k, j_k), (i_1, j_k)$$

is a sequence of indices to the elements in f. If we can connect the indices one to another with straight lines, we get a loop. If there exists a set Q over Σ such that

$$fc(j_1) = g_1 c(j_1), \quad \text{where } g_1 = O[c(j_1): i_1(Q) \to i_2(Q)](f)$$

$$fr(i_2) = h_2 r(i_2), \quad \text{where } h_2 = O[(r(i_2): j_1(Q) \to j_2(Q)](f)$$

$$\vdots$$

$$fc(j_k) = g_k c(j_k), \quad \text{where } g_k = O[c(j_k): i_k(Q) \to i_1(Q)](f)$$

$$fr(i_1) = h_1 r(i_1), \quad \text{where } h_1 = O[(r(i_1): j_k(Q) \to j_1(Q)](f)$$

then there is an ambiguous loop, denoted by

$$\{(i_1, j_1), (i_2, j_1), (i_2, j_2), \ldots, (i_k, j_k), (i_1, j_k); Q\}$$

For example, Figures 3.8(a) and (b) have the same normal 2D string. This ambiguity comes from the dotted loop.

Theorem 3.2. A symbolic picture f is ambiguous under the normal 2D string if and only if there exists an ambiguous loop in f.

Proof. Suppose the ambiguous loop is

$$\{(i_1, j_1), (i_2, j_1), (i_2, j_2), \ldots, (i_k, j_k), (i_1, j_k); Q\}$$

We will prove that f and g have the same 2D string, where

$$g = O[c(j_1): i_1(Q) \to i_2(Q)]O[c(j_2): i_2(Q) \to i_3(Q)] \ldots O[c(j_k): i_k(Q) \to i_1(Q)]$$

Since the 2D string is a row by row and column by column representation, if two pictures have the same row strings and the same column strings, they have

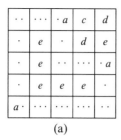 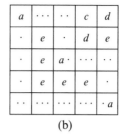

	$\cdots$	$\cdot$ a	c	d
$\cdot$	e	$\cdot$	d	e
$\cdot$	e	$\cdot\cdot$	$\cdots$	$\cdot$ a
$\cdot$	e	e	e	$\cdot$
$a\cdot$	$\cdots$	$\cdots$	$\cdots$	$\cdot\cdot$

(a)

a	$\cdots$	$\cdot\cdot$	c	d
$\cdot$	e	$\cdot$	d	e
$\cdot$	e	$a\cdot$	$\cdots$	$\cdot\cdot$
$\cdot$	e	e	e	$\cdot$
$\cdot\cdot$	$\cdots$	$\cdots$	$\cdots$	$\cdot$ a

(b)

Figure 3.8 Ambiguous loop.

the same 2D string. Therefore, we only need to check if f and g have the same segment strings.

The pictures f and g may only differ at rows $i_1, i_2, \ldots, i_k$ and columns $j_1, j_2, \ldots, j_k$, and it is straightforward to prove f and g cannot be different at those rows and columns. If f and g are different, but with the same 2D string, we can prove that there must exist an ambiguous loop in f. First, we search f and g column by column at the same time. Let $f(i_1, j_1)$ be the first element found to be different from g. Without loss of generality, suppose $f(i_1, j_1)$ is nonempty and $g(i_1, j_1)$ is empty. We conclude that there must exist a $f(i_2, j_1)$ such that $f(i_2, j_1) = \{\cdot\}$, $g(i_2, j_1) = f(i_1, j_1)$ and $fc(j_1) = g_c(j_1)$. We can repeat this procedure until we have found an ambiguous loop. Since the picture is finite, this procedure will terminate. □

3.6. SUMMARY

This chapter presents a methodology for symbolic picture representation by 2D strings. Table 3.2 summarizes the important properties of various 2D string representations.

Table 3.2 Summary of 2D string representation.

String type	Normal 2D string	Absolute 2D string	Reduced 2D string	Augmented 2D string
Example string	$(a : bc < a,\ a : ba < c)$	$(a : b = c < a =,$ $a : b = a < c =)$	$(abc < a,\ aab < c)$	$(abc < a,\ 124 < 3)$
Picture to string	procedure **2Dstring**	procedure **2Dstring**	procedure **2Dstring**	procedure **2DstringA**
String to picture	procedure **2Dpicture**	procedure **2Dpicture**	procedure **2Dpicture**	procedure **2Dpicture**
String match	procedure **2Dmatch**	procedure **2Dmatch**	procedure **2Dmatch**	procedure **2DmatchA**
Ambiguity condition	Theorem 3.2 ambiguous loop	unambiguous	Theorem 3.1 $(a < a,\ a < a)$ subsequence	unambiguous

REFERENCES

Chang, S. K. (1971) The reconstruction of binary patterns from their projections. *Communications of the ACM* **14**, 21–5.

Chang, S. K. and Clarisse, O. (1984) Interpretation and construction of icons for man–machine interface in an image information system. *IEEE Proceedings on Languages for Automation*, pp. 38–45.

Chang, S. K., Shi, Q. Y. and Yan, C. W. (1987) Iconic indexing by 2D strings. *IEEE Transactions on Pattern Analysis and Machine Intelligence* **9**, 413–28.

Drakopoulos, J. and Constantopoulos, P. (1989) An exact algorithm for 2D string matching. Technical Report 21, Institute of Computer Science, Foundation for Research and Technology, Heraklion, Greece.

Petrakis, E. G. M. (1993) Image representation, indexing and retrieval based on spatial relationships and properties of objects. Ph.D. thesis. Technical report FORTH-ICS/TR-075, Institute of Computer Science, Foundation for Research and Technology, Heraklion, Greece.

4

Generalized 2D Strings

As discussed in Chapter 3, a *symbolic picture* or simply a *picture* is a grid where some of the slots are filled by picture objects. The symbolic projections originally were defined for a fixed-sized grid, where all objects are point-like objects. When we deal with objects of variable sizes and shapes, the 2D strings become insufficient to represent the complex relationships among them.

This chapter presents the concept of the generalized 2D string. In section 4.1 we discuss why variable-sized grids are needed. The segmentation problem is introduced in section 4.2. The formal definition of generalized 2D strings is given in section 4.3, followed by the definition of empty space objects in section 4.4. Section 4.5 introduces the different types of projections, which will allow the rotation of objects. Finally, section 4.6 gives a classification of the different types of 2D and 3D strings proposed so far by various researchers. The important variations of 2D strings will be further discussed in subsequent chapters.

4.1. VARIABLE-SIZED GRIDS

The symbolic projection technique introduced in Chapter 3 assumes a fixed-sized grid. An immediate generalization is to allow for variable-sized grids. In other words, the objects can have variable size and shape, but we still reduce them to point-like objects, so that each object is represented by, for example, its centroid. The grid lines are drawn between the point-like objects, so that the "slots" are variable in size.

As an example, the picture in Figure 4.1 can be represented by the 2D string $(u, v) = (a < b = c, a = b < c)$. The symbol "$<$" denotes the left–right spatial relation in string u, and the below–above spatial relation in string v. The symbol "$=$" denotes the spatial relation "at approximately the same spatial location as". Since the three objects are regarded as point-like objects, and the x-coordinates of the centroid of b and c are within a nearness threshold, the two objects are regarded as at approximately the same horizontal location. Therefore, the 2D string representation can be seen to be the symbolic

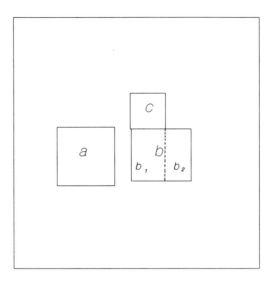

Figure 4.1 Symbolic projections of a picture consisting of objects with different size.

projections of the point-like objects in a picture along the vertical and horizontal directions.

In the above representation, the spatial relational operator "=" can be omitted, so that the 2D string is more efficiently represented by $(u, v) = (a < bc, ab < c)$. If the symbolic picture is given, we can take the symbolic projections to obtain the 2D string (u, v). Conversely, if the (u, v) is given, we can reconstruct a picture having symbolic projections (u, v), although the reconstruction may not be unique. Efficient algorithms for picture reconstruction and similarity retrieval have been developed (Chang *et al.*, 1987).

The two basic spatial relational operators can be augmented by other operators. The edge-to-edge local operator, denoted by the symbol "|", can be used when two objects are in direct contact either in the left–right or in the below–above direction (Jungert, 1988). Figure 4.2 illustrates the edge-to-edge relationship and the corresponding string representation. As to be discussed in Section 4.2 and illustrated in Figure 4.4, the edge-to-edge operator can be used advantageously to segment an object into connected subobjects.

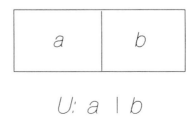

Figure 4.2 An example of the edge-to-edge operator.

4.2. IMAGE SEGMENTATION BY CUTTING LINES

The three spatial operators, "<", "=" and "|", are the most basic spatial operators. The edge-to-edge operator allows us to represent the touching of two objects. For example, if we are to describe the relationships between the two objects b and c in Figure 4.1, we can use the 2D string $(bc, b|c)$.

However, the above representation still does not describe fully the relationships between b and c. With the edge-to-edge operator, we can further segment an object into its constituent parts. This is accomplished by introducing cutting lines. Going back to Figure 4.1, the object b can be segmented into b_1 and b_2, by drawing a cutting line as shown. Now the relationships between b and c are expressed by $(b_1c|b_2, b|c)$.

In the above we use b_1 and b_2 to denote the two parts of b. If we use b to represent either the entire object b, or a part of it, then the 2D string becomes $(bc|b, b|c)$.

As another example, for objects that encompass other objects, such as the objects a and b shown in Figure 4.3(a), we can describe their relations by regarding them to be in the same slot. Using a variable-sized grid, the size of the slots can be defined arbitrarily.

The 2D string for Figure 4.3(a) is $(ab < c, a < bc)$. Notice the minimum enclosing rectangles of objects a and b overlap. This is an indication that segmentation may be necessary.

If the object b is segmented into objects x and y, as shown in Figure 4.3(b), the 2D string representation becomes $(x < ay < c, xa < yc)$.

Planar objects (such as lakes, towns, etc.) can be segmented into primitive constituent objects. Linear objects (such as roads, rivers, etc.) can be segmented into line segments. Point objects, of course, need not be further segmented.

This approach for *hierarchical 2D string encoding* is outlined on page 52:

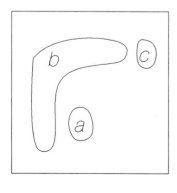

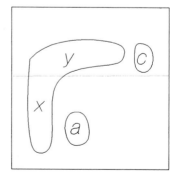

(a)   **(b)**

Figure 4.3 Encompassing objects (a) and their segmentation (b).

Procedure **2Dindex**(picture, u, v)
begin
   /*object recognition*/
   recognize objects in the picture;
   find minimum enclosing rectangle (MER) of each object;
   /*segmentation*/
   while the MERs overlap
     begin
       segment overlapping objects into constituent objects;
       find MER of each segmented object;
     end
   /*now all objects are disjoint*/
   find centroid of each object;
   find 2D string representation using Procedure **2Dstring**
end

The segmentation of an object can be done by drawing cutting lines, as shown in Figures 4.1 and 4.3(b). A systematic way of drawing the cutting lines will be presented in section 4.3. More sophisticated cutting mechanisms will be presented in Chapter 5.

4.3. DEFINITION OF GENERALIZED 2D STRING

Based upon the previously introduced three spatial operators, a systematic way of drawing the cutting lines is as follows: First, the extremal points are found in both the horizontal and vertical directions. An example is illustrated in Figure 4.4. Next, vertical and horizontal cutting lines are drawn through these extremal points. This technique gives a natural segmentation of planar objects into the constituent parts.

   With the cutting mechanism, we can also formulate a general representation,

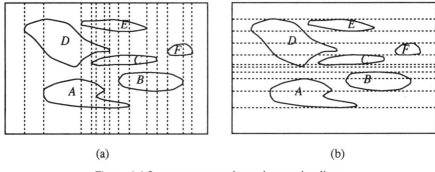

(a)   (b)

Figure 4.4 Image segmentation using cutting lines.

encompassing the other representations based upon different operator sets. This consideration leads to the formulation of a generalized 2D string system (Chang, Jungert and Li, 1989).

A *generalized 2D string system* is a five-tuple $(\Sigma, C, E_{\text{op}}, \mathbf{e}, "\langle, \rangle")$, where Σ is the vocabulary; C is the *cutting mechanism*, which consists of cutting lines at the extremal points of objects; $E_{\text{op}} = \{<, =, |\}$ is the set of *extended spatial operators*; $\mathbf{e}$ is a special symbol which can represent an area of any size and any shape, called the *empty-space object*; and "$\langle, \rangle$" is a pair of operators which is used to describe local structure.

The cutting mechanism defines how the objects in an image are to be segmented, and also makes it possible for the local operator "$\langle, \rangle$" to be used as a global operator to be inserted into the original 2D strings. The symbolic picture in Figure 4.4 has cutting lines as shown by dotted lines. The *generalized 2D string* representation is as follows:

u: $D \mid A\,\mathbf{e}\,D \mid A\,\mathbf{e}\,D\,\mathbf{e}\,E \mid A\,\mathbf{e}\,C\,\mathbf{e}\,D\,\mathbf{e}\,E \mid A\,\mathbf{e}\,A\,\mathbf{e}\,C\,\mathbf{e}\,D\,\mathbf{e}\,E \mid A\,\mathbf{e}\,C\,\mathbf{e}$

   $D\,\mathbf{e}\,E \mid A\,\mathbf{e}\,C\,\mathbf{e}\,E \mid A\,\mathbf{e}\,B\,\mathbf{e}\,C\,\mathbf{e}\,E \mid B\,\mathbf{e}\,C\,\mathbf{e}\,E \mid B\,\mathbf{e}\,C \mid B \mid B\,\mathbf{e}\,F \mid F$

v: $A \mid A\,\mathbf{e}\,B \mid B < D \mid D\,\mathbf{e}\,C \mid D\,\mathbf{e}\,F \mid D \mid D\,\mathbf{e}\,E$

4.4. THE EMPTY SPACE OBJECT

In the above, the symbol $\mathbf{e}$ represents "empty space". The term "empty space" was first introduced by Lozano-Perez, to support the full description of a "room" (Lozano-Perez, 1981). Here we generalize Lozano-Perez's concept and use the special symbol $\mathbf{e}$ to represent empty areas of any size and any shape. Therefore, the expression $A\,\mathbf{e}\,B$ can be rewritten as AB, and the generalized 2D strings can be simplified:

u: $D \mid A D \mid A D E \mid A C D E \mid A A C D E \mid A C D E \mid A C E \mid A B C E \mid B C E \mid$

   $B C \mid B \mid B F \mid F$

v: $A \mid A B \mid B < D \mid D C \mid D F \mid D \mid D E$

Furthermore, the expression $B < D$ in the v-string can also be written as $B \mid \mathbf{e} \mid D$, indicating objects B and D are not touching.

The special empty-space symbol $\mathbf{e}$ and operator-pair "$\langle, \rangle$" provide the means to use generalized 2D strings to substitute for other representations. In other words, we can transform another representation into the generalized 2D string and conversely.

In Chapter 14, when we discuss the σ-tree spatial data model, a new type of 3D generalized empty space object $\mathbf{v}$ will be introduced. The volumetric 3D empty space object $\mathbf{v}$ is similar to $\mathbf{e}$ except that it is three-dimensional.

4.5. PROJECTION TYPES

Symbolic projection is a qualitative method in which various types of projections against the coordinate axis are performed as a means for generation of syntactic descriptions of space in order to identify a large number of object relations. Since the projections of the objects play such an important role it is necessary to have an understanding of the various types of projections that exist and are used in the method. For this reason those types that are most frequently used will be presented.

In the original approach by Chang *et al.*, projections perpendicular to the *x*- and *y*-coordinate axis were used. Furthermore, the original approach was characterized by projections of the centroids of the minimum bounding enclosing rectangle of the objects. In later approaches this was changed because that type of projection leads to an object-view that was inherently point oriented, giving rise to limitations in the way the syntactic object structures were interpreted by humans and by machines. An improved projection type that came to generalize the original approach is the *interval projection type*, which projects the extremal points of the object, that is the projection of the characteristic attributes corresponding to the minimum enclosing rectangles, discussed earlier in this chapter. This projection technique makes it easier to determine the relations among overlapping object projections. For this reason it was possible to determine a much richer set of object relations than could be identified using centroid projections. Centroid and interval projections are illustrated in Figures 4.5(a) and (b) respectively where the objects are represented with their minimal enclosing rectangles.

Slope projection is an extension of Symbolic Projection that came of the necessity to locate hidden or "badly located" points in space. It has also been used for determination of qualitative directions and distances. However, this technique must be handled with special restrictions since it may give rise to inconsistencies in the syntactic space descriptions. The method may in some cases be uneconomical since it may require extra computations. The slope

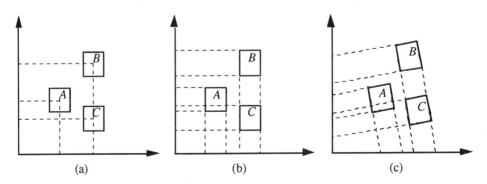

Figure 4.5 Centroid projections (a), interval projections (b) and slope projections (c).

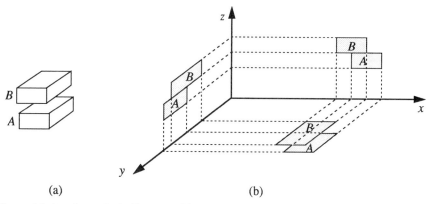

Figure 4.6 A universe including two objects (a) and their corresponding projections onto the $y–x$, $z–x$ and $y–z$ planes (b).

projections are discussed in depth in Chapter 10, and a simple example of this projection type is shown in Figure 4.5(c).

The projection types that have been discussed so far are the most commonly used, especially in 2D applications. Another type that also is used in 2D applications is the polar projection type, which is strongly related to the slope projections. This approach will be discussed further in Chapter 5.

In Section 2.6 the σ-tree was introduced to support generalization of symbolic projection into higher dimensions. In such applications further extensions on how the projections should be viewed, as well as how they should be performed are necessary. One approach is illustrated in Figure 4.6. This type is more oriented towards projections that correspond to surfaces generated from 3D objects. A consequence of this is that the number of projection directions is increased as can be seen in Figure 4.6(b), where there are three surface projections, one in each plane in the coordinate system, i.e. the $x–y$-, $x–z$- and $y–z$-planes. Normally the surface projections can be reduced into normal 2D projections, two for each plane in the coordinate system, as will be seen in Chapter 13.

4.6. A TAXONOMY OF 2D AND 3D STRINGS

Table 4.1 provides a taxonomy of different type of 2D and 3D strings proposed so far by various researchers; it enables us to study the various approaches systematically.

As can be seen from Table 4.1, the original 2D string has been extended in various ways both in two dimensions and in three dimensions. The important types will be discussed in subsequent chapters. References can be found at the end of each chapter, as well as in the general bibliography at the end of the book. The *N-string* is discussed in Chou *et al.* (1994), where analysis and experimental results are reported on the storage efficiency of the N-string for

Table 4.1. A taxonomy of 2D and 3D strings.

Type	Proposed by	Main features and where discussed
2D B-string	S. Y. Lee, M. C. Yang and J. W. Chen	It does not require cutting lines to partition the objects. By using the ranks of symbols, the spatial relationships can be derived (section 8.5 of Chapter 8). Interval projection strings are similar to B-strings.
2D C-string	S. Y. Lee and F. J. Hsu	It uses a rich set of spatial operators so that a symbolic picture can be represented by 2D C-strings economically using very few cutting lines. It supports similarity retrieval and spatial reasoning (section 5.3 of Chapter 5).
Index 2D C-string	G. Petraglia, M. Sebillo, M. Tucci and G. Tortora	It is able to maintain information about different subparts of the same object, so that different types of matching can be defined, and rotations of symbolic picture become possible (section 5.4 of Chapter 5).
2D G-string	E. Jungert and S. K. Chang	It uses three spatial operators and the cutting lines to produce the 2D G-string representation for pictures containing objects of all sizes and shapes (section 4.3 of this chapter).
2D H-string	S. K. Chang and Y. Li	It uses hierarchical spatial operators to combine quad-tree with 2D string (section 8.1.1 of Chapter 8).
2D N-string	Annie Y. H. Chou, C. C. Chang and W. P. Yang	It is based upon quad-tree and equivalent to 2D H-string in postfix notation, but is storage-space efficient.
2D T-string	Y. G. Sun	It uses spatial operators to represent topological, ordering and auxiliary (surrounding) relations. It supports spatial reasoning and has been applied to geographic information systems (section 6.5 of Chapter 6).
Generalized projection strings	E. Jungert	It extends symbolic projection from the observer's point of view, so that it can support spatial reasoning (Chapter 9).
3D relations	A. Del Bimbo, M. Campanai and P. Nesi	It uses pairwise relations of objects in the scene to characterize and retrieve pictures (section 6.7 of Chapter 6).
3D string	Y. Li and S. K. Chang	It generalizes 2D string to 3D string, with corresponding concepts of the reconstruction of pictures (section 13.1 of Chapter 13).

encoding symbolic pictures into quad-tree-like codes, and the time efficiency in carrying out the transformations. Because the N-string is very similar to the H-string, it will not be discussed separately.

REFERENCES

Chang, S. K., Shi, Q. Y. and Yan, C. W. (1987) Iconic indexing by 2D strings. *IEEE Transactions on Pattern Analysis and Machine Intelligence* **9**, 413–28.

Chang, S. K., Jungert, E. and Li, Y. (1989) Representation and retrieval of symbolic pictures using generalized 2D strings. *Proceedings of the SPIE Visual Communications and Image Processing Conference*, pp. 1360–72.

Chou, A. Y. H., Chang, C. C. and Yang, W. P. (1994) Symbolic indexing by N-strings. Technical Report, Department of Computer and Information Science, National Chiao Tung University, Hsinchu, Taiwan.

Jungert, E. (1988) Extended symbolic projections as a knowledge structure for spatial reasoning and planning. In *Pattern Recognition*, ed. J. Kittler, pp. 343–51. Springer-Verlag, Berlin.

Lozano-Perez, T. (1981) Automatic planning of manipulator transfer movements. *IEEE Transactions on Systems, Man and Cybernetics* **11**, 681–98.

5

Local Operators and Spatial Relations

Local operators play a fundamental role in symbolic projection for identification of spatial relations between objects that are in close relation to each other, and various sets of operators have been defined for various types of applications. Many of the local operators have similarities to the temporal operators defined by Allen (1983). The different operator sets will be discussed and illustrated further in this chapter. The requirements of certain operators make it necessary to split the images orthogonally to the coordinate axis. Such splittings, generally called *cuttings*, play a fundamental role in some approaches to Symbolic Projection. These cuttings are discussed in some detail. However, there exist other types of spatial relations that do not entirely relate to the operators by Allen, for example the orthogonal relations in section 5.5. Orthogonal relations correspond to a set of relations that are basically used to identify complex relations where the objects involved are represented with minimal bounding rectangles that are either completely or partly overlapping. Besides the above techniques for identification of spatial relations by means of local operators, some other methods are introduced which are partly related to Symbolic Projection in that they can also be used for spatial reasoning as well.

Sections 5.1 and 5.2 are both concerned with extensions to the original operator set. In sections 5.3 and 5.4, two further and more sophisticated cutting mechanisms are introduced as complements to the extended operator sets. The orthogonal relations follow in section 5.5, while in section 5.6 some other methods, partly related to Symbolic Projection, are presented.

5.1. EXTENSIONS OF THE ORIGINAL OPERATOR SET

In the original approach to Symbolic Projection a limited set of relational operators, $\{=, <\}$ was introduced. Later, it turned out that this small set was not sufficient. The main reason for this insufficiency is that objects were regarded as point objects and hence the method was less useful for more complicated problems. A solution to this problem was to expand the number of operators in the original set. The first attempt in this direction was made by Jungert (1988), and Chang and Li (1988) proposed another. Jungert's operator set was mainly intended for local reasoning and was later used for path-finding, while Chang and Li's work was intended for logical description of multi-resolution structures similar to, for instance, quad-trees. Further extensions followed, of which some of the more important are discussed in sections 5.1, 5.2 and 5.3.

Jungert's extension to the original set of relational operators is strongly related to the temporal intervals described in Allen (1983). This is not surprising, since in Symbolic Projection each projection string corresponds to a single dimension and this is also the case in the temporal intervals described by Allen. For this reason it is easy to see that the relational operators describing completely and partly overlapping intervals must be present in any extension to Symbolic Projection. Allen defines 13 different relations, while Jungert's extension just contains 8. However, a closer look at the two sets shows that Allen's set includes the "inverses" of the relations described by Jungert. From a practical point of view there is hence no real difference between the two sets. Thus given a set of pictorial objects and a set of spatial relational operators, a 2D string over the complete set of objects is defined as

$$i_1 x_1 i_2 x_2 \ldots i_n V : \ldots i_1 x_3 \ldots i_2 x_4 \ldots i_m$$

where $i_1, i_2, \ldots i_m, \ldots, i_n$ are members of the set of pictorial objects and $x_1, x_2, \ldots$ are members of the set of spatial relational operators. The pictorial objects can be identified either by their names or by any other type of symbols. The set of operators is $\{\backslash, \%, \vDash, \dashv, /, =, <, \sim\}$ where the operators are used to specify the spatial relations between the pictorial objects. As before, u: is the label denoting the string of the symbolic projections along the x-axis while v: corresponds to the symbolic projections along the y-axis. The order of the objects along each coordinate axis is preserved in the 2D string; this correspondence must always be maintained. The order in the v-string is a permutation of the objects in the u-string. The definitions of the operators are found in Table 5.1 below, and the relations are also illustrated in Figure 5.1.

The extension to symbolic projection given here was developed to performing local spatial reasoning, i.e. the motivation for extending the original set of spatial relational operators was primarily not to give a complete and formalized image description but rather to identify the relations between a small number of objects local to a given area. Among the operators in the extended set the "|" (edge-to-edge) operator was also discussed in Chapter 4, where it was shown that this operator can be used for global reasoning as well.

The % operator has turned out to be less useful since it is a bit fuzzier than the others, but nevertheless it should not be left out since there are applications

Table 5.1 The definitions of the extended operator set according to Jungert

---

$A = B ::= \text{centroid}(A) = \text{centroid}(B)$ and $\text{length}(A) = \text{length}(B)$
$A \vDash B ::= \min(A) = \min(B)$ and $\text{length}(A) < \text{length}(B)$
$A < B ::= \max(A) < \min(B)$
$B \dashv A ::= \max(B) = \max(A)$ and $\text{length}(B) > \text{length}(A)$
$A \backslash B ::= \min(A) < \min(B)$ and $\min(B) < \max(A)$ and $\max(A) < \max(B)$
$B \% A ::= \min(B) < \min(A)$ and $\max(A) < \max(B)$
$B / A ::= \max(B) < \max(A)$ and $\min(A) < \max(B)$ and $\min(B) < \min(A)$
$A \mid B ::= \max(A) = \min(B)$

---

where it will be needed. Among the others two pairs can be identified, i.e. $\{/, \backslash\}$ and $\{\vDash, \dashv\}$. These two pairs are mirror images of one another. In Figure 5.1 the operators of the extended set along the x-direction are clearly self-describing and need no further presentation besides the definitions given in Table 5.1.

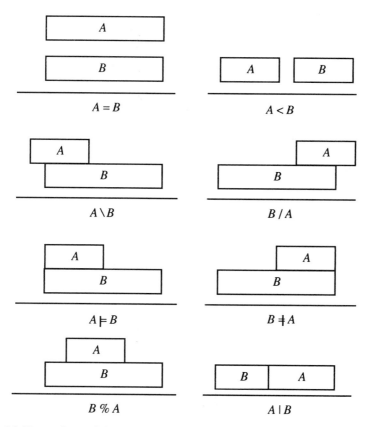

Figure 5.1 Illustrations of the extended set of spatial relational operators projected in one dimension.

5.2. FURTHER EXTENSIONS TO THE OPERATOR SET

The extension to Symbolic Projection in section 5.1 was mainly developed for local spatial reasoning, and is not the only one in existence. In section 5.2.1 a split of the edge-to-edge operator called *refined edge-to-edge* will be demonstrated (Chang and Jungert, 1991; Chang and Li, 1988; Chang *et al.*, 1990). This split was originally defined to represent the images as multiresolution hierarchies similar to quad-trees. In section 5.2.2 a further extension proposed by Lee and Hsu (1990, 1991) will be discussed that can be applied on a global level and used for similarity retrieval. This aspect of the approach will be discussed in Chapter 6.

5.2.1. Refined Edge-to-Edge Operators

A symbolic picture or simply a picture is an image where some of the slots are filled by picture objects. In pictorial information retrieval, the goal is often to retrieve pictures satisfying certain spatial relations such as, for example, "find the tree to the left of the house." To specify such spatial relations, two spatial relational operators were originally proposed. However, as has already been discussed, this set of basic spatial relational operators must be enlarged. The local edge-to-edge operator denoted by "|" can be used when two objects are in direct contact either in the left–right or in the below–above direction.

The set of the three spatial operators $\{<, =, |\}$ corresponds to the minimal set of spatial operators. From Section 4.2, it is clear that it can be used to describe any images in terms of Symbolic Projections completely, although this may not always be efficient or optimal. It turns out that a further extension of the edge-to-edge operator is possible; this is illustrated in Figure 5.2 where the edge-to-edge operator has been replaced by three local operators. Using these operators the original $\langle u \cdot v \rangle$ strings of a symbolic picture can be transformed such that, they can represent even more complex spatial relations among the pictorial objects in an image. For example, the $\langle u, v \rangle$-strings for the picture shown in Figure 5.3 can be transformed from $(a < bc, ab < c)$ into $(a < bc, ab \vdash c)$. The use of the refined edge-to-edge operators will be discussed in section 1 of Chapter 8.

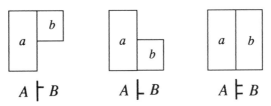

Figure 5.2 The refined edge-to-edge operators.

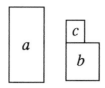

Figure 5.3 A simple picture suitable for description with the refined edge-to-edge operators.

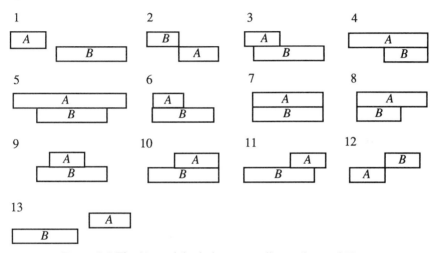

Figure 5.4 The 13 spatial relations according to Lee and Hsu.

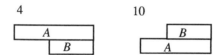

Figure 5.5 A spatial relation and its inverse according to Lee and Hsu.

5.2.2. The Spatial Operators According to Lee and Hsu

The extension suggested by Lee and Hsu is in fact an extension of the operator set suggested by Jungert that was described in section 5.1. The difference between Jungert's and Lee and Hsu's sets is that the latter also contains all possible inverses and for that reason their set is identical to Allen's (1983) set of temporal operators. Lee and Hsu's set can be seen in Figure 5.4.

An important question is whether the inverse spatial operators are really necessary or whether they can be left out. The answer is that they clearly are necessary in Allen's time-dependent application which is one-dimensional, while Symbolic Projection is concerned with two or three dimensions. In the original work by Chang *et al.* the order of the projected objects was of no importance, i.e. the *u*-string becomes the same in both alternatives in Figure 5.5 which correspond to relations 4 and 10 in Lee and Hsu's set. Hence, *AB* is equal

Table 5.2 The definitions of the characteristic spatial operators

Notation	Condition	Meaning
$A < B$	$\text{end}(A) < \text{begin}(B)$	A disjoint B
$A = B$	$\text{begin}(A) = \text{begin}(B)$ and $\text{end}(A) = \text{end}(B)$	A equals B
$A \mid B$	$\text{end}(A) = \text{begin}(B)$	A edge-to-edge B
$A \% B$	$\text{begin}(A) < \text{begin}(B)$ and $\text{end}(A) > \text{end}(B)$	A contains B (different bounds)
$A [B$	$\text{begin}(A) = \text{begin}(B)$ and $\text{end}(A) > \text{end}(B)$	A contains B (same begin)
$A] B$	$\text{begin}(A) < \text{begin}(B)$ and $\text{end}(A) = \text{end}(B)$	A contains B (same end)
A/B	$\text{begin}(A) < \text{begin}(B)$ and $\text{end}(A) < \text{end}(B)$	A partly overlap B

to *BA*. Taking this into consideration, the inverse functions are not always needed since the difference in order between the objects, *A* and *B*, is mirrored in the *v*-string. Consequently, inserting the inverses in the set of spatial relational operators is the same as including information from the orthogonal string, i.e. from the *y*-axis. Whether this should be done or not can be debated, but there may be applications where this should be allowed as well. Table 5.2 illustrates the notation and meaning as well as the conditions of the operators.

Lee and Hsu have taken their reasoning mechanism even further by taking the pairwise relations into account, i.e. by combining binary relations along both coordinate axes. Hence, it has been possible to identify 169 different types of spatial relations in two dimensions, which can be seen in Table 5.3.

The "contain" and "belong" relations require further explanation, since for these two cases one of the objects is entirely inside (overlapping) the other object. For this reason the *overlapping* part is illustrated with a lighter pattern in the table to avoid having two different relations that look the same. For instance, the relations $=]$ and $]^* =^*$ would otherwise be identical.

5.3. THE SPARSE CUTTING MECHANISM (LEE AND HSU)

The 2D C-string method introduced by Lee and Hsu (1992), is clearly an important extension of the method for local spatial reasoning introduced by Jungert (1988). Their main application is similarity retrieval, for which an improved cutting mechanism has also been developed. The motivation for the new cutting mechanism is that the original, discussed in Chapter 4, led to too many object slices. Lee and Hsu's cutting mechanism is more economical with respect to the number of cuttings required in a specific image.

Cutting according to Lee and Hsu is performed between partly overlapping objects. More specifically, cutting is performed in such a way that one of the overlapping objects is split into two parts, that is, cutting takes place at the end

Table 5.3 The 169 types of spatial relations in two dimensions according to Lee and Hsu

Disjoint (48)			Join (40)			Partial overlap (50)			Contain (16)	Belong (16)
< <	/ <*	/* <*	| |	% |*	|* %	/ /	= /*	] [*	= =	=* =*
< |	/<	/* <	|/	/* |	|* [*	/]	[/	] %*	=]	=*]*
< /	] <*	/* <*	|]	[* |*	|* =	/ %	[/*	]]*	= %	=* %*
<]	% <	<* <	| %	= |	|* [	/ [*	%* /	% [*	= [	=* [*
< %	% <*	<* |	| [*	= |*	|* %*	/ =	%* /*	% %*	] =	]* =*
< [*	/* <	<* /	| =	[|	|*]*	/ [	]* /	%]*	]]	]*]*
< =	/* <*	<*]	| [	[|*	|* /*	/ %*	]* /*	[[*	] %	]* %*
< [	= <	<* %	| %*	%* |*	|* /*	/]*	/* /	[%*	] [	]* [*
< %*	= <*	<* [*	|]*	%* |*	/ /*	/*]	[ ]*	% =	%* =*
<]*	[<	<* =	| /*	]* |	[/	/* %	[*]	% [	%* [*
< /*	[<*	<* [	| |*	/* |*	] /*	/* [*	[* %	%]	%* %*
< |*	%* <	<* %*	/ |	/* |	% /	/* =	[* [	% [	%* [*
< <*	%* <*	<*]*	/ |*	/* |*	% /*	/* [	%*]	[=	[* =*
/<	]* <	<* /*	] |	/* |	/* /	/* %*	%* %	[ ]	[*]*
| <*	]* <*	<* |*	] |*	/* /	[* /*	[* /*	%* [	[%	[* %*
/<	/* <	<* <*	% |	|*]	= /	/* /*	]*]	[[	[* [*
						]* [	]* %		

of the first ending object in a pair of overlapping objects. The object that is not split is called the *dominant* or *dominating object*. A simple illustration of this cutting mechanism can be found in Figure 5.6. The arrangement in Figure 5.6(a) can thus be represented as $A\,]\,B\,]\,C\,|\,B\,[\,C$; compare this with the original cutting mechanism of the general string type which gives $A\,|\,A=B\,|\,A=B=C\,|\,B=C\,|\,B$. A comparison shows that Lee and Hsu's cutting mechanism requires just a single cutting for the objects in Figure 5.6(a), while the original approach requires no fewer than four. The number of relations is different as well. Lee and Hsu require four operators in the

(a)   (b)

Figure 5.6 *A* is dominating both *B* and *C* (a), *B* is dominating *C* where both *B* and *C* are contained in *A* (b).

example, while the general string type requires eight. This is due to the different cutting mechanisms. Furthermore, the end-bound point of a dominating object does not partition another other object if the latter contains the former, as is illustrated in Figure 5.6(b). In this example object *B* is dominating *C* but both *B* and *C* are contained in *A*. Thus the latter object is not cut and the resulting string becomes *A* % (*B* | *C* | *C*).

Figure 5.7 shows a more complex example that illustrates Lee and Hsu's cutting mechanism. Here there are three cuttings along the *x*-axis and just a single one along the *y*-axis. This should be compared to the original cutting mechanism that would have required 12 (respectively 10) cuttings. From Figure 5.7 the following strings can now be determined:

$$u: \quad D \,]\, A \,]\, E \,]\, C \,|\, A = C = E \,]\, B \,|\, B = (C\,[\,E < F)\,|\, F$$

$$v: \quad A \,]\, B \,|\, B < D \,]\, (C \,|\, F < E)$$

In the example just 13 segmented subobjects are created along the *x*-axis, compared to the general string type which gives 26. According to Lee and Hsu their cutting mechanism is always more economical than the original for overlapping objects and at least as economical for nonoverlapping objects.

The algorithm for the cutting mechanism, which includes the creation of the

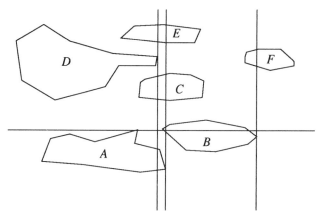

Figure 5.7 An illustration of the cutting mechanism according to Lee and Hsu.

2D C-strings, can now be described. Assume the objects $s_1, s_2, \ldots, s_n$ which are recognized in an image and then enclosed within minimal bounding rectangles. Then the following notation is used for these objects:

p_{ib}^x corresponds to the begin-bound point of s_i along the x-axis

p_{ie}^x corresponds to the end-bound point of s_i along the x-axis

p_{ib}^y corresponds to the begin-bound point of s_i along the y-axis

p_{ie}^y corresponds to the end-bound point of s_i along the y-axis.

Then the 2D C-string can be constructed from the symbolic picture f by first applying the cutting algorithm. Then the u- and v-strings are constructed individually since their cuttings are orthogonal. Hence the cutting algorithm can be described as follows.

Procedure **Cutting**$(s_1, s_2, \ldots, s_n)$

begin

/* output from this procedure is a 2D C string */

/* step 1. Sort all the begin-bounds and end-bounds p_{ib}^x and p_{ie}^x of s_i for $i = 1, 2, \ldots, n$. */

/* step 2. For the sorted points, group the points with the same value into a same value list */

/* step 3. For each same value list loop from step 4 to step 9 otherwise put out the 2D C string. */

*/ step 4. Check whether there are any end bound points in the list if not go to step 9. */

*/ step 5. Find the dominating objects from the objects in the given end-bound list cuttings are performed at the end bound points of the dominating objects. */

*/ step 6. Cut those objects whose begin bounds are within the range of the dominating objects. The cutting is split into three phases.

(1) The objects partly overlapping with the dominant object are segmented. According to the begin-bound point values of latter objects, from largest to the smallest, the same begin bound objects are chained by "="-operators. The larger value list is merged into the smaller value list by "]" operators.

(2) The dominating objects, if more than one, are chained by "="-operators. If there further begin bound values which are equal, then the dominating list is extended with that value with a "["-operator, otherwise the dominating list is merged into the former object by an "%"-operator.

(3) If the containing object in (1) and (2) contains other former objects, the latter object will be connected to the former by "<" or "|" depending on whether the "edge" flag was set in step 8. */

*/ step 7. If there are no further unfinished objects situated before or at the same position as the dominating objects put the dominating object into the 2D C-string. */

*/ step 8. The remaining sub-part of segmented objects in step 6(1) are viewed as

the new objects with the begin-bound as the location of cutting line and the corresponding "edge" flag is marked. */
*/ step 9. Collect the begin bound points of these new objects. */
end

5.4. POLAR AND CONCENTRIC CUTTINGS

The cutting mechanism introduced by Lee and Hsu can be applied to polar representations as well. This has been demonstrated by Petraglia *et al.* (1995). The main purpose of their work was to identify a mapping between images and indices that is normalized, that is a mapping that avoids information loss and permits recognition of similarities between images with respect to given positions in the images. Thus normalization with respect to rotation implies the use of a polar cutting mechanism, which will be described in this section.

An image can be associated with a pair of iconic indices. The iconic 2R string index is defined with respect to a given object where polar axes originate from its *rotation center* which coincides with the centroid of the given object, i.e. the centroid of the minimal bounding rectangle of the object. It is clear that any object centroid in the image can be chosen as the rotation center. From a practical viewpoint the object closest to the center of the image is best suited to this task. For each object, four boundary points can be identified, i.e. from the pair of lines starting at the rotation center and which correspond to the edges of the radial segment of the object, and the concentric circles that are the tangents of the object with respect to the rotation center. Subsequently, the sector-direction (s-direction) is referred to a clockwise movement and circle-direction (c-direction) to a movement from the origin and outwards. This is illustrated in Figure 5.8.

Using this cutting mechanism, every spatial relation can be described in terms of conditions on the beginning and end points along both the s- and c-directions. Hence these projections correspond to a type of interval projection. The set of spatial relations defined from these two types of cuttings is defined and illustrated in Figure 5.9 and 5.10 for the 2R and 2D string operators in the

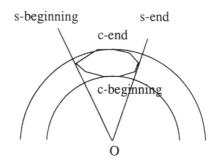

Figure 5.8 The start and end points along the s- and c-directions.

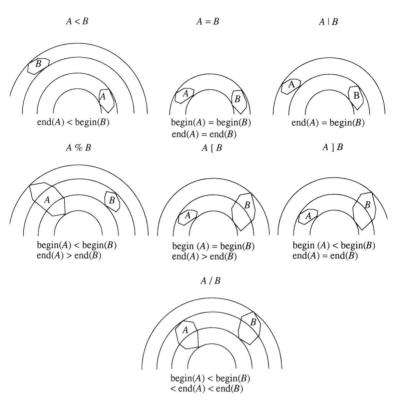

Figure 5.9 The definitions of the spatial operators of the 2R string along the c-direction.

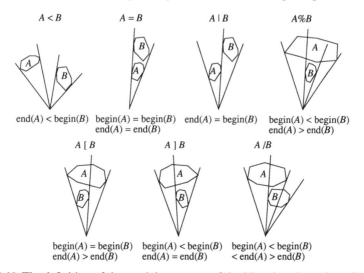

Figure 5.10 The definition of the spatial operators of the 2R string along the s-direction.

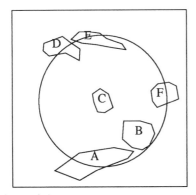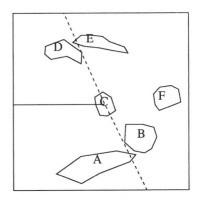

Figure 5.11 The 2R string cutting mechanism; the c-direction (a) and the s-direction (b).

c- and s-directions respectively. For example, the disjoint relation "$<$" along the c-direction is satisfied, that is $A < B$, if the end point of A precedes the beginning of B when moving from the origin towards the edge of the image.

For the above, a traditional cutting mechanism was assumed. However, this type of cutting approach can be refined for the 2R strings by analogy to the one proposed by Lee and Hsu which was discussed in section 5.3. The main difference between the two is that the cutting lines must relate to the origin, i.e. the rotation center. The projections are made with respect to a reference radius and the rotation center. The string that represents the projection in the s-direction repeats its first operand (object or subpart); in other words, it has the same operand as its first and last entities. Figure 5.11 illustrates the cutting and the generation of the string pair from each one of the orthogonal directions:

$$c: \quad B_1 \,]\, A_1 \,]\, D_1 = F_1 \,]\, E_1 \,|\, D_2 \,[\, A_2 \,[\, F_2 \,[\, E_2$$

$$s: \quad D_1 \,]\, E_1 \,|\, E_2 < F_1 < B_1 \,]\, A_1 \,|\, A_2 < D_1$$

5.5. THE ORTHOGONAL RELATIONS

The original approach to symbolic projection only allowed two types of spatial relations, left–right and below–above. A way of representing more complex spatial relations is first to split the object into segments from which all left–right and below–above relations among the components can be identified. Simple spatial relations can then be transformed into complex ones. Such simple spatial relations, which will be discussed subsequently, are called *orthogonal relations* (Chang and Jungert, 1986, 1990).

For objects where the minimum bounding rectangles are available, three types of topological relations between the rectangles can be identified:

- nonoverlapping rectangles
- partly overlapping rectangles
- completely overlapping rectangles.

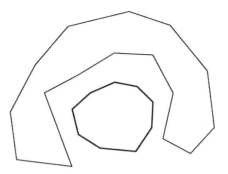

Figure 5.12 Two objects with overlapping MBRs.

The first alternative, where the rectangles do not overlap, is trivial and will normally not cause any problems because the object relations are simple. The other two might sometimes cause problems especially when one of the objects partly surrounds the other. Figure 5.12 demonstrates a problem of this type. The fundamental issue here is to find a method that easily describes the relations between the objects. The method is called *orthogonal relations*, because it deals with spatial relations that are orthogonal to each other.

The basic idea is to regard one of the objects as a *point of view object* (PVO). The PVO is projected on the other object in at most four directions (north, east, south, west). Hence, at least one or at most four subparts of the other object can be "seen" in the projection directions from the PVO. A part of the object that is actually "seen" is always in that direction where the two MBRs overlap, partly or completely. This is illustrated in Figures 5.13 and 5.14. The subobjects will in the next step of the method be regarded as point objects, which correspond to the centroids of those rectangles that enclose the subobject. It is a fairly simple operation to identify and generate these points if the objects are represented in RLC. Thus in the next step the objective is to identify the relations between the POV and the subobjects seen by the symbolic projection intervals. It is also of interest to observe that the orthogonal relations are established by a procedure that has similarities with a cutting mechanism, that is, it is correct to view this as

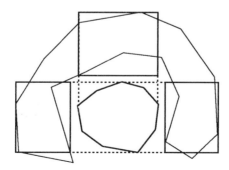

Figure 5.13 The PVO and its corresponding orthogonal relations.

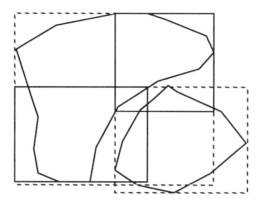

Figure 5.14 The PVO and its corresponding orthogonal relations for partly overlapping
  MBRs.

local cutting mechanism that is concerned only with a pair of neighboring
objects.

Each one of the subobjects constitutes a subobject orthogonal to the PVO,
and, since the subobjects are regarded as points, a sparse description of the
original object is generated. From this viewpoint, it does not matter whether the
objects are of extended or linear type. However, it is of importance that the
subobjects are interpreted correctly. Figure 5.15(a) shows a correct interpretation of a north and a west segment, while the interpretation of the same element
in Figure 5.15(b) is erroneous. The natural interpretation is to look clockwise.
Hence, nine different combinations can be identified:

2 points: N–E, E–S, S–W, W–N
3 points: N–E–S, E–S–W, S–W–N, W–N–E
4 points: N–E–S–W

No other interpretations are allowed. It is also possible to regard the elements in
between the orthogonal ones, but this is not necessary since enough information
is available anyway. See section 5.5.2 for further discussions of the technique.

The technique of finding orthogonal relations is described in the following
algorithm.

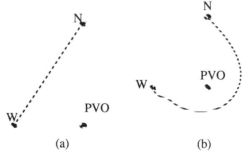

(a)   (b)

Figure 5.15 A correct (a) and an erroneous (b) interpretation of orthogonal relations.

Procedure **Ortho**(x, y)
begin
    /*this procedure finds the orthogonal relations of object x with respect to object y*/
    /*find the minimum enclosing rectangle of x and y*/
    find Mer (x), find Mer (y);
    /*find the four relational objects of object y intersecting with the extensions of object x*/
    y-W = W-extension (Mer (x)), Mer (y);
    y-E = E-extension (Mer (x)), Mer (y);
    y-N = N-extension (Mer (x)), Mer (y);
    y-S = S-extension (Mer (x)), Mer (y);
    return ($\{y$-W, y-E, y-N, y-S$\}$);
end

Figure 5.16 Three steps of a segmentation example which is based on the orthogonal relation technique where the original objects (a) are used to identify the orthogonal objects (b) and then the latter are expanded into segments (c).

5.5.1. Segmentation by Orthogonal Relations

The technique of finding orthogonal relations can be applied as a segmentation process to objects in an image. First, the image is pre-processed and the objects are recognized. Then, for each object x, the corresponding orthogonal relational objects in y are found. If the number of orthogonal objects is less than 2, the minimum bounding rectangles (MBRs) of objects x and y are disjoint. Therefore, no further processing is needed. If the number of orthogonal relational objects is greater than or equal to 2, then they will be added to the list Rel (y). After all object pairs have been processed, then for each object y a list of orthogonal objects Rel (y) has been accomplished. The object y can then be segmented into a number of segments corresponding to the number of members in Rel (y). The reference point of each segment will become the center point of each orthogonal relational object. The symbolic projection strings $\langle u, v \rangle$ can then be obtained from this symbolic picture, where each segment is regarded as a separate object. The algorithm for generation of the orthogonal segments can be described as follows.

Procedure **OrthoSegment**(f, u, v)
begin
    /* object recognition */
    recognize objects in the picture f;
    /* initialization */
    for each object x
    Rel (x) is set to empty;
    /* find orthogonal relations */
    for each pair of objects x and y
    begin
    find Ortho (x, y) /*orthogonal relations of object x with respect to object y*/;
    if $|$ Ortho $(x, y)| > 1$ then Rel $(y) =$ Rel $(y) \cup$ Ortho (x, y);
    find Ortho (y, x) /*orthogonal relations of object y with respect to object x*/
    if $|$ Ortho $(y, x)| > 1$ then Rel $(x) =$ Rel $(x) \cup$ Ortho (y, x);
    end
    /* segmentation */
    for each object x
    segment x into objects Rel (x);
    /*2D string encoding */
    apply procedure **2Dstring** (f, m, u, v, n);
end

An example of the segmentation technique is given in Figure 5.16(a)–(c). The original image contains objects a, b, and c, where a and b are point objects, as shown in Figure 5.16(a). After the procedure Ortho has been applied, it is found that Ortho $(a, c) = \{x, y, z\}$, and Ortho $(b, c) = \{w, y\}$. Therefore, object c contains four relational objects x, y, z, and w, as shown in Figure 5.16(b). The segmentation result is illustrated in Figure 5.16(c). The object c is thus split into four segments, x, y, z, and w. To obtain these segments, each orthogonal relational object is "grown" until it meets its orthogonal relational neighbors. The center of each relational object is used as the reference point of each segment. In this way, all orthogonal spatial relations are obtained and preserved.

The 2D string encoding of the picture in Figure 5.16 is $\langle u, v \rangle = \langle x < az < w < by, z < bw < xya \rangle$. To simplify the encoding, the orthogonal relational objects that are adjacent and belong to the same object must be merged. For example, if two orthogonal relational objects are merged into one segment, the reference point of either object can be used, or alternatively the centroid of the merged object can be used as its new reference point, provided that all orthogonal spatial relations are still valid. For example, if the objects w and y are merged into y the encoding becomes $\langle u, v \rangle = \langle x < az < y < b, z < b < xya \rangle$. All orthogonal spatial relations are thus preserved.

5.5.2. A KNOWLEDGE-BASED APPROACH TO SPATIAL REASONING

Symbolic projection and its application to orthogonal relations provide a means that can be used as a basis for spatial reasoning. This will be illustrated in a number of examples below. From the 2D string representation, spatial relations can be derived without any loss of information. Furthermore, from these spatial relations even more complex spatial relations can be derived. Therefore, by combining the 2D string representation with a knowledge-based system, flexible means of spatial reasoning and image information retrieval and management can be provided. The knowledge-based approach to spatial reasoning is illustrated by means of the two examples below.

Example 5.1. An island outside a coastline

Figure 5.17 illustrates the orthogonal relations. The 2D symbolic projections are then

$$u: \quad C_1 < C_2 i$$
$$v: \quad C_2 < C_1 i$$

The following rule can now be applied:

if $u: r_1 < r_2 p$ and $v: r_2 < r_1 p$
then
fact: (south rp) (west rp)
text: "The object $\langle p \rangle$ is partly surrounded by the object $\langle r \rangle$ on its south and west sides"

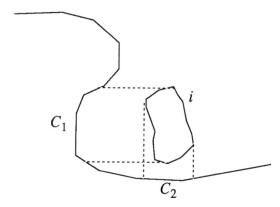

Figure 5.17 An island and its corresponding orthogonal relations in a coastline, for example.

When applying this rule to the example, r will be substituted by C and p by i, that is, (south Ci) and (west Ci). These facts are now stored in the fact database. The textual part could be presented to the user.

Example 5.2. A forest near a lake

In Figure 5.18, L_1, L_2, and L_3 illustrate the orthogonal relational objects which are part of a lake. The 2D symbolic projections are

$$u: \quad L_3 < FL_1 < L_2$$
$$v: \quad L_3 FL_2 < L_1$$

This is accompanied by the following rule:

if $u: r_1 < pr_2 < r_3$ and $v: r_1 pr_3 < r_2$
then
fact: (west rp) (north rp) (east rp)
text: "The object $\langle p \rangle$ is partly surrounded by the object $\langle r \rangle$ on its west, north and east side"

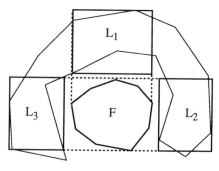

Figure 5.18 The orthogonal relations between a forest and a lake.

In this example, p is substituted by forest F and r by lake L. By successively applying rules that correspond to each of the basic orthogonal relation types, it is possible to identify all object-to-object relations. For example, the conclusions in Example 5.2 are (west L F), (north L F), and (east L F).

5.5.3. Visualization of Symbolic Pictures

The knowledge-based approach to spatial reasoning can be expanded further. In the examples in Section 5.5.2, it was demonstrated how spatial relationships between various spatial objects can be inferred from the projection strings. In this section, it will be demonstrated how spatial relationships can be inferred from symbolic descriptions of images.

Consider the following problem: given the 2D string representation of an image, can the original image be recreated from the projection strings in symbolic form? An algorithm for reconstruction of a symbolic picture from its 2D string representation can be found in Chang and Jungert (1990), where the objects are assumed to be point objects that are pieces of segmented objects. Somehow these segmented pieces must be reconstructed (and reconnected). Of course, the RLC description can be retrieved from the image database directly and manipulated to visualize the objects. This is, however, time-consuming. An alternative is to apply connection rules to reconnect the segments from a reconstructed symbolic picture. Hence, by successively applying the types of rules discussed in Section 5.5.2 on more complex structures, supplemented with some further rules, it will be possible to generate sequences of fact that can be used to connect the orthogonal relational objects. This method is especially useful when the orthogonal relational objects are expanded into segments. Example 5.3 shows the rules and assertions needed to reconstruct a description of an image.

Example 5.3

The problem is to describe an image and generate an approximate visualization. The image contains three objects, X, Y and Z. X is a linear object, and Y and Z are point objects. From the original image the following 2D symbolic projection strings have been generated using their corresponding orthogonal relations.

X-to-Y projections

$$u: \quad X_2 < X_3 Y X_1$$
$$v: \quad X_3 < X_2 Y < X_1$$

X-to-Z projections

$$u: \quad X_5 < X_4 Z$$
$$v: \quad X_4 < X_5 Z$$

Y-to-Z projections

$$u: \quad Y < Z$$
$$v: \quad Z < Y$$

In the X-to-Y projection strings, one of the basic rules that describes the applied relations is:

if $u: r_2 < r_3 p r_1$ and $v: r_3 < r_2 p < r_1$
then
fact: (south rp) (west rp) (north rp)
text: "The object $\langle p \rangle$ is surrounded by $\langle r \rangle$ on its south, west and north side"

Thus it can be concluded that:

(south $X\ Y$),   (west $X\ Y$)   and (north $X\ Y$)

The relation X-to-Z uses the same basic rule as in Example 5.1 in section 5.5.2. Hence the following statements are found:

(west $X\ Y$)   and   (south $X\ Y$)

Finally, for the relation between Y and Z, none of the basic orthogonal relation rules are useful because this relation is even more primitive. The relations between Y and Z are of the nonoverlapping rectangle type. Hence the following rule can be used:

if $u: r < p$ and $v: p < r$
then
fact: (northwest rp)
text: "The object $\langle r \rangle$ is to the northwest of $\langle p \rangle$"

It is easy to see that eight further rules of similar type can be identified. The result of the execution of this rule gives:

(northwest $Y\ Z$)

The next step is to verify whether Y and Z reside on the same side of X or whether they are on different sides. In the latter case a rule from which

(between $X\ Y\ Z$)

can be concluded must be identified. However, this is not the case here. This implication is only mentioned because there must be some rules available that describe relations of this kind. Instead, the following rule is applied:

if (west $r\ p$) and (west $r\ q$)
then (west $r\ p\ q$)

from which

$$(\text{west } X \ Y \ Z)$$

is inferred.

By using rules of the type identified so far, it will be possible to infer the assertions that can be used to describe the image. However, it is obvious that general information concerning the types of the objects and the relative distances between them must be available as well when eventually describing the image. The solution to this problem is illustrated in Figure 5.19(a) where it is assumed that X is a road and Y and Z are buildings. Figures 5.19(b) and 5.19(c), finally, illustrate two similar examples where the images have been reconstructed from the symbolic descriptions. In these two latter cases the projection strings are given as well as the corresponding symbolic pictures. Observe also, that although the forest in both examples is of extended object type, the orthogonal relational objects that are of interest are still handled as point objects. It is left to the reader to identify the remaining rules that are needed to solve these two problems and eventually reconstruct the images from the projection strings.

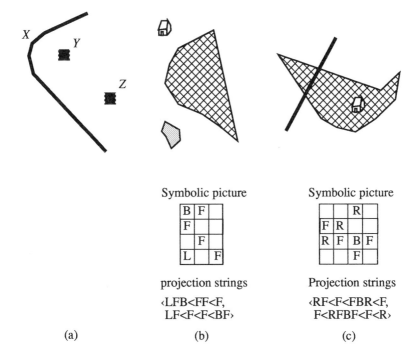

Figure 5.19 The result of connecting orthogonal relational objects: (a) of the road X and the neighboring buildings Y and Z; (b) a building, a forest and a lake; (c) a road, a forest and a building.

5.6. PARTLY RELATED METHODS

There are techniques partly related to Symbolic Projection in that they have a qualitative representation, and for this reason a couple of such techniques are presented here from that perspective. In Chapter 10 some further techniques on qualitative reasoning are presented.

5.6.1. Directed Regions (Tang and Ma)

Tang and Ma's approach (1991) is based on what is somewhat ambiguously called a cutting mechanism. Instead, the method should be regarded as a way of splitting the image space into regions or frames. Below this splitting method is a partition technique, which forms the basis for the reasoning in a method only partly related to Symbolic Projection, i.e. in this approach no global projected strings are used although projection intervals are defined for both the x- and y-coordinate axes. The method is hence concerned with local reasoning; in other words, the purpose is to identify various types of binary object relations in a local environment.

Partition of the space into regions can be described as follows. A central region corresponding to a minimum bounding rectangle of an object, also called the *orthogonal circumscribed rectangle*, is first defined. The central frame thus generated is called the central directed region. There are also eight frames surrounding the central one. The eight surrounding frames are open-ended except on the side facing the central region and the sides common to the surrounding frames. The partition approach is called a *directed system* and the

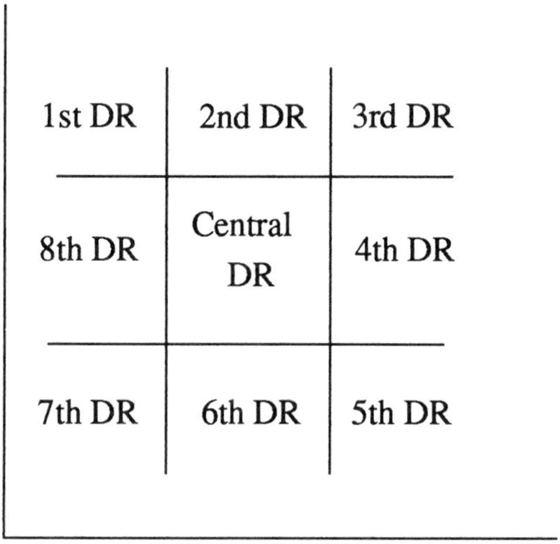

1st DR	2nd DR	3rd DR
8th DR	Central DR	4th DR
7th DR	6th DR	5th DR

Figure 5.20 The partition of a direction system in nine direction regions.

subspaces (frames) are called *directed regions* (DR). An example of the division of the space is presented in Figure 5.20.

For this direction system, 11 different object relationships can be identified. The definition of these relationships is as follows.

Definition. Given two closed intervals A corresponding to $[a_1, a_2]$ and B corresponding to $[b_1, b_2]$, the following spatial relations can be identified between the closed intervals:

$$
\begin{aligned}
A = B & \quad a_1 = b_1 < a_2 = a_1 \\
A < B & \quad a_2 \leqslant b_1 \\
A > B & \quad b_2 \leqslant a_1 \\
A \subset B & \quad a_1 > b_1 \quad \text{and} \quad a_2 < b_2 \\
A \supset B & \quad a_1 < b_1 \quad \text{and} \quad a_2 > b_2 \\
A \,|{\subset}\, B & \quad a_1 = b_1 < a_2 < b_2 \\
A \,{\subset}|\, B & \quad b_1 < a_1 < a_2 = b_2 \\
A \,|{\supset}\, B & \quad a_1 < b_1 < a_2 = b_2 \\
A \,{\supset}|\, B & \quad a_1 = b_1 < b_2 < a_2 \\
A \,|{\times}\, B & \quad a_1 < b_1 < a_2 < b_2 \\
A \,{\times}|\, B & \quad b_1 < a_1 < b_2 < a_2
\end{aligned}
$$

The relationship with Allen's operators is obvious. The main difference is that two relations have been merged, resulting in 11 operators instead of 13. Tang and Ma have redefined "less than" such that the second relation above $a_2 \leqslant b_1$ is a merge of $a_2 < b_1$ (less than) and $a_2 = b_1$ (touch or edge-to-edge in Symbolic Projection). The same redefinition is made for $A > B$ in the above list of operators. Clearly, this way of redefining the operators induces limitations.

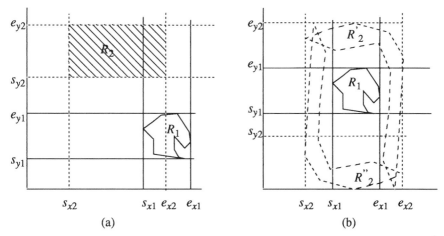

(a) (b)

Figure 5.21 Determination of object relations in a directed system where (a) corresponds to a certain relationship while (b) is ambiguous.

The reasoning method described by Tang and Ma suffers from the same problems as many other reasoning methods that are based on orthogonal views of space. This is illustrated in Figure 5.21, where the object R_1 corresponds to the object inside the central DR. In Figure 5.21(a) the relation between the objects can easily be determined while in Figure 5.21(b) the relations illustrates an ambiguous situation where the relationship cannot easily be determined. A solution to this problem is to apply the same method as for the orthogonal relations, as discussed in section 5.5. This is obviously true for both objects R_2' and R_2''.

5.6.2. Topological Relations

Topological relations play an important role in the work by Egenhofer (1994). This approach has recently attracted considerable interest and will for many reasons become an important approach within qualitative spatial reasoning. It is discussed further in Chapter 10, but the basic elements of the approach are introduced here. Another work that is related in many ways to Egenhofer's is that of Smith and Park (1992). However, neither of these two methods is related to Symbolic Projection except that all three can be used for qualitative spatial reasoning and thus contains qualitative elements. The basic topological relations suggested by Egenhofer are illustrated in Figure 5.22. There are eight altogether, and the following notation is used:

disjoint(A, B)
meet(A, B)
equal(A, B)
covers(A, B)
coveredBy(B, A)

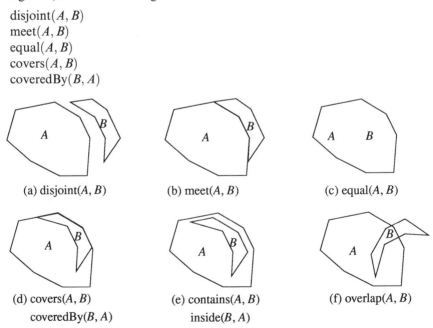

(a) disjoint(A, B) (b) meet(A, B) (c) equal(A, B)

(d) covers(A, B) (e) contains(A, B) (f) overlap(A, B)
coveredBy(B, A) inside(B, A)

Figure 5.22 The topological relations defined by Egenhofer.

contains(A, B)
inside(B, A)
overlap(A, B)

These eight topological relations are all rotation invariant. Egenhofer is not concerned with their underlying definitions contrary to Symbolic Projection which, however, is not a rotation invariant method. Among these relations, "covers" and "coveredBy" as well as "contains" and "inside" are inverses of each other. From the illustrations in Figure 5.22 it does not, at first, look as if Egenhofer's relations have much resemblance to those discussed so far. However, if they are considered as a 2D generalization of Allen's approach the relationship becomes clear.

In Egenhofer's method the usual concept of point-set topology with open and closed sets is assumed. Thus the interior of a set A, denoted by A°, is the union of all open sets in A. The closure of A, denoted by $\bar{A}$, is the intersection of all closed sets of A. The complement of A with respect to the embedding space $\mathbb{R}^n$, denoted by A^{-}, corresponds to the sets of all points of $\mathbb{R}^n$ that are not contained in A. The boundary of A, denoted ∂A, is the intersection of the closure of A and the closure of the complement of A. Clearly, the union of ∂A, A° and A^{-} corresponds to $\mathbb{R}^n$.

$$I_0(A, B) = \begin{bmatrix} \varnothing & \varnothing & \neg\varnothing \\ \varnothing & \varnothing & \neg\varnothing \\ \neg\varnothing & \neg\varnothing & \neg\varnothing \end{bmatrix} \qquad I_1(A, B) = \begin{bmatrix} \varnothing & \varnothing & \neg\varnothing \\ \varnothing & \neg\varnothing & \neg\varnothing \\ \neg\varnothing & \neg\varnothing & \neg\varnothing \end{bmatrix}$$

(a) disjoint(A, B) (b) meet(A, B)

$$I_2(A, B) = \begin{bmatrix} \neg\varnothing & \varnothing & \varnothing \\ \varnothing & \neg\varnothing & \varnothing \\ \varnothing & \varnothing & \neg\varnothing \end{bmatrix} \qquad I_3(A, B) = \begin{bmatrix} \neg\varnothing & \varnothing & \varnothing \\ \neg\varnothing & \varnothing & \varnothing \\ \neg\varnothing & \neg\varnothing & \neg\varnothing \end{bmatrix}$$

(c) equal(A, B) (d) inside(A, B)

$$I_4(A, B) = \begin{bmatrix} \neg\varnothing & \varnothing & \varnothing \\ \neg\varnothing & \neg\varnothing & \varnothing \\ \neg\varnothing & \neg\varnothing & \neg\varnothing \end{bmatrix} \qquad I_5(A, B) = \begin{bmatrix} \neg\varnothing & \neg\varnothing & \neg\varnothing \\ \varnothing & \varnothing & \neg\varnothing \\ \varnothing & \varnothing & \neg\varnothing \end{bmatrix}$$

(e) coveredBy(A, B) (f) contains(A, B)

$$I_6(A, B) = \begin{bmatrix} \neg\varnothing & \neg\varnothing & \neg\varnothing \\ \varnothing & \neg\varnothing & \neg\varnothing \\ \varnothing & \varnothing & \neg\varnothing \end{bmatrix} \qquad I_7(A, B) = \begin{bmatrix} \neg\varnothing & \neg\varnothing & \neg\varnothing \\ \neg\varnothing & \neg\varnothing & \neg\varnothing \\ \neg\varnothing & \neg\varnothing & \neg\varnothing \end{bmatrix}$$

(g) covers(A, B) (h) overlap(A, B)

Figure 5.23 The specifications of the topological relations between two point sets in 2D.

Identification of the topological relations between two point sets, A and B, can be made by the nine possible intersection combinations of the boundary, interior and complement of A and B. All these intersections can be represented in a 3×3 matrix, subsequently called the *9-intersection matrix*, that is:

$$I(A, B) = \begin{bmatrix} A^\circ \cap B^\circ & A^\circ \cap \partial B & A^\circ \cap B^- \\ \partial A \cap B^\circ & \partial A \cap \partial B & \partial A \cap B^- \\ A^- \cap B^\circ & A^- \cap \partial B & A^- \cap B^- \end{bmatrix}$$

Let the parts **a** and **b** be arbitrary elements of $\{A^\circ, \partial A, A^-\}$ and $\{B^\circ, \partial B, B^-\}$, respectively. An index, such as $\mathbf{a}_i$ or $\mathbf{b}_i$, will indicate corresponding elements of the two sets A and B, i.e. either both interiors, both boundaries or both exteriors. The notion $I[\mathbf{a}, \mathbf{b}]$ will be used for a particular intersection. For instance, $I[\partial, ^\circ]$ thus describes the boundary–interior intersection in the matrix, i.e. $\partial A \cap B^\circ$.

By deriving the 9-intersection matrix for a pair of objects, the topological relations between the objects can be determined; to which standard inference rules about point sets can be applied. The 9-intersections for the eight different topological relations between two point sets are shown in Figure 5.23.

The reasoning technique for determination of transitive object relations will be discussed in Chapter 10.

Acknowledgement

The major contribution of M. Egenhofer in Section 5.6.2 is acknowledged.

REFERENCES

Allen, J. F. (1983) Maintaining knowledge about temporal intervals. *Communications of the ACM* **26**, 832–43.

Chang, S. K. and Jungert, E. (1986) A spatial knowledge structure for image information systems using symbolic projections. *Proceedings of the Fall Joint Computer Conference, Dallas*, pp. 79–86.

Chang, S. K. and Jungert, E. (1990) A spatial knowledge structure for visual information systems. In *Visual Languages and Applications*, ed. T. Ichikawa, E. Jungert and R. R. Korfhage, pp. 277–304. Plenum Press, New York.

Chang, S. K. and Jungert, E. (1991) Pictorial data management based upon the theory of Symbolic Projection. *Journal of Visual Languages and Computing* **2**, 195–215.

Chang, S. K. and Li, Y. (1988) Representation of multi-resolution symbolic and binary using 2DH strings. *Proceedings of the IEEE Workshop on Languages for Automation*, pp. 190–5.

Chang, S. K., Jungert, E. and Li, Y. (1990) The design of pictorial databases based upon the theory of symbolic projections. In *Design and Implementation of Large Spatial Databases*, ed. A. Buchmann, O. Gunther and T. R. Smith, pp. 303–23. Springer-Verlag, Berlin.

Egenhofer, M. (1994) Deriving the combination of binary topological relations. *Journal of Visual Languages and Computing* **5**, 133–49.

Jungert, E. (1988) Extended symbolic projections as a knowledge structure for spatial reasoning and planning. In *Pattern Recognition*, ed. J. Kittler, pp. 343–51. Springer-Verlag, Berlin.

Lee, S.-Y. and Hsu, F.-J. (1990) 2D C-string: A new spatial knowledge representation for image database systems. *Pattern Recognition* **23**, 1077–87.

Lee, S.-Y. and Hsu, F.-J. (1991) Picture algebra for spatial reasoning of iconic images represented in 2D C-string. *Pattern Recognition Letters* **12**, 425–35.

Lee, S.-Y. and Hsu, F.-J. (1992) Spatial reasoning and similarity retrieval of images using 2D C-string knowledge representation. *Pattern Recognition* **25**, 305–18.

Petraglia, G., Sebillo, M., Tucci, M. and Tortora, G. (1996) A normalized index for image databases. In *Intelligent Image Database Systems*, ed. S. K. Chang, E. Jungert and G. Tortora. World Scientific, Singapore.

Smith, T. R. and Park, K. K. (1992) Algebraic approach to spatial reasoning. *International Journal of Geographical Information Systems* **6**, 177–92.

Tang, M. and Ma, S. D. (1991) A new method for spatial reasoning in image databases. *Proceedings of the 2nd Working Conference on Visual Database Systems*, pp. 37–48, Budapest.

6

Applications to Image Information Retrieval

The 2D string representation is an efficient way to represent symbolic pictures, allowing an effective means for queries on image databases, spatial reasoning, visualization and browsing. At the same time, we note that the performance of the 2D string iconic indexing depends on the abstraction from segmented images to symbolic pictures. Many researchers have looked for good abstraction techniques from iconic images to symbolic representation. See for example Sties *et al.* (1976), Tanimoto (1976), Klinger *et al.* (1978), Shapiro and Haralick (1979), Chang and Liu (1984). The iconic indexing approach should be combined with pattern recognition so that iconic indices can be automatically created. The active index technique will be presented in Chapter 12.

We now describe some typical applications of the theory of Symbolic Projection to image information retrieval. These are based upon works by researchers from many different countries and are indicative of the diversity of potential applications. In section 6.1, we first discuss image database queries. An intelligent image database system based on the 2D string pictorial data structure is described in section 6.2. This system supports spatial reasoning, flexible image information retrieval, visualization, and traditional image database operations. The pictorial data structure based upon 2D strings provides an efficient means for iconic indexing in image database systems, and spatial reasoning. Section 6.3 presents an application of the 2D C-string to similarity retrieval, where the problem of matching images containing many objects is transformed into a problem of 2D C-string matching. Section 6.4 presents an application to a CAD database of ships, where 3D matching is accomplished through string matching using a 3D voxel model. Section 6.5 illustrates an application to GIS, and section 6.6 demonstrates applicability to similarity retrieval of Chinese characters. Section 6.7 describes 3D image database querying, and Section 6.8 presents an application to medical image retrieval.

6.1. IMAGE DATABASE QUERIES

Although image information systems primarily deal with image data, many query languages developed for image information systems are command languages, or commands plus expressions (Joseph and Cardenas, 1988). Some of them follow a SQL-like syntax. For example, here is a PSQL query (Rossopoulos *et al.*, 1988):

> SELECT STATE_NAME, STATE_REGION, AREA(STATE_REGION)
> FROM STATE
> WHERE POPULATION >500,000;

The GRIM_DBMS graphical image database management system supports a query language with fuzzy measures (Rabitti and Stanchev, 1989):

> RETRIEVE IMAGES (hospital_building/0.9)
> CONTAINING
> ((double_bedroom/1.0) AND (number_of_doors $\geq$ 2));

On the other hand, in one of the early image information systems, the advantage in expressing queries by pictorial examples was already recognized. IMAID supports query-by-pictorial-example (QPE), but the queries are specified using tables, in the style of query-by-example (QBE) (Chang and Fu, 1980).

Since then, considerable progress has been made in the query-by-example approach. In IIDS, to be described in section 6.2, a pictorial query can be expressed directly as a picture, which is then converted into 2D strings for matching against the iconic index of the image database (Chang *et al.*, 1988). This query-by-example approach, combined with direct manipulation interface, can support querying by image content in 2D or 3D. The user can point at an image object and ask for all images "like-this", i.e. all images having similar features. Therefore, the user interface must be supported by algorithms for similarity retrieval.

In many applications, there is also a growing need for querying and visualizing images of different modality. The user should also be allowed to switch between interaction paradigms, and/or combine interaction paradigms. For example, the user can use iconic approach to specify "like-this" queries, tabular approach to specify conditional queries, and graphs such as the entity–relationship diagram to specify more complicated queries (Chang *et al.*, 1992a). Such multiparadigmatic visual queries will be discussed in Chapter 11.

6.2. AN INTELLIGENT IMAGE DATABASE SYSTEM (CHANG *ET AL.*)

IIDS is an intelligent image database system which uses a relational database management system to manage textual data and 2D string data associated with an image (Chang *et al.*, 1988). The architecture of IIDS is illustrated in Figure 6.1. The IDBM subsystem consists of a relational database management system and an image manipulation module augmenting the DBMS. The spatial reasoning module supports processing of queries which correspond to predicates. 2D strings are converted to atoms of the form "east of XY", "north of XY", "same x position XY", and "same y position XY". The user interface takes the user's query and maps it on to the matching predicate. For binary relations, the user specifies one, two or none of the objects in the relation. The spatial reasoning module checks to see if the relation is true (both objects specified) or finds all possible solutions (one or none of the objects specified).

The visualization module reconstructs the symbolic picture from the 2D string representation, then using the spatial reasoning and connection rules in the knowledge base KB shown in Figure 6.1, a visualized picture can be obtained. The 2D string iconic indexing module of IIDS is an implementation of the methods described in Chapter 4. This module is controlled by image databases management (IDBM) subsystem whose core is the relational database management system. In IIDS, the user is prompted to construct a 2D string for the image.

We now describe a sample session using IIDS. The intelligent image database system allows the transfer of pictures and data from one system to another. This exchange takes place when the user enters the system. The next step takes us to the database portion of IIDS. At this level, the user is presented with a screen that is divided into three major sections. These sections are the DISPLAY WINDOW, the MESSAGE CENTER, and the MENU WINDOW. The MENU WINDOW presents to the user a series of options at various levels.

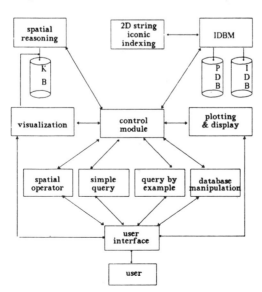

Figure 6.1 Architecture of IIDS.

Selection of one menu item will have either of the following results: an expanded submenu may appear, or a subprocess may be invoked. The MESSAGE CENTER is used for echoing the user's choices, displaying error messages, and providing special instructions or information. The DISPLAY WINDOW provides an area for printing the results of a user's queries. An example of these windows and the top-level menu is shown in Figure 6.2.

At the top level, the user is presented with a set of choices. The user may select DATABASE MANIPULATION, QUERY, VISUALIZATION, COMMUNI-CATIONS, and INFORMATION. The DATABASE MANIPULATION choice will allow the user to store and retrieve pictures. QUERY allows the user to ask questions about the pictures stored in the database. VISUALIZATION will allow the display of various queries and COMMUNICATIONS will allow for the transfer of information between systems. Finally, INFORMATION will provide help and hints about the system.

In the DATABASE MANIPULATION submenu, the user can execute the following:

- Activate Next Image, which displays the current camera image on the video screen.
- *Write Image*, which saves the current image on the video screen to the disk.
- *Load Image*, which retrieves images from the disk.
- *Delete Image*, which removes a picture from the disk.

Before items are written to the disk, the user can give a name to the picture and specify various keywords that denote the picture. The picture is stored on the disk according to its frame number and the frame numbers are automatically updated by the system.

The main focus of the system is in the QUERY submenu. The data that is retrieved is a string representation of the picture along with the name and related keywords. In QUERY BY NAME mode, the user can enter the name of

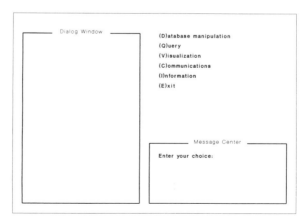

Figure 6.2 Top-level menu of IIDS user interface.

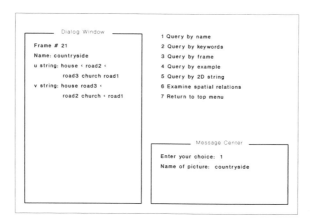

```
 _____ Dialog Window _____         1 Query by name

 Frame # 21                              2 Query by keywords

 Name: countryside                       3 Query by frame

 u string: house ‹ road2 ‹               4 Query by example

           road3 church road1            5 Query by 2D string

 v string: house road3 ‹                 6 Examine spatial relations

           road2 church ‹ road1          7 Return to top menu



                                         _____ Message Center _____

                                         Enter your choice:  1

                                         Name of picture:  countryside
```

Figure 6.3 Query-by-name and the retrieved symbolic picture.

a picture, and the system will retrieve all pictures that have that name (names are not unique). In QUERY BY KEYWORDS mode, the user can enter one or more keywords and the system will retrieve all pictures that contain the user's specified keywords. In QUERY BY FRAME mode, the user can enter a specific frame number. The system will return only one picture since the frame number is the unique key for that picture.

Figure 6.3 shows an example of the user selecting QUERY BY NAME and the results of a successful retrieval. In that figure, the results are shown in the display window. The user may want to compare Figure 6.3 with the picture shown in Figure 6.4. In the picture, there is a house in the left-hand corner, a

Figure 6.4 Retrieved image corresponding to the symbolic picture of Figure 6.3.

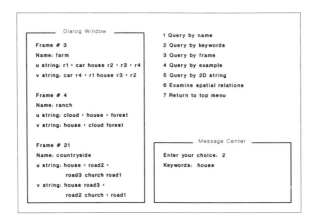

Figure 6.5 Query-by-keys and the retrieved symbolic pictures.

church appears on the right-hand side and a road loops by the church. When the picture was written to the database, it was given the name "countryside" and the keywords "house", "church", and "road" were entered. In addition, the picture was encoded with a 2D string.

If we consider road as three separate objects – road 1, the part of the road above the church; road2, the part of the road to the left of the church; and road3, the part of the road below the church – then there are five major objects in the picture. The objects appear in the picture from left to right as follows: the house is to the left; road2 is in the middle; road3, the church and road1 are on the right. Therefore the *u*-string is encoded as house < road2 < road3 church road1. If we now examine the picture from the bottom up, the house and road3 are at the bottom; road2 and the church are in the middle; and road1 is at the top. So the *v*-string is encoded as house road3 < road2 church < road1.

Figure 6.5 is an example of QUERY BY KEYWORDS. Here the user has

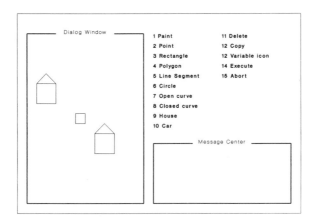

Figure 6.6 Query-by-example.

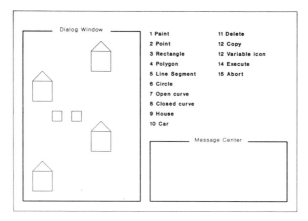

Figure 6.7 Retrieved symbolic picture for the query of Figure 6.6.

selected to find all the pictures that have the keyword "house" associated with them. Figure 6.5 shows the results of such a query, and the display center shows that there are three frames with that keyword.

With QUERY BY EXAMPLE, another major submenu is encountered. Here, a series of icons is presented to the user. An example of this is shown in Figure 6.6. This submenu allows the user to place the selected icon at any position on the screen. If the icon is placed near a boundary, it is properly clipped. Such choices as HOUSE, CIRCLE, and RECTANGLE will draw a house, circle, and rectangle, respectively. Once the desired icons are placed on the screen, the user can EXECUTE the display window. When such an execution takes place, the icons and their relative positions are converted into their corresponding 2D strings and the query is then made into the database.

Figure 6.6 shows a sample query. Here the user has posed the query: "Find all pictures that have a car with buildings to the northwest and to the southeast". The car is represented by a small rectangle in this situation. Figure 6.7 shows a

Figure 6.8 Retrieved image corresponding to the symbolic picture of Figure 6.6.

response to that query. Note that the buildings and the car in the response are in the same relative position as they were in the query, but the positions of objects in the response do not have to be exactly the same as the positions of the objects in the query. The reader should also compare Figures 6.6 and 6.7 with Figure 6.8. Figure 6.8 shows the actual picture that is stored in the database. Note that the orientation of the cars in the picture is not the same as the orientation of the car in the query or of the cars in the response. It is relative position of the objects that we are concerned with.

By performing QUERY BY 2D STRING, the user can specify the relative positions of various objects. Retrieval will then match against those 2D strings which are in the database.

In the EXAMINE SPATIAL RELATIONS mode, the user can then make queries about the retrieved image. Such questions as EAST OF (X, Y) and SURROUNDED BY (X, Y) can be asked, so that object X which is east of (or surrounded by) object Y can be determined. The system allows for either or both of the variables to be instantiated, so that the query can find everything that is east of something, or confirm that "the house is east of the car".

The menu shown in Figure 6.9 is the submenu associated with EXAMINE SPATIAL RELATIONS. Here the user has chosen "X east of all", which means that the system will find all the objects that object X is east of. The response to the query contains two lines: church is east of road2, and church is east of the house. The reader may want to compare these results with the picture in Figure 6.4.

We can extend the query to a *composite picture query* such as: "Find those pictures with object A to the east of object B, where object B contains objects C and D, and object C is to the north of object D." To deal with such composite picture queries (Chang *et al.*, 1992b), it is assumed that the picture is represented by a generalized 2D string without nesting operators "$\langle , \rangle$". The generalized

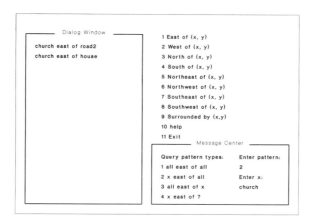

Figure 6.9 Examining spatial relationships.

2D string representation of a picture query may have arbitrarily nested expressions.

An *elementary object* is an object without nested expressions (an object without local structure). A *composite object* is an object with nested expressions (an object with local structure). The *level* of an object is the number of nested parentheses of an object. Since each composite object may contain several (elementary or composite) objects, the query may have several nesting levels. Figure 6.10 is an example of a query with several composite objects.

The **composite_2DmatchA** algorithm is the recursive extension of the **2Dmatch** algorithm described in Chapter 3. The difference is that the multiple levels of the composite picture query must first be considered. Moreover, each matching element now becomes a set of matching elements for a composite object. The recursive technique is used to deal with the multiple levels, but the path-finding method can be used to find the path between the composite objects. In the algorithm **2DmatchA**, each object is an indivisible object which has a unique rank value, and two objects are either totally overlapping or totally disjoint. However, when we consider the rank value of two composite

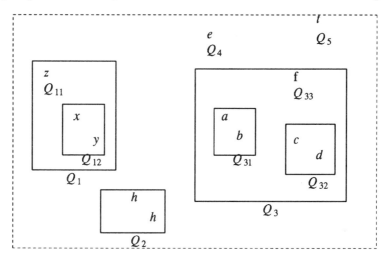

Figure 6.10 The query picture shows the nesting of the composite objects. The objects Q_1 has subobjects Q_{11} and Q_{12}, for example. The picture is a two-level conceptual pictorial query.

$$Q = \{Q_1, Q_2, Q_3, Q_4, Q_5\}$$
$$Q_1 = \{Q_{11}, Q_{12}\} \qquad Q_2 = \{Q_{21}, Q_{22}\} \qquad Q_3 = \{Q_{31}, Q_{32}, Q_{33}\}$$
$$Q_4 = \{e\} \qquad Q_5 = \{t\}$$
$$Q_{11} = \{z\} \qquad Q_{12} = \{x, y\}$$
$$Q_{21} = Q_{22} = \{h\}$$
$$Q_{31} = \{a, b\} \qquad Q_{32} = \{c, d\} \qquad Q_{33} = \{f\}$$

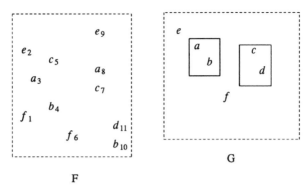

F

Figure 6.11 The picture F stored in pictorial database indexed by 2D string method. The query picture G is a conceptual pictorial query, with one level of nesting.

objects, we have an uncertain rank value. That makes the definition of type-2 subpicture unnecessary.

Figure 6.11 illustrates the picture and the two-level composite picture query. The picture has objects $f_1, e_2, a_3, b_4, c_5, f_6, c_7, a_8, e_9, b_{10}, d_{11}$, where the same symbol represents the same type of object. The picture query has objects a, b, c, d, e, f.

Applying the recursive matching algorithm, the following *matching pools* are obtained:

$$\text{matching\_pool}(\{e\}) = \{2, 9\}$$
$$\text{matching\_pool}(\{a, b\}) = \{3\text{-}4, 3\text{-}10, 8\text{-}10\}$$
$$\text{matching\_pool}(\{f\}) = \{1, 6\}$$
$$\text{matching\_pool}(\{c, d\}) = \{5\text{-}11, 7\text{-}11\}$$

where the notation 3-4 indicates a solution composed of objects in the picture whose indexes are 3 and 4, respectively. The dash between a sequence of numbers indicates there exists a path such that every pair of nodes satisfies the type-*i* condition. Invoking recursively the algorithm **Composite_2DmatchA**

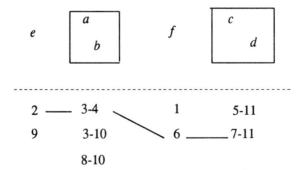

Figure 6.12 The figure shows the result of the **Composite_2DmatchA**. The path 2 – 3-4 – 6 – 7-11 means that we get a solution set $\{e_2, a_3, b_4, f_6, c_7, d_{11}\}$.

for the second level, the solution 2-3-4-6-7-11 is found, corresponding to objects $e_2, a_3, b_4, f_6, c_7, d_{11}$ that match the picture query. Figure 6.12 illustrates this situation. Thus there is a subpicture which matches the query picture.

6.3. SIMILARITY RETRIEVAL

A technique for similarity retrieval based on symbolic projection has been developed by Lee and Hsu (1992). A cornerstone of the technique is their sparse cutting mechanism, which was discussed in section 5.3. Here too, the variation of the projection strings called the *2D C-string* is used. Similarity retrieval, generally speaking, is an operation unique to image database systems that has no counterpart in any other type of database system. The purpose is to retrieve images for similarities that may or may not be present. Traditional approaches to this problem are measured on the basis of the maximum-likelihood or minimum distance criterion. A method of the minimum distance type, which uses 1D strings, is described in Fu (1982) where the minimum distance criterion between two strings is defined in terms of the minimum number of errors occurring when transforming one of the strings into the other. Lee and Hsu have in their approach to similarity retrieval, adopted and modified a technique where the images are represented in terms of 2D C-strings. The means for determination of various degrees of similarity is based on the idea of subpicture matching given in the original approach to symbolic projection, that is the type-0, type-1 and type-2 2D substrings (see Chapter 3).

A string γ is a type-*i* 1D subsequence of string α, if the following rule is applied: if γ is contained in α, and $a_1 w_1 b_1$ is a substring of γ, and a_1 matches a_2 in α and b_1 matches b_2 in α then

(type-0): $r(b_2) - r(a_2) \geqslant r(b_1) - r(a_1)$ or $r(b_1) - r(a_1) = 0$
(type-1): $r(b_2) - r(a_2) \geqslant r(b_1) - r(a_1) > 0$ or $r(b_2) - r(a_2) = r(b_1) - r(a_1) = 0$
(type-2): $r(b_2) - r(a_2) = r(b_1) - r(a_1)$

where $r(s)$ is defined as 1 plus the number of "$<$" preceding the symbol s. Let $\langle u, v \rangle$ and $\langle u', v' \rangle$ be the 2D string representations of picture sf and f' respectively. $\langle u', v' \rangle$ is a type-*i* 2D subsequence of $\langle u, v \rangle$ if u' is a type-*i* 1D subsequence of u, and v' is a type-*i* 1D subsequence of v. then, f' is a type-*i* subpicture of f. Therefore, a picture f' is a type-*i* subpicture of f if $\langle u', v' \rangle$ is a type-*i* 2D subsequence of $\langle u, v \rangle$. Thus, the picture matching problem becomes that of 2D string subsequence matching. However, the original 2D string representation must be modified since it does not permit representation of overlapping objects. The 2D C-string is therefore used instead, but this requires a specifically adopted similarity retrieval technique for this string type. According to the ranks of symbols in 2D C strings the spatial relationships between two segmented objects can be inferred, but this requires a new definition of the type-i 2D subsequences, which is defined accordingly.

Definition 6.1. Picture f' is a type-i unit picture of f, if f' is a picture containing the two objects A and B, represented as u: $Ar_{AB}^{u'} B$, v: $Ar_{AB}^{v'} B$, and A and B are also contained in f, and the relations between A and B in f are represented as u: $Ar_{AB}^{u} B$, and v: $Ar_{AB}^{v} B$, then

> (type-0): category (r_{AB}^{u}, r_{AB}^{v}) = category $(r_{AB}^{u'}, r_{AB}^{v'})$
> (type-1): (type-0) and $(r_{AB}^{u} = r_{AB}^{u'}$ or $r_{AB}^{v} = r_{AB}^{v'})$
> (type-2): $r_{AB}^{u} = r_{AB}^{u'}$ and $r_{AB}^{v} = r_{AB}^{v'}$

where Category $(r^{u}, r^{u'})$ denotes the relation category of the spatial relationship as shown in Table 5.2 in chapter 5. The pair (A, B) is called a type-i similar pair and is denoted by p.

Definition 6.2. A picture f_1 is a type-i 2D similar picture to f_2, if there exists a picture f', such that f' is a type-i unit picture of f_1, and a type-i unit picture of f_2. f' is called the type-i common unit picture of f_1 and f_2.

Definition 6.3. The set of type-i similar pairs between f_1 and f_2 is $P = \{p_1, p_2, \ldots, p_n\}$. The picture f'' constructed from P is a type-i common subpicture of f_1 and f_2.

Definition 6.4. If f'' is the type-i common sub-picture of f_1 and f_2, then the type-i similar degree between f_1 and f_2 is the number of symbols of f'', and f_1 is said to be type-i similar to f_2.

Definition 6.5. f_1 is the type-i most similar picture of f_2, if f_1 is type-i similar to f_2 with the largest similar degree.

The 2D C-string similar retrieval algorithm according to Lee and Hsu can now be described as follows.

Input: two 2D C-string representations of pictures f_1 and f_2.
Output: the type-i similar sub-picture of f_1 and f_2.
1. Construct the association graphs for the two pictures f_1 and f_2.
2. The matched objects between f_1 and f_2 form the set vertices.
3. For each type-i similar pair, there is an edge connecting the object pair.
4. Find the cliques of the association-graph. A complete subgraph containing the maximum number of vertices, i.e. the maximal complete subgraph or clique constitutes the type-i common sub-picture between f_1 and f_2.
5. The clique with type-i largest similar degree is the type-i most similar subpicture of f_1 and f_2.

Take the pictures f_1 and f_2 in Figure 6.13 as an example. There are eight pictorial objects in each of these two pictures. The 2D C-strings of the pictures

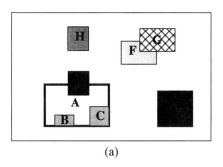 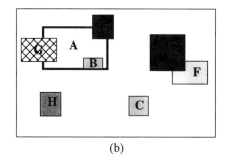

(a)   (b)

Figure 6.13 An example of similarity retrieval with two input images f_1 (a) and f_2 (b).

become:

$$f_1: \quad u: \quad A \,]\, (B \,]\, E = H \,|\, E = H \,|\, C) < F \,]\, G \,|\, D \,[\, G$$
$$v: \quad A = (D \,[\, C \,[\, B \,|\, E) \,|\, E \,|\, F \,]\, G \,|\, H \,[\, G \tag{6.1}$$

$$f_2: \quad u: \quad G \,]\, H \,]\, A \,|\, A \,]\, (B \,]\, E \,|\, E) \,|\, E < C \,|\, D \,]\, F \,|\, F$$
$$v: \quad H = C < F \,]\, A = B = D \,]\, G \,|\, A = (D \,[\, G \,|\, E) \,|\, E \tag{6.2}$$

By applying the reasoning rules for the determination of the similarities the spatial relationships among objects can be inferred by the ranks of the symbols. α_1 and β_1 (α_2 and β_2) stand for two symbolic objects in the picture $f_1(f_2)$ respectively. Thus:

$$f_1: \quad (\alpha_1 r_1^u \beta_1) = \{A \% B, A \,]\, C, A < D, A \% E, A < F, A < G, A \% H, B < C,$$
$$B < D, B / E, B < F, B < G, B / H, C < D, C \,|^* E,$$
$$C < F, C < G, C \,|^* H, D <^* E, D \,|^* F, D /^* G, D <^* H,$$
$$E < F, E < G, E = H, F / G, F <^* H, G <^* H\}$$

$$(\alpha_1 r^{v_1} \beta_1) = \{A \,[\, B, A \,[\, C, A \,[\, D, A / E, A < F, A < G, A < H, B \,[^* C,$$
$$B \,[^* D, B < E, B < F, B < G, B < H, C \,[^* D, C < E,$$
$$C < F, C < G, C < H, D \,|\, E, D < F, D < G, D < H, E \,|\, H,$$
$$E < G, E < F, F / G, F \,|\, H, G / H\} \tag{6.3}$$

from which the category (r_1^u, r_1^v) can be deduced:

(contain *AB*),	(contain *AC*)
(disjoint *AD*),	(partovlp *AE*)
(disjoint *AF*),	(disjoint *AG*)
(disjoint *AH*),	(disjoint *BC*)
(disjoint *BD*),	(disjoint *BE*)
(disjoint *BF*),	(disjoint *BG*)
(disjoint *BH*),	(disjoint *CD*)
(disjoint *CE*),	(disjoint *CF*)

$$(\text{disjoint } BG), (\text{disjoint } CH)$$
$$(\text{disjoint } DE), (\text{disjoint } DF)$$
$$(\text{disjoint } DG), (\text{disjoint } DH)$$
$$(\text{disjoint } EF), (\text{disjoint } EG)$$
$$(\text{disjoint } EH), (\text{disjoint } FG)$$
$$(\text{disjoint } FH), (\text{disjoint } GH) \tag{6.4}$$

f_2: $(\alpha_2 r_2^u \beta_2) = \{A \% B, A < C, A < D, A / E, A < F, A < G, A /^* H, B < C,$
$\qquad B < D, B / E, B < F, B <^* G, B <^* H, C | D, C <^* E, C < F,$
$\qquad C <^* G, C <^* H, D <^* E, D / F, D <^* G, D <^* H, E <^* F,$
$\qquad E <^* G, E <^* H, F <^* G, F <^* H, G] H \}$

$(\alpha_2 r_2^v \beta_2) = \{A [B, A <^* C, A [D, A / E, A / F, A \% G, A <^* H, B <^* C,$
$\qquad B [^* D, B < E, B /^* F, B / G, B <^* H, C < D, C < E, C < F,$
$\qquad C < G, C = H, D | E, D /^* F, D] G, D <^* H, E <^* F, E |^* G,$
$\qquad E <^* H, F / G, F <^* H, G <^* H \} \tag{6.5}$

and category (r_2^u, r_2^v):

$$(\text{contain } AB), \quad (\text{disjoint } AC)$$
$$(\text{disjoint } AD), \quad (\text{partovlp } AE)$$
$$(\text{disjoint } AF), \quad (\text{partovlp } AG)$$
$$(\text{disjoint } AH), \quad (\text{disjoint } BC)$$
$$(\text{disjoint } BD), \quad (\text{disjoint } BE)$$
$$(\text{disjoint } BF), \quad (\text{disjoint } BG)$$
$$(\text{disjoint } BH), \quad (\text{disjoint } CD)$$
$$(\text{disjoint } CE), \quad (\text{disjoint } CF)$$
$$(\text{disjoint } CG), \quad (\text{disjoint } CH)$$
$$(\text{disjoint } DE), \quad (\text{partovlp } DF)$$
$$(\text{disjoint } DG), \quad (\text{disjoint } DH)$$
$$(\text{disjoint } EF), \quad (\text{disjoint } EG)$$
$$(\text{disjoint } EH), \quad (\text{disjoint } FG)$$
$$(\text{disjoint } FH), \quad (\text{disjoint } GH) \tag{6.6}$$

According to the definition of type-i similar picture for $i = 0, 1, 2$, the

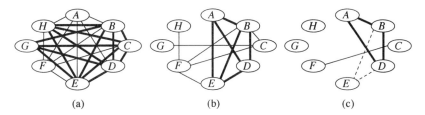

Figure 6.14 Type-0 similar pairs between f_1 and f_2 (a), type-1 similar pairs between f_1 and f_2 (b) and type-2 similar pairs between f_1 and f_2.

following set of type-i similar pairs of the pictures f_1 and f_2 can be identified:

type-0: $(AB), (AD), (AE), (AF), (AH), (BC), (BD), (BE), (BF),$
$(BG), (BH), (CD), (CE), (CF), (CG), (CH), (DE), (DG),$ (6.7)
$(DH), (EF), (EG), (EH), (FH), (GH)$

type-1: $(AB), (AD), (AE), (BC), (BD), (BE), (BF), (CF), (CG),$
$(DE), (EF), (FH)$ (6.8)

type-2: $(AB), (AD), (BD), (BE), (CF), (DE)$ (6.9)

From Figure 6.14(a)–(c), the maximal complete subgraphs of type-0, type-1 and type-2 are found respectively. The corresponding type-i longest common subpicture of f_1 and f_2 is constructed from the following object set:

type-0: $\{B, C, D, E, F, H\}$
type-1: $\{A, B, D, E\}$
type-2: $\{A, B, D\}$ or $\{B, D, E\}$

In this example, f_1 is type-0, type-1, type-2 similar to f_2 with similar degree of 6, 4 and 3, respectively.

6.4. CAD DATABASE (HILDEBRANDT AND TANG)

A CAD database for ships supports queries to retrieve ships with "two cranes on the hull with a superstructure behind them and a mast, radar and funnel on the superstructure." To process such queries Hildebrandt and Tang (1992) applied the symbolic projection technique in 3D to symbolic voxel models.

CAD data is usually stored in one of two forms: the boundary representation or the constructive solid geometry form. In the boundary representation an object is segmented into nonoverlapping faces. Each face is modelled by bounding edges and edges by end vertices, so the object is modelled by a tree of depth 3. In constructive solid geometry primitives such as cylinders, boxes and cones are combined and modified by operations such as union, intersection, difference, rotation and scale. The database catalog of the CAD database is assumed to contain a simple voxel model generated from the CAD data for queries.

GIS databases typically store a collection of co-registered two-dimensional

Figure 6.15 Voxel ship model.

images of certain properties such as brightness, spot height and slope, together
with vector data such as roads, river and contours. Layers of raster data can be
interpreted directly as images and grouped together to give voxel data. Vector
data would have to be first converted into low resolution raster data and then

```
Level : 10
b b b b b b b b b b b b b b b b b b b b b b
b b b b b b b b b b b b b b b b b b b b b b
b b b b b b b b b b b b b b b b b b b b b b
b b b b b b b b b b b b b b b b b b b b b b
b b b b b b b m b b b b b b b b b b b b b b
b b b b b b b m b b b b b b b b b b b b b b
b b b k b b b m b b b b b b b k b b b b b
b b b k b b b m b b b b b b b k b b b b b
b b b k b b b m b b f f b b b k b b b b b
b b b k b b b m b b f f b b b k b b b b b
b b b k b b s s s s s s s b k b b b b b
b b b k b b s s s s s s s b k b b b b b
b h h h h h h h h h h h h h h h h b b b b
b h h h h h h h h h h h h h h h h b b b b
b h h h h h h h h h h h h h h h h b b b b
b h h h h h h h h h h h h h h h h b b b b
b b b b b b b b b b b b b b b b b b b b b b
b b b b b b b b b b b b b b b b b b b b b b
b b b b b b b b b b b b b b b b b b b b b b
b b b b b b b b b b b b b b b b b b b b b b
```

Figure 6.16 Slice of voxel model: h = hull, s = superstructure, k = kingpost, m = mast
and f = funnel.

```
pattern:
Level : 0        Level : 1
k m f k          k m f k
k s s k          k s s k
h h h h          h h h h
```

Figure 6.17 A query pattern.

used as image or voxel data. The simplified 2D representation could then be used on each band in the GIS to index spatial information, and it may be possible to use the 3D form to index spatial relations between all bands.

The simple voxel model encodes a 3D object into slices. An example is illustrated in Figure 6.15, which shows a 20×20 voxel model. Such models could be used to index a collection of CAD models and would allow 3D spatial queries. A slice through this model is shown in Figure 6.16, where the letters "h", "s", "k", "m" and "f" denote "hull", "superstructure", "kingpost", "mast" and "funnel", respectively. In Figure 6.17, a $3 \times 4 \times 2$ pattern of a type-1 query is shown. A single type-1 match returned is indicated by the arrows in Figure 6.18. This search pattern would be used with a query such as "Find ship with two kingposts above hull with superstructure in between them and mast followed by funnel above superstructure."

A graphical user interface was constructed for a prototype ship database application (Figure 6.19). For textual information associated with a ship, typical database forms were employed. To enter and display the spatial relations

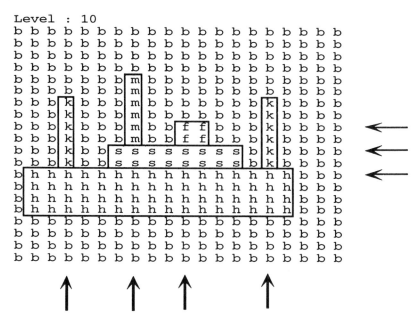

Figure 6.18 A single type-1 match.

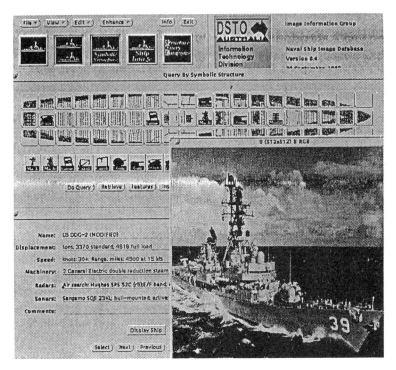

Figure 6.19 Graphical user interface combining textual and symbolic query with image retrieval.

Figure 6.20 Query-by-symbolic-structure form.

in the database, a graphical interface was employed where icons representing objects could be placed on the gridded outline of a hull, viewed from above (Figure 6.20). From this input, the required spatial relations could be determined and placed in a relational table to perform the database query. In addition to type-*i* queries, the system also supports pairwise relations matching so that similar patterns can be found more efficiently.

6.5. GEOGRAPHICAL INFORMATION SYSTEMS (GIS) (SUN)

The application of the theory of Symbolic Projection to GIS was studied by Yuguo Sun (1993).

Sun generalized the 2D G-string to the 2D T-string. The 2D T-string is able to represent three different types of qualitative spatial relations: *topological relations*, *ordering relations* and *auxiliary relations*. The topological relations describe local spatial relations, such as "equal", "disjoint", "meet", "edge", "contain" and "partial overlap". The ordering relations are the two basic global spatial relations "<" and "|". The auxiliary relations include "surround", "partially surround", and "quasi-partially surround". The cutting mechanism basically follows the cutting mechanism for the G-string, but refined by additional rules so that only the necessary cutting lines important to one of

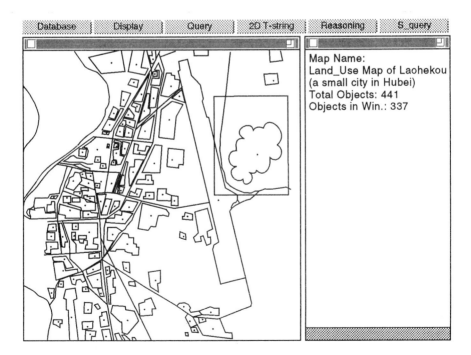

Figure 6.21 Land use map.

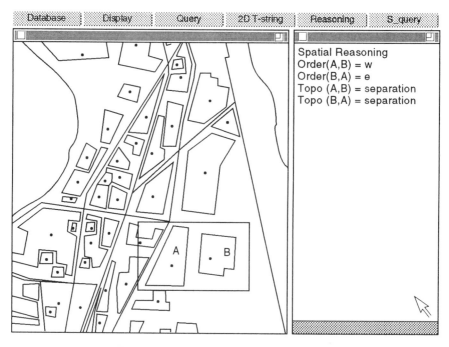

Figure 6.22 Example of spatial reasoning.

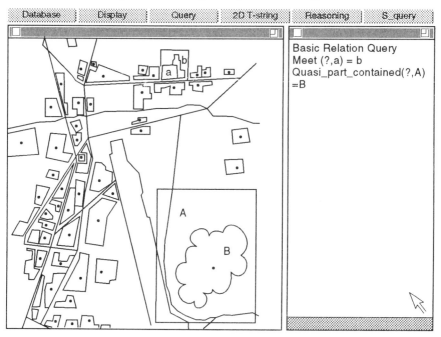

Figure 6.23 Basic spatial relation query.

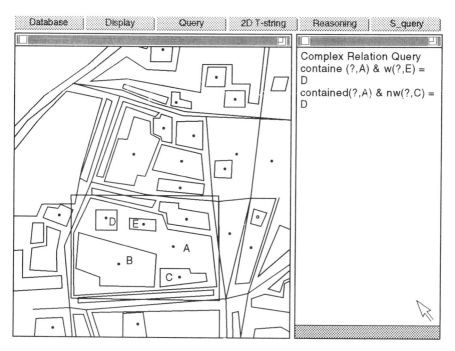

Figure 6.24 Complex spatial relation query.

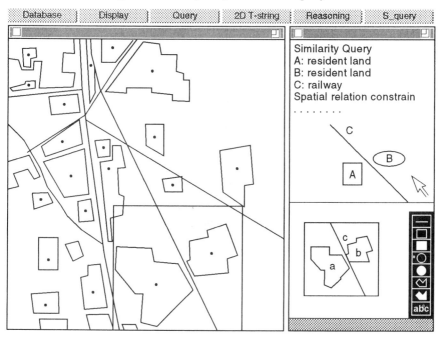

Figure 6.25 Similarity query.

the above types of relations will be drawn. Techniques for constructing the 2D T-string from the symbolic picture, and for spatial reasonings, have been developed. Since they are similar to the techniques described in Lee and Hsu (1992), the details will not be presented here.

An experimental spatial relations retrieval system based upon the 2D T-string was implemented for geographical information retrieval. The system supports spatial reasoning, basic spatial relation query, complex spatial relation query and similarity query. The geographical data is the land use map of Laohekou City in Hubei Province of China. The land use map is shown in Figure 6.21. The map contains 441 objects, and 337 of them are displayed in the window area. Figure 6.22 is an example of spatial reasoning. The user selects a rectangular area of interest. The ordering relations ("A is to the west of B", and "B is to the east of A") and topological relations ("A and B are separated spatially") can be derived from the 2D T-string and displayed in the window area on the right.

Figure 6.23 illustrates a basic spatial relation query. The system can find out object B meets object A, and object B is quasi-part-contained in A.

Figure 6.24 illustrates a complex spatial relation query. The system can find out object D is contained in A and to the west of E, and object D is contained in A and to the northwest of C.

Figure 6.25 illustrates a similarity query. The query is shown in the upper right window, where A and B are residential land, and C is the railway, and their approximate spatial relations are as shown. The result is displayed in the lower right window.

For similarity retrieval, time-consuming graph matching is required. However, in practical applications, the targets are restricted to a prespecified small window area, and the constraints include not only spatial constraints, but also constraints on the objects' attribute values such as shape, color, etc. Therefore, similarity retrieval can be computed in a reasonable time.

6.6. RETRIEVAL OF SIMILAR CHINESE CHARACTERS (CHANG AND LIN)

Although many methods have been proposed to solve it, the problem of retrieving spatially similar Chinese characters still remains. There are several

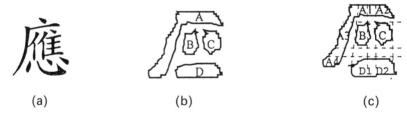

(a)   (b)   (c)

Figure 6.26 The original image of a Chinese character (a), the symbolic picture (b) and the segmented symbolic picture (c).

motivations for considering the retrieval of similar Chinese characters. First, it can be useful in learning Chinese characters. The structurally similar Chinese characters can be retrieved and presented to students, so that they can remember the components of the characters and their meanings. Second, similarity retrieval is also useful for Chinese character recognition, because it is capable of clustering similar characters.

Chang and Lin (1994) applied Symbolic Projection theory to Chinese character retrieval, by regarding the Chinese character as a symbolic picture. As illustrated in Figure 6.26(a), the original image corresponds to a Chinese character. A pattern recognition algorithm can be applied to segment the image into four major components A, B, C and D, as illustrated in Figure 6.26(b). The technique of orthogonal relations described in Section 5.5 (Chang and Jungert, 1990) can then be applied to discover the important orthogonal relations and convert the symbolic picture into the 2D string. As illustrated in Figure 6.26(c), the following orthogonal relations are discovered:

Ortho-relation(B, A) = {A1, A3}
Ortho-relation(C, A) = {A2, A3}
Ortho-relation(B, D) = {D1}
Ortho-relation(C, D) = {D2}

Therefore, A is segmented into four pieces, and D is segmented into two pieces. The 2D string is (A4 < A3 < A1 B D1 < A2 C D2, A4 D1 D2 < A3 B C < A1 A2). Given a Chinese character, it can be transformed into the 2D string. We can then match it against the 2D strings of other Chinese characters. By using type-0, type-1 or type-2 matching and a cost algorithm, we can find somewhat similar, partially similar or completely similar Chinese characters.

In a related application, 2D strings have been applied to the retrieval and recognition of handwritten signatures (Sethi and Han, 1995).

6.7. 3D IMAGE DATABASE QUERYING (DEL BIMBO *ET AL.*)

An extension of 2D-strings to deal with 3D imaged scenes was proposed in Del Bimbo *et al.* (1993). The approach relies on the consideration that 2D iconic queries and 2D string-based representations are effective for the retrieval of images representing 2D objects or very thin 3D objects, but they might not allow an exact definition of spatial relationships for images representing scenes with 3D objects. In fact, in this case an incorrect representation of the spatial relationships between objects may result due to two distinct causes. First, 2D icons cannot reproduce scene depth. 2D icon overlapping can be used only to a limited extent since it affects the understandability of the query. Second, as demonstrated by research in experimental and cognitive psychology, the mental processes of human beings simulate physical world processes. Computer-generated line drawings representing 3D objects are regarded by human beings as 3D structures and not as image features, and they imagine spatial transformations directly in 3D space.

Therefore, an unambiguous correspondence is established between the iconic query and image contents, if the spatial relationships referred to are those between the objects in the scene represented in the image, rather than those between the objects in the image. The dimensionality of data structures associated with icons must follow the dimensionality of the objects in the scene represented in the image. A 3D structure should be employed for each icon to describe a 3D scene. An example is illustrated in Figure 6.27. Representations of images are derived considering 3D symbolic projections of objects in the 3D imaged scene. Thirteen distinct operators, corresponding to the interval logic operators, distinguish all the possible relationships between the intervals corresponding to the object projects on each axis.

Retrieval systems employing this ternary representation of symbolic projections have been expounded in Del Bimbo *et al.* (1993, 1994). In these approaches, the user reproduces a 3D scene by placing 3D icons in a virtual space and sets the position of the camera in order to reproduce both the scene and the vantage point from which the picture was taken. A spatial parser

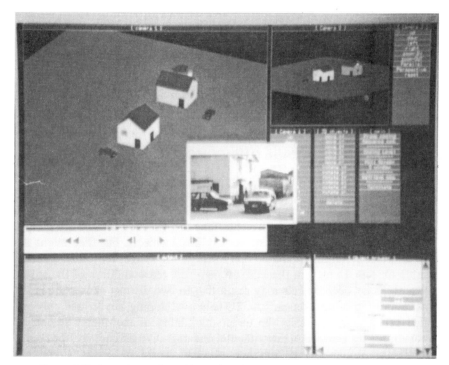

Figure 6.27 Querying a three-dimensional scene using pairwise 3D relations.

translates the visual specification into the representation language and retrieval again is reduced to a matching between symbolic strings.

6.8. MEDICAL IMAGE DATABASE SYSTEM (KOSTOMANOLAKIS *ET AL.*)

2D strings have been used in recognizing fungi in medical research (Dorf *et al.*, 1994). This section describes the incorporation of 2D strings in a medical image database system.

Radiological examinations are extremely important in health care. X-ray film is the medium conventionally used for medical image archival purposes. A picture archiving and communication system (PACS) is a computer system that supports digital image handling in a hospital environment. Facilities typically provided by a PACS include image entry, archiving, communication, presentation, etc. A PACS is connected by high-speed network with the hospital information system (HIS) in order to handle textual patient data together with images.

Such new environments open a whole new world of possibilities for the utilization of medical images in the clinical environment, including computer assisted diagnosis, radiotherapy planning, surgery planning, medical training, etc. Medical image indexing and retrieval by content, in particular, play a special role in this new setting.

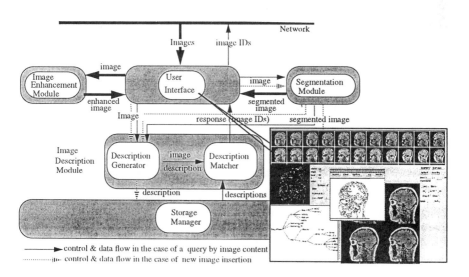

Figure 6.28 The architecture of the I^2C system.

I^2C is an image database system which has been developed as a platform for the design, implementation and evaluation of medical image indexing and retrieval by content schemes (Kostomano *et al.*, 1993). The main concept in the design of I^2C are image classes and image description types. An image class encapsulates algorithms for the organization, processing and indexing of the images in it. When a request for retrieval is placed with I^2C, it is directed to the appropriate class. The concept of the image description type encapsulates all the details of an indexing and retrieval by content scheme, including the use of 2D strings. The architecture of the I^2C system is illustrated in Figure 6.28. This system allows the user to define regions of interest (ROI) on the query image, and adjust the relative importance of different regions as well as their characteristics. The user can draw a sketch, for example, and adjust the search parameters, to direct the image retrieval process.

The issue of designing an active image information system, to monitor user-defined regions of interest (also called hot spots), will be discussed further in Chapter 12.

REFERENCES

Chang, C. C. and Lin, D. C. (1994) Utilizing the concept of longest common subsequence to retrieve similar Chinese characters. *Computer Processing of Chinese and Oriental Languages* **8**, 177–91.

Chang, N. S. and Fu, K. S. (1980) Query-by-pictorial example. *IEEE Transactions on Software Engineering* **6**, 519–24.

Chang, S. K. and Jungert, E. (1990) A spatial knowledge structure for visual information systems. In *Visual Languages and Applications*, ed. T. Ichikawa, E. Jungert and R. R. Korfhage, pp. 277–304. Plenum Press, New York.

Chang, S. K. and Liu, S. H. (1984) Indexing and abstraction techniques for pictorial databases. *IEEE Transactions on Pattern Analysis and Machine Intelligence* **6**, 475–84.

Chang, S. K., Yan, C. W., Arndt, T. and Dimitroff, D. (1988) An intelligent image database system. *IEEE Transactions on Software Engineering, Special Issue on Image Databases*, pp. 681–8.

Chang, S. K., Costabile, M. F. and Levialdi, S. (1992a) A framework for intelligent visual interface design for database systems. *Proceedings of the International Workshop on Interfaces to Database Systems (IDS92)*, pp. 377–391, Glasgow.

Chang, S. K., Lee, C. M. and Dow, C. R. (1992b) A 2D-string matching algorithm for conceptual pictorial queries. *Proceedings of the SPIE/IS&T 1992 Symposium on Electronic Image: Science and Technology, Conference on Image Storage and Retrieval Systems*, pp. 47–58, San Jose, CA.

Del Bimbo, A., Campanai, M. and Nesi, P. (1993) A three-dimensional iconic environment for image database querying. *IEEE Transactions on Software Engineering* **19**, 997–1011.

Del Bimbo, A., Vicarion, E. and Zingoni, D. (1994) A spatial logic for symbolic description of image contents. *Journal of Visual Language and Computing* **5**, 267–286.

Dorf, M. L., Mahler, A. F. and Lehmann, P. F. (1994) Incorporating semantics into 2-D strings. *Proceedings of the 22nd Annual ACM Computer Science Conference*, Phoenix, AZ, ed. D. Cizmar, pp. 110–15.

Fu, K. S. (1982) *Syntactic Pattern Recognition and Applications*. Prentice-Hall, Englewood Cliffs, NJ.

Hildebrand, J. W. and Tang, K. (1992) Symbolic two and three dimensional picture retrieval. *Workshop on Two and Three Dimensional Spatial Data: Representation and Standards*, Australian Pattern Recognition Society, Perth, WA.

Joseph, T. and Cardenas, A. F. (1988) Picquery: A high level query language for pictorial database management. *IEEE Transactions on Software Engineering* **14**, 630–8.

Klinger, A., Rhode, M. L. and To, V. T. (1978) Accessing image data. *International Journal of Policy Analysis Information Systems* **1**, 171–89.

Kostomanolakis, S., Lourakis, M., Chronaki, C. *et al.* (1993) Indexing and retrieval by pictorial content in the I^2C image database system. Technical report, Institute of Computer Science, Foundation for Research and Technology, Heraklion, Greece.

Lee, S.-Y. and Hsu, F.-J. (1992) Spatial reasoning and similarity retrieval of images using 2D C-string knowledge representation. *Pattern Recognition* **25**, 305–18.

Rabitti, F. and Stanchev, P. (1989) GRIM_DBMS: A GRaphical IMage DataBase Management System. *Visual Database Systems*, pp. 415–30. IFIP.

Rossoupoulos, N., Faloutsos, D. and Leifker, D. (1988) An efficient pictorial database system for PQSL. *IEEE Transactions on Software Engineering* **14**, 639–50.

Sethi, I. K. and Han, K. (1995) Use of local structural association for retrieval and recognition of signature images. *SPIE Proceedings of Storage and Retrieval for Image and Video Databases III*, San Jose, CA, ed. W. Niblack and R. C. Jain, Vol. 2420, pp. 125–34.

Shapiro, L. G. and Haralick, R. M. (1979) A spatial data structure. Technical report CS 79005-R, Department of Computing Science, Virginia Polytechnic Institute and State University.

Sties, M., Sanyal, B. and Leist, K. (1976) Organization of object data for an image information system. *Proceedings of the 3rd International Joint Conference on Pattern Recognition*, pp. 863–9.

Sun, Yuguo (1993) PhD thesis, *Description of Topological Spatial Relations and Representation of Spatial Relations Using 2D T-String*, Wuhan Surveying Technology University, Wuhan, China.

Tanimoto, S. L. (1976) An iconic/symbolic data structuring scheme. In *Pattern Recognition and Artificial Intelligence*, pp. 452–71. Academic Press, New York.

PART II

ADVANCED THEORY WITH APPLICATIONS TO SPATIAL REASONING

7

An Image Algebra

The requirements on abstract or high-level data structures for manipulation and transformation of symbolic images normally emerge from both algorithmic and heuristic types of applications. In particular, there is an obvious interest in integration of such data structures both in image database systems and in visual-type database systems. There is also an increasing interest in using qualitative methods as means for both cognitive modelling and iconic indexing. This chapter describes an image algebra which is based on Symbolic Projection including a limited extension of the original set of relational operators. In the algebra a generalized *empty space object* is also introduced, which supports general descriptions and manipulations of images.

The background of the image algebra in section 7.1 is followed by a discussion of the basic operators and the partition technique in section 7.2. The empty space object and the laws of the generalized empty space are introduced in the sections 7.3 and 7.4 respectively. The object manipulation rules are discussed in section 7.5 while section 7.6 is concerned with unification, section 7.7 with overlapping objects, section 7.8 with operator precedence, and, finally, some concluding remarks are made in section 7.9.

7.1. BACKGROUND

Many alternative approaches to pictorial data structures have been developed. Among these hierarchical structures of various kinds such as the quad-tree have been subject to particular interest. However, most such approaches demonstrate a lack of generality, in that they normally do not support reasoning for determination of various types of spatial relations. These structures are for the most part intended for storage of spatial data. However, a few attempts have been made to overcome these problems. Winston (1984) describes an approach called "natural constraints" intended mainly for description of the contours of objects in an image. Winston's method requires much preprocessing, which again is a drawback. Picture grammars (Fu, 1982) can also be employed advantageously in certain applications. Kundu (1988) has developed another hierarchical structure for the description of 2D planes where the tree is of irregular type. However, there is often a need to transform the different representations, so there can be a good match. The image algebra of Jungert and Chang (1989, 1993) avoids many of the drawbacks of other approaches.

7.2. BASIC GLOBAL OPERATORS AND THE PARTITION TECHNIQUE

In Chapter 4, it was shown that "<", "=" and "|" are the three most basic spatial operators. In the image algebra, the same set of relational operators is used. Along with the general set of relational operators, a method to partition the image into smaller sections has been developed, that is the cutting mechanism, which plays an important role in the algebra. Hence, an image is partitioned according to the rules that were introduced in Chapter 4 and then refined into more sophisticated cutting techniques in Chapter 5. However, some obvious consequences will occur when an image is partitioned in this way. For instance, a local extreme point of an object is either a concave or a convex object-point. A line corresponding to a cut partitions all objects that are crossed by that line. The number of cuts parallel to each coordinate axis might not be the same, since a cut is only made orthogonal to the coordinate axis to which the extreme point can be directly projected. The original partitioning technique is illustrated in Figure 7.1.

The 2D-strings corresponding to the partitions made in Figure 7.1 are then generated in accordance with the principles outlined in Chapter 3, using the operator set defined for the generalized projection strings. Thus, from Figure 7.1 the following projection strings can be obtained:

$$u: \quad A < B \mid CB \mid C \mid CC \mid C$$
$$v: \quad C \mid BC \mid ABC \mid AB \mid B$$

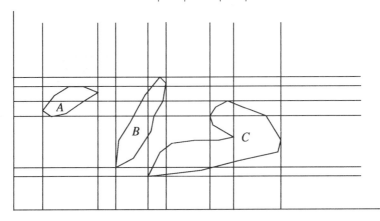

Figure 7.1 Illustrations of the cutting technique used in the image algebra.

When a sequence *ABC* occurs, it means that for the corresponding row (column) there are three pieces of the objects *A*, *B*, and *C* present, in that order. There may or may not be gaps between them. When we write *CC*, it means that in this row (column) there are two pieces of the same object, and there is always a gap between the two subobjects if they are present in the same row or column. This information is useful when, for instance, trying to reconstruct a symbolic picture from given projection strings.

7.3. THE EMPTY SPACE OBJECT

The concept of "empty space" is introduced not only to describe the complete image but also to make it possible to describe space outside the objects of interest. The term "empty space" was first used in this sense by Lozano-Perez (1981) partly for the same purpose as here, i.e. to build up a structure which can be used to describe the space of an image. Lozano-Perez used a quad-tree related structure for this purpose. There are also other examples where the empty space is used, directly or indirectly. The quad-trees (Hunter and Steiglitz, 1979) may also include blank quadrants, which can be part of the background. In Jackins and Tanimoto (1980) oct-trees may have an octant labelled "VOID" (which is empty) that is either transparent or white. However, Lozano-Perez considered the "empty space" to be objects similar to other real objects. This is also the notion adopted in the image algebra.

Besides, the empty space objects, Lozano-Perez introduced FULL and MIXED spaces as well. The empty space outside the objects are solid rectangles representing the free space outside so-called MIXED cells. Here, however, the empty space is divided into smaller areas as well, but not necessarily into rectangles. The shape of the empty space blocks are the result of the partition of the image. Hence, they are linear along the cutting line and along any of the coordinate axes. But when it touches an object, the empty space gets a shape corresponding to the shape of the object. However, when used as a knowledge structure the useful heuristic approach is to regard the empty space objects as rectangles. This was also done by Holmes and Jungert using a somewhat different approach that will be discussed in section 3 of Chapter 10.

An empty space area is defined as:

$$\mathbf{e}_{i,j}^s = [T_{i-1}^s, T_i^s]_j \quad \text{for } i = 1, \ldots, n \quad \text{and} \quad j = 1, \ldots, m$$

where $T_{i-1}^s = x_{i-1}$ and $T_i^s = x_i$ for $s = u$, $T_{i-1}^s = y_{i-1}$ and $T_i^s = y_i$ for $s = v$. x_i is the coordinate of a vertical cutting line, while y_i is the coordinate of a horizontal cutting line, n is the number of cutting lines, and m the number of e-objects in the corresponding interval. Therefore, $\mathbf{e}_{i,j}^s$ refers to the jth piece of empty space between the two cutting lines T_{i-1}^s and T_i^s: $T_0^u = x_0$ corresponds to the y-axis and $T_0^v = y_0$ corresponds to the x-axis. Hence, an empty space object is defined as the interval between two consecutive partition lines (cutting lines)

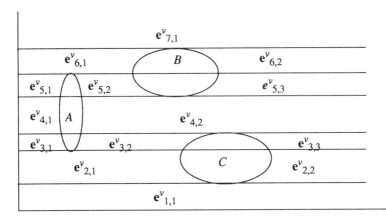

Figure 7.2 The empty space areas corresponding to the projections along the y-axis.

along any of the coordinate axis. The two other sides could be either the coordinate axis and an object or two objects. Figure 7.2 shows three objects A, B and C and the partition of the area between the objects. Each empty space area is considered an object of its own.

The partition in Figure 7.2 gives the following 2D-string along the y-axis:

$$v: \quad \mathbf{e}^v_{1,1} \mid \mathbf{e}^v_{2,1} \, C \, \mathbf{e}^v_{2,2} \mid \mathbf{e}^v_{3,1} \, A \, \mathbf{e}^v_{3,2} \, C \, \mathbf{e}^v_{3,3} \mid \mathbf{e}^v_{4,1} \, A \, \mathbf{e}^v_{4,2} \mid \mathbf{e}^v_{5,1} \, A \, \mathbf{e}^v_{5,2} \, B \, \mathbf{e}^v_{5,3} \mid \mathbf{e}^v_{6,1} \, B \, \mathbf{e}^v_{6,2} \mid \mathbf{e}^v_{7,1}$$

The situation that occur for the u-string projections is analogous and corresponds to Figure 7.3, from which we obtain the following string:

$$u: \quad \mathbf{e}^u_{1,1} \mid \mathbf{e}^u_{2,1} \, A \, \mathbf{e}^u_{2,2} \mid \mathbf{e}^u_{3,1} \mid \mathbf{e}^u_{4,1} \, B \, \mathbf{e}^u_{4,2} \mid \mathbf{e}^u_{5,1} \, C \, \mathbf{e}^u_{5,2} \, B \, \mathbf{e}^u_{5,3} \mid \mathbf{e}^u_{6,1} \, C \, \mathbf{e}^u_{6,2} \mid \mathbf{e}^u_{7,1}$$

Any portion of the free space can be identified by means of set theoretical expressions. This is demonstrated in Figure 7.4(a) and (b). The shaded area **a** in Figure 7.4(a) is calculated from

$$\mathbf{a} = \mathbf{e}^u_{2,2} \cap \mathbf{e}^v_{6,1}$$

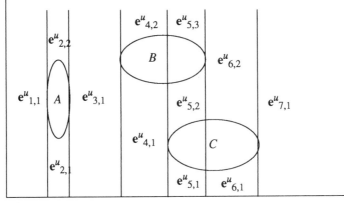

Figure 7.3 The U-string partition of the empty space for Figure 7.2.

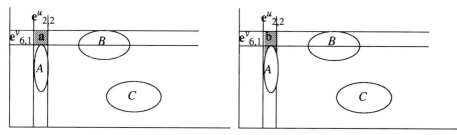

Figure 7.4 Two examples illustrating arbitrary subareas of empty space.

The area **b** in Figure 7.4(b) corresponds to:

$$\mathbf{b} = \mathbf{e}_{2,2}^u \cap ((\mathbf{e}_{4,1}^v \,|\, \mathbf{e}_{5,1}^v \,|\, \mathbf{e}_{6,1}^v) \cup (\mathbf{e}_{4,2}^v \,|\, \mathbf{e}_{5,2}^v))$$

Any empty space defined by the cutting lines can be represented this way. The example from Figure 7.4(b) shows also that the shape of the identified area does not necessarily have to be rectangular.

The strings can also be stripped with respect to the empty objects, e.g.

$$u: \quad A < B \,|\, CB \,|\, C$$
$$v: \quad C \,|\, AC \,|\, A \,|\, AB \,|\, B$$

Hence, the following rule is applicable:

$$s: \langle \text{unstripped string} \rangle \Rightarrow s: \langle \text{stripped string} \rangle$$

The stripping rules will be discussed in detail in the next section.

Empty space objects $\mathbf{e}_{i,j}^u$ can also be generalized such that no differentiation is made between them. Hence, the expressions from Figures 7.2 and 7.3 can be transformed into

$$u: \quad \mathbf{e} \,|\, \mathbf{e}A\mathbf{e} \,|\, \mathbf{e} \,|\, \mathbf{e}B\mathbf{e} \,|\, \mathbf{e}C\mathbf{e}B\mathbf{e} \,|\, \mathbf{e}C\mathbf{e} \,|\, \mathbf{e}$$
$$v: \quad \mathbf{e} \,|\, \mathbf{e}C\mathbf{e} \,|\, \mathbf{e}A\mathbf{e}C\mathbf{e} \,|\, \mathbf{e}A\mathbf{e} \,|\, \mathbf{e}A\mathbf{e}B\mathbf{e} \,|\, \mathbf{e}B\mathbf{e} \,|\, \mathbf{e}$$

where **e** is the generalized empty space object. Hence, **e** indicates empty space either between two objects or between one object and the edge of the image, without considering the size, the shape or the position of the empty space. These 2D strings can, of course, also be stripped, and as will be demonstrated below, the empty space object is an important aspect of the algebra on image manipulation. The empty space object **e** is therefore a natural extension of the language based on the global operator set which was defined in Chapter 4.

7.4. LAWS OF THE GENERALIZED EMPTY SPACE

Several laws apply to 2D-strings that contains the **e**-object. The most basic ones are four which emphasize mostly the effects of stripping. These laws work only in one direction, i.e. they cannot easily be reversed, and are analogous to what was shown about stripping above.

The laws of stripping

(i) $s: \mathbf{e}\,a \Rightarrow s: a$

(ii) $s: \mathbf{e}\,|\,a \Rightarrow s: a$

(iii) $s: a\,|\,\mathbf{e} \Rightarrow s: a$

(iv) $s: a\,\mathbf{e} \Rightarrow s: a$

In the above laws, "$\Rightarrow$" means "replaced by" and a is an arbitrary object not equal to the empty object $\mathbf{e}$. For example, law (i) says that a string $\mathbf{e}\,a$ can be replaced by a, because the empty space object $\mathbf{e}$ can be stripped.

Since $\mathbf{e}$ is a generalized expandable empty space object, expressions of the following type will never be permitted:

$$\mathbf{e} < a \quad \text{or} \quad a < \mathbf{e}$$

because this would imply that there are some other object between $\mathbf{e}$ and a, but such an object can never be stripped off without violating the basic laws of Symbolic Projection.

A further consequence of the $\mathbf{e}$-object is the law

(v) $a\,|\,\mathbf{e}\,|\,b \Leftrightarrow a < b$

where the notation "$\Leftrightarrow$" means that the replacement works in both directions and does not violate what was said about the "$<$" relations above.

Since the $\mathbf{e}$-objects do not correspond to any particular size or shape they can be split or merged in any arbitrary direction. Hence, four further laws that reshape $\mathbf{e}$-objects can be identified.

The laws of reshaping

(vi) $s: \mathbf{e}\,a\,|\,\mathbf{e} \Leftrightarrow s: \mathbf{e}\,(a\,|\,\mathbf{e})$

(vii) $s: a\,\mathbf{e}\,|\,\mathbf{e} \Leftrightarrow s: (a\,|\,\mathbf{e})\,\mathbf{e}$

(viii) $s: \mathbf{e} \Leftrightarrow s: \mathbf{e}\,\mathbf{e}$

(ix) $s: \mathbf{e}\,\mathbf{e} \Leftrightarrow s: \mathbf{e}\,|\,\mathbf{e}$

Some other transformations exist but they are derivable from the above, e.g.

$$s: \mathbf{e}\,\mathbf{e} \Leftrightarrow s: \mathbf{e}\,|\,\mathbf{e}$$

7.5. OBJECT MANIPULATION LAWS

The laws on the generalized e-object can be extended and applied to objects of any kind. The two most basic laws are *distributive laws* for compound objects. These laws are similar to distributive rules in algebra, for example, $ab + ac \Leftrightarrow a(b + c)$. The laws are given in their general form as follows:

(x) $s:$ $a_1 a_2 \ldots a_n \,|\, a_1 a_{n+1} \ldots a_m \,|\, \ldots \,|\, a_1 a_{m+p} \ldots a_k$
  $\Leftrightarrow a_1 (a_2 \ldots a_n \,|\, a_{n+1} \ldots a_m \,|\, \ldots \,|\, a_{m+p} \ldots a_k)$

(xi) $s:$ $a_1 a_2 \ldots a_{n-1} a_n \,|\, a_{n+1} \ldots a_m a_n \,|\, \ldots \,|\, a_{m+p} \ldots a_k a_n$
  $\Leftrightarrow (a_2 \ldots a_{n-1} \,|\, a_{n+1} \ldots a_m \,|\, \ldots \,|\, a_{m+p} \ldots a_k) a_n$

In both laws expressions within parentheses are considered *compound objects*, i.e. objects composed from other objects or subobjects. This assumption is general throughout the image algebra.

The laws (x) and (xi) can be given a visual interpretation which shows the applicability of these laws. For example, according to (x):

$$u: \quad ab \,|\, ac \,|\, ad \Leftrightarrow u: a(b \,|\, c \,|\, d) \tag{7.1}$$

This example is visualized in Figure 7.5.

$$u: ba \,|\, ca \,|\, da \Leftrightarrow u: (b \,|\, c \,|\, d)a \tag{7.2}$$

This is a transformation according to the law (xi). The visual interpretation of this example is similar to the one in Figure 7.5, although in (7.2) the object a is on top of b, c and d, instead of being below them as in Figure 7.5. From this example, it should be clear that the laws (x) and (xi) are applicable under the assumption that the objects forming a compound object share the same cutting lines. Moreover, implicit in both (x) and (xi) is the assumption that

$$(a_1 \,|\, \ldots \,|\, a_1) \Rightarrow a_1 \tag{7.3}$$

Since both laws are bidirectional this will not cause any harm, but taken from a broader perspective (7.3) could imply loss of information, as will be illustrated below.

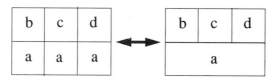

Figure 7.5 Visualization of the transformation in (7.1).

Another law that is in a way a generalization of (x) and (xi) is:

(xii) s: $a_1 \ldots a_{i-1} a_i a_{i+1} \ldots a_m \mid a_{m+1} \ldots a_{i-1+m} a_i a_{i+1+m} \ldots a_{2m} \mid \ldots$
$\mid a_{nm+1} \ldots a_{i-1+nm} a_i a_{i+1+nm} \ldots a_{(n+1)m} \mid$
$\Leftrightarrow (a_1 \ldots a_{i-1} \mid a_{m+1} \ldots a_{i-1+m} \mid \ldots \mid a_{n,m+1} \ldots a_{i-1+nm})$
$a_i(a_{i+1} \ldots a_m \mid a_{i+1+m} \ldots a_{2m} \mid \ldots \mid a_{i+1+nm} \ldots a_{(n+1)m})$

An example of law (xii) is:

$$u: AaC \mid BaC \mid BaD \Leftrightarrow u: (A \mid B \mid B) a (C \mid C \mid D) \qquad (7.4)$$

Here $(A \mid B \mid B)$ can be transformed into $(A \mid B)$ and $(C \mid C \mid D)$ into $(C \mid D)$ but in both these cases information is lost, because

$$u: (A \mid B \mid B) a (C \mid C \mid D) \Rightarrow u: (A \mid B) a (C \mid D) \Rightarrow u: AaC \mid BaD \neq AaC \mid BaC \mid BaD$$
$$(7.5)$$

This is a result that obviously does not correspond to the original expression in (7.4). Hence there is a serious risk of losing information if (7.3) is applied. This is especially apparent in (xii). The obvious and visual transformation of the expression in (7.4) can be seen in Figure 7.6.

The next law to be introduced is (xiii), which is a simplification of rule (xii):

(xiii) $(a_1 \mid \ldots \mid a_i \mid \ldots \mid a_n)(b_1 \mid \ldots \mid b_i \mid \ldots \mid b_n) \Leftrightarrow a_1 b_1 \mid \ldots \mid a_i b_i \mid \ldots \mid a_n b_n$

This law can be illustrated by the following example:

$$u: (A \mid A \mid C)(D \mid B \mid B) \Leftrightarrow u: AD \mid AB \mid CB \qquad (7.6)$$

What actually happens in law (xiii) is that the interpretation of the image is changed from a row view into a column view. This aspect will be discussed further in section 7.7. However, the law (xiii) can be generalized further. When the law is concerned with only a pair of rows where the compound object, including the subobjects $a_1, \ldots, a_n$, corresponds to the first row, while $b_1, \ldots, b_n$ corresponds to the second. This limitation is overcome, as can be seen, in law (xiv) below:

(xiv) $a_{11} a_{12} \ldots a_{1n} \mid \ldots \mid a_{i1} a_{i2} \ldots a_{in} \mid \ldots \mid a_{n1} a_{n2} \ldots a_{nn}$
$\Leftrightarrow (a_{11} \mid \ldots \mid a_{i1} \mid \ldots \mid a_{n1})(a_{12} \mid \ldots \mid a_{i2} \mid \ldots \mid a_{n2}) \ldots$
$(a_{1n} \mid \ldots \mid a_{in} \mid \ldots \mid a_{nn})$

The example in (7.7) shows a special case of law (xiv), where the risk of losing information is apparent as well:

$$u: \quad (A \mid A \mid B \mid B \mid C)(F \mid G \mid G \mid H \mid H) \Leftrightarrow AF \mid AG \mid BG \mid BH \mid CH \qquad (7.7)$$

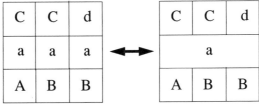

Figure 7.6 Visual interpretation of the transformation in (7.4).

In the above, only the u-strings have been considered. It is, however, clear from the general representation of the laws that projections along both axes are applicable. Furthermore, in the examples and in the laws the assumptions have been made that the transformations can be carried out without taking information from the orthogonal object string into account. This is, however, not always the case. Assume, for instance, the cases in (7.8) and (7.9):

$$u: \quad AB \mid A \tag{7.8}$$

$$v: \quad A \mid BA \tag{7.9}$$

According to law (x) the u-string can be rewritten as

$$u: \quad A(B \mid) \tag{7.10}$$

But this does not look appropriate, because the meaning of $(B \mid)$ is not defined in the algebra and by investigating the v-string it becomes more likely that

$$u: \quad A(B \mid A) \tag{7.11}$$

because it can easily be shown that A has the shape of a rotated letter "L".

There are two additional laws that have already been mentioned in the above discussion but have not explicitly been pointed out. These laws are related to the reshaping laws that are applied to the empty space laws, thus:

(xv) $s\!: a \mid \ldots \mid a \Rightarrow s\!: a$

(xvi) $s\!: a \ldots a \Rightarrow s\!: a$

As pointed out earlier the elimination of the "|" operator may lead to loss of information, therefore law (xv) has to be handled with care. The inverse of both rules is not completely clear, since they will require creation of new partition lines.

7.6. UNIFICATION

Earlier it was demonstrated how loss of information in algebraic expressions can occur and thus may cause serious problems. However, there is one aspect of the algebra that can contribute to the solution of this problem. This aspect is called *unification* since it contributes to the transformation of all different orthogonal strings into a single unified string along an arbitrarily chosen coordinate axis. The unified string comprises all available information. Unification is illustrated as follows. From Figure 7.6 the following algebraic strings can be created:

$$u: \quad AaC \mid BaC \mid BaD \tag{7.12}$$

$$v: \quad AB \mid a \mid CD \tag{7.13}$$

If law (xiv) is applied to the u-string in (7.12) above, it will be transformed into

$$u: \quad (A\,|\,B\,|\,B)(a\,|\,a\,|\,a)(C\,|\,C\,|\,D) \tag{7.14}$$

This expression is equivalent to

$$u: \quad (A\,|\,B\,|\,B) = (a\,|\,a\,|\,a) = (C\,|\,C\,|\,D) \tag{7.15}$$

As mentioned already in Section 7.4, the interpretation of this transformation is that a column structure, as in (7.12), is transformed into a row structure (7.15). Now, since a row in the u-string is equivalent to a column in the v-string then, for instance, the first row $u(1)$ corresponds to the first column $v(1)$:

$$u(1): \quad A\,|\,B\,|\,B \Leftrightarrow v(1): AB \tag{7.16}$$

Hence, the v-string is actually equal to ABB although it includes less information than the u-string. Nevertheless it is easy to see that a row in the u-string can be transformed into a column of a v-string by simply interchanging the "|" operators with "=" operators and vice versa. Now, consider the relationships between the rows in the u-string and the relationships between the columns of a v-string. From this it follows that the "=" relations of the u-string correspond to the "|" relations in the v-string. Consequently, a row-oriented u-string can be transformed into a v-string by just substituting the "=" operators with "|" operators and the "|" operators with "=" operators. Relation (7.14) thus turns into (7.17), where u' corresponds to a string of v-type:

$$u': \quad ABB\,|\,aaa\,|\,CCD \tag{7.17}$$

Observe that the transformed string in (7.17) actually contains more information than the original v-string (7.13). Another illustration of this is found in example (7.8).

$$u: \quad AB\,|\,A \Rightarrow u': (A\,|\,B)A \Rightarrow u': AA\,|\,BA \tag{7.18}$$

In (7.18) there is no need to transform the u-string into a row-oriented expression before it is unified. A direct transformation is permitted, and here as well the new v-string contains more information than the original. Apparently, there is now enough information available to identify the L-shaped object in the image by just looking at the new v-string.

Again, consider Figure 7.6, and the strings given in (7.4) and (7.5):

$$u: \quad AaC\,|\,BaC\,|\,BaD \tag{7.19}$$

$$v: \quad AB\,|\,a\,|\,CD \tag{7.20}$$

From (7.19) the corresponding v-string transformation is

$$u': \quad (A\,|\,a\,|\,C)(B\,|\,a\,|\,C)(B\,|\,a\,|\,D) \Rightarrow ABB\,|\,a\,|\,CCD \tag{7.21}$$

and from (7.20) the transformed v-string becomes:

$$v': \quad (A\,|\,B)a(C\,|\,D) \Rightarrow AaC\,|\,BaD \tag{7.22}$$

Expression (7.21) contains more information from (7.20) while expression (7.22) contains less than (7.19). The string containing most information is the one with the largest number of subobjects and that information is kept in the unification step. In the example above, the string containing the most information is the string corresponding to (7.19).

Now, the main question is when should it be permissible to unify a given image along an arbitrary coordinate axis. It is not sufficient to say that it should be done from the string corresponding to axis that contains most information. The answer is that *unification* can be done in any direction if and only if all cutting lines parallel to both the *x*- and the *y*-axes are considered. This way the image can be looked at as a set of "superpixels" subsequently called *frames*. In a sense, this is a limitation of the method, since strings minimized with respect to the number of subobjects are not permitted if all information should be retained.

7.7. OVERLAPPING OBJECTS

In the majority of the examples considered so far only the **MBR** of the objects have been used. In many cases such rectangles will overlap, hence there must be a way to deal with the projection strings for such rectangular objects. An example of two overlapping rectangles or objects, *A* and *B*, is given in Figure 7.7.

So far, we have no way to express overlap in the projection strings. A way of dealing with this problem in the symbolic algebraic expressions is:

$$u: \quad A \mid B\{BA\}A \mid B \tag{7.23}$$

$$v: \quad B \mid A\{AB\}B \mid A \tag{7.24}$$

Here, the overlapping portions of the rectangles *A* and *B* are denoted by $\{AB\}$ or $\{BA\}$. The order of the objects or subobjects that are situated inside the parentheses is irrelevant.

The above strings are equivalent to stripped strings which cannot be unified, according to the observations made in section 7.6, since at least two frames are excluded in the expressions. The excluded frames are all part of the empty space. Hence, the empty space that surrounds the objects must be considered in the

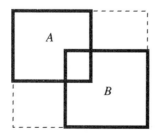

Figure 7.7 An illustration of two overlapping rectangles.

expressions to allow unification. Taking the empty space into consideration the
expressions (7.23) and (7.24) are transformed into:

$$u: \quad \mathbf{e}AA \mid B\{BA\}A \mid BB\mathbf{e} \tag{7.25}$$

$$v: \quad \mathbf{e}BB \mid A\{AB\}B \mid AA\mathbf{e} \tag{7.26}$$

The part of the empty space in these two expressions which is taken into
consideration is inside the dashed rectangle in Figure 7.7. The use of the partial
empty space here is just for reasons of simplicity, but obviously in certain
applications the use of a partial empty space could be motivated for efficiency
reasons, for example.

It is easy to show that unification of both (7.25) and (7.26) works here as well
and in particular for (7.25), the resulting unified string becomes:

$$u': \quad (\mathbf{e} \mid A \mid A)(B \mid \{BA\} \mid A)(B \mid B \mid \mathbf{e}) \tag{7.27}$$

A further derivation yields

$$u': \quad (\mathbf{e} \mid A \mid A)(B \mid \{BA\} \mid A)(B \mid B \mid \mathbf{e}) \Rightarrow \mathbf{e}BB \mid A\{BA\}B \mid AA\mathbf{e} \tag{7.28}$$

As can be seen, $\{BA\}$ is handled as a single subobject and, as a whole, the strings
behave in the same manner as was already pointed out in the section on
unification. A further implication concerning the "overlap object" is that a
cutting must take place along its boundaries otherwise the projection strings will
not become complete, in other words they will lack information. This can be
seen in (7.23) and (7.24) where parts of both A and B include frames that have
been merged.

7.8. OPERATOR PRECEDENCE

In conventional algebra, precedence among operators is specified rigorously.
Precedence has to be considered here as well, although the situation in this
symbolic algebra is somewhat different than for conventional algebra. The main
difference between the symbolic and the conventional algebra is that here we are
talking about objects, which have spatial extension, and their relations in a
space that is at least two-dimensional.

The discussion is first concerned with expressions of the type given in section
7.6 on unification, illustrated, for instance, in expression (7.26). Precedence is
independent of the direction of the actual coordinate axis along which the
objects have been projected. Therefore, the following discussion concerns any
arbitrary string, i.e. the s-string. In this discussion the "$=$" operator, which
normally is dropped, must be present to clarify all aspects of the operator
precedence in the image algebra.

$$s: \quad (A \mid B \mid B) = (a \mid a \mid a) = (C \mid C \mid D) \tag{7.29}$$

In expression (7.29) the subexpressions within parentheses can be considered
as units, which correspond to rows parallel to the projection axis. Hence,

precedence in (7.29) depends more on the parentheses and less on the operators. However, if the expression is unified with respect to some other projection direction, it becomes:

$$s': \quad A = B = B\,|\,a = a = a\,|\,C = C = D \tag{7.30}$$

Here the units are of the type:

$$X_1 = X_2 \ldots = X_n \tag{7.31}$$

In these units the ties between the subobjects are stronger than in units of the type:

$$Y_1\,|\,Y_2\,|\, \ldots \,|\,Y_n \tag{7.32}$$

This is true except when parentheses are present, as in (7.29). As reflected in (7.31) and (7.32), the "=" operator has higher precedence than the "|" operator. Observe that the parentheses in (7.29) cannot be removed, because this would change the semantic meaning of the expression.

The units which are discussed above are either columns (7.31) or rows (7.32) with respect to the direction of the projection. Hence, they are not objects in the ordinary sense; instead they are a collection of subobjects of various types, i.e. frames. A consequence of this discussion is that in (7.29) rows are units within parenthesis built up by subobjects separated by "|" operators. Columns are units built up by subobjects separated by "=" operators.

7.9. REMARKS

The main aspects of the described symbolic image algebra, intended for manipulation and transformation of objects present in images, is based on symbolic projection and contains:

- a set of three elementary and powerful relational operators
- generalized empty space objects
- means for description of multi-level data structures (see Chapter 8).

From these fundamental aspects, a number of algebraic laws have been defined which make it possible to manipulate and transform symbolic images in arbitrary ways. The technique can be used to create a knowledge structure and also for the development of graph structures (see Chapter 8) which can be applied to heuristic navigation (see Chapter 10). Hence, the algebra can serve as a framework for symbolic image processing.

A further important application for the image algebra is qualitative spatial reasoning: for example, to plan for the shortest path between two arbitrary points in a map of free space and obstacles. The image algebra can be applied to convert symbolic projections of a map into a connectivity graph in a way that is similar to the technique of Holmes and Jungert which will be discussed in Chapter 10. Another application would be to find out the degree of similarity between two images or objects present in the images. The images must in that case first be transformed and represented in terms of symbolic projections: then the laws of the image algebra can be applied and string matching techniques can subsequently be performed to determine the similarities in the images or between the objects.

REFERENCES

Fu, K. S. (1982) *Syntactic Methods in Pattern Recognition*. Academic Press, New York.
Hunter, G. M. and Steiglitz, K. (1979) Linear transformation of pictures represented by quad-trees. *Computer Graphics and Image Processing* **10**, 289–96.
Jackins, C. L. and Tanimoto, S. L. (1980) Oct-trees and their use in representing three-dimensional objects. *Computer Graphics and Image Processing* **14**, 249–70.
Jungert, E. (1989) Symbolic expressions within a spatial algebra: unification and impact upon spatial reasoning. *Proceedings of the IEEE Workshop on Visual Languages*, 157–162. Rome.
Jungert, E. and Chang, S. K. (1989) An algebra for symbolic image manipulation and transformation. In *Visual Database Systems*, ed. T. L. Kunii, pp. 301–17. North-Holland, Amsterdam.
Jungert, E. and Chang, S. K. (1993) An image algebra for pictorial data manipulation. *Computer Vision, Graphics and Image Processing: Image Understanding* **58**, 147–60.
Kundu, S. (1988) The equivalence of the subregion representation and the wall representation for a certain class of rectangular dissections. *Communications of the ACM* **31**, 752–63.
Lozano-Perez, T. (1981) Automatic planning of manipulator transfer movements. *IEEE Transactions on Systems, Man and Cybernetics* **11**, 681–98.
Winston, P. H. (1984) *Artificial Intelligence*. Addison-Wesley, Reading, MA.

8

Transformations

Means for transforming of various types of projection strings into other types of projection strings are examples of useful operations. Generation of various types of hierarchical string types, as well as various types of rotations and transformations – for instance, the transformation of projection strings into a type of connectivity graph – are other examples. It will be shown in this chapter that it is not always possible to transform projection strings of one type into another. However, there exists a number of useful transformation types of which hierarchies and rotations are among the most important.

The descriptions of projection string hierarchies in section 8.1 are followed by the introduction of the generalized 2D string type and those local operators that can be applied to them in section 8.2. Rules for transformation of algebraic string expressions into connectivity graphs are introduced in section 8.3. Section 8.4 contains descriptions of two different string rotation methods. The problem of rotation invariance is introduced in section 8.5, which also presents some outlines to its solution. A technique for compacting algebraic strings is illustrated in section 8.6.

8.1. HIERARCHIES

In pictorial retrieval, queries concerned with object hierarchies are common. So far, two alternative hierarchical approaches to pictorial retrieval based on symbolic projections have been described in the literature. The first one uses a set of spatial operators and is proposed by Chang *et al.* (1990) and Chang and Jungert (1991). The second approach is based on the image algebra that was discussed in Chapter 7 and by Jungert and Chang (1989, 1993).

8.1.1. Hierarchical spatial operators

An augmentation of the operator set $\{=, <, |\}$ which is used for general string representation can be used to describe object hierarchies. This is done by splitting and extending the meaning of the edge-to-edge operator. Thus, three

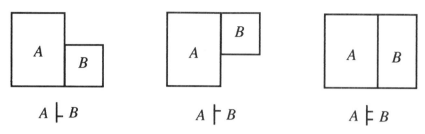

$$A \vdash B \qquad\qquad A \vdash B \qquad\qquad A \vDash B$$

Figure 8.1 Refined edge-to-edge operators.

operators, illustrated in Figure 8.1, have been described to accomplish descriptions of object hierarchies. This subset of operators is called the *refined edge-to-edge operator set* and was introduced in section 5.2.1.

By using the refined set of edge-to-edge operators the original $\langle u, v \rangle$ strings of a symbolic picture can be extended so that they can represent more complex spatial relations among picture objects.

The augmented or refined operators in Figure 8.1 can now be used to define a larger set of hierarchical spatial operators, so that a symbolic picture can be represented hierarchically. In a way, this combines the quad-tree representation and the 2D string representation. The set

$$\Downarrow = \{\downarrow, \vdash, \vdash, \vDash\} \tag{8.1}$$

is defined as the *down-level operators*. The down-level operators define the local spatial relations among picture objects in terms of subdividing picture blocks into quadrants. Similarly, the *up-level* operators can be defined as:

$$\Uparrow = \{\uparrow, \dashv, \dashv, \dashv\} \tag{8.2}$$

The up-level operators and the down-level operators must be used as pairs. Therefore, the down-level operators are similar to left parentheses, and the up-level operators are similar to right parentheses. $\downarrow$, $\vdash$, $\vdash$ and $\vDash$ have the opposite meaning to $\uparrow$, $\dashv$, $\dashv$ and $\dashv$, respectively. Thus the operators in the two sets are pairwise equivalent. When subdividing a picture block into quadrants, 16 different pairs of down-level and up-level operators can be obtained as illustrated in Figure 8.2.

The last operator pair in Figure 8.2 corresponds to the case where no further subdivision can be made. An illustration of this hierarchical spatial operators on a symbolic picture P which results in a symbolic string representation called a *2D H-string* is shown in Figure 8.3.

The four quadrants are represented by Q_1 (up-left), Q_2 (down-left), Q_3 (up-right) and Q_4 (down-right). The operator pairs in Figure 8.2 can now be used to encode a picture recursively. The picture P is first encoded as $\Downarrow Q_1 Q_2 Q_3 Q_4 \Uparrow$, and then each quadrant is encoded in the same manner. The encoded 2D H-string of P then becomes:

$$\text{2D H}(P) = \vDash \vDash DC \vdash f \uparrow \dashv \vDash BA \uparrow W \vdash \downarrow a \dashv \vdash bdc \dashv E \dashv \dashv \tag{8.3}$$

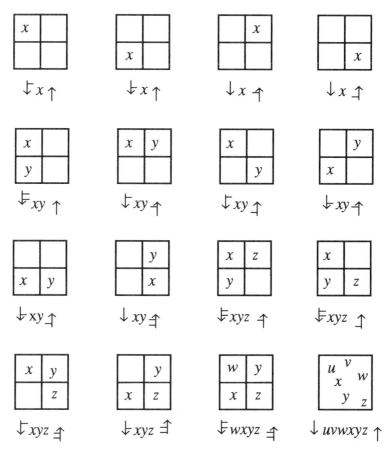

Figure 8.2 Sixteen pairs of down-level and up-level operators.

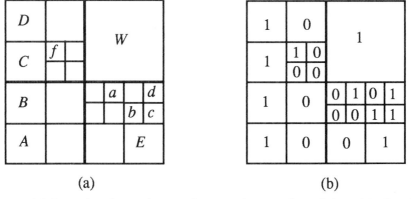

(a)   (b)

Figure 8.3 Example of two images that can be transformed into 2D H-string representation.

Similarly, the binary picture P shown in Figure 8.3(b) can be represented by:

$$2D \ H(P) = \; \models \; \models 11 \; \curlyvee 1 \uparrow \dashv \models 11 \uparrow 1 \curlyvee \downarrow 1 \dashv \curlyvee 111 \dashv 1 \dashv \dashv \qquad (8.4)$$

8.1.2. The algebraic multilevel structure

Consider the symbolic image in Figure 8.4(a), from which the following 2D-string along the x-axis can be obtained:

$$u: \quad AF \mid BGH \mid CGI \mid DK \qquad (8.5)$$

This string can then, according to law (xii), (see Chapter 7) be transformed into

$$u: \quad AF \mid (B \mid C)G(H \mid I) \mid DK \qquad (8.6)$$

and finally through law (xiv) it becomes

$$u: \quad (A \mid (B \mid C) \mid D)(F \mid G(H \mid I) \mid K) \qquad (8.7)$$

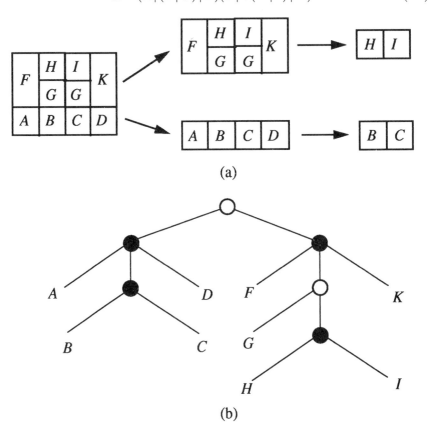

(a)

(b)

Figure 8.4 (a) A compound object; (b) Example of algebraic compound object degeneration.

In this example it is easy to see that in the final expression a hierarchical structure is obtained; this is also illustrated in Figure 8.4(b). Using this technique to determine compound objects and their hierarchical structures, any multilevel structure can be obtained. The multilevel approach is illustrated in the example in Figure 8.5, where the image is transformed into a structure analogous to a quad-tree. The view of the quad-tree itself can easily be obtained from the figure on the right in Figure 8.6. The 2D-strings of the image have the basic representation:

$$u: \quad Ce \,|\, CeB \,|\, eB \,|\, e \,|\, De \,|\, De\,A \tag{8.8}$$

$$v: \quad CeD \,|\, Ce \,|\, e \,|\, eBe\,A \tag{8.9}$$

The structures in (8.8) and (8.9) do not include any subobject levels. Thus in accordance with the usual technique for quad-tree construction the image is first split into four subrectangles of equal size, then in the next step the subrectangles are split again in the same way if required and so on. The procedure when transforming the 2D-strings is similar, except that at each division new compound objects are created on lower levels in the multilevel structure.

The first division step is

$$u: \quad (u_1 \,|\, u_2) = u\!: \;(Ce \,|\, Ce\,B \,|\, e\,B) \,|\, (e \,|\, De \,|\, De\,A) \tag{8.10}$$

$$v: \quad (v_1 \,|\, v_2) = v\!: \;(CeD \,|\, Ce \,|\, e) \,|\, (e \,|\, eBe\,A) \tag{8.11}$$

where

$$u_1 = (Ce \,|\, Ce\,B \,|\, e\,B) \tag{8.12}$$

$$u_2 = (e \,|\, De \,|\, De\,A) \tag{8.13}$$

$$v_1 = (CeD \,|\, Ce \,|\, e) \tag{8.14}$$

$$v_2 = (e \,|\, eBe\,A) \tag{8.15}$$

This first step is fairly straightforward, although it is obvious that the generation of this multilevel structure requires some kind of low-level search while the program is performing the transformation.

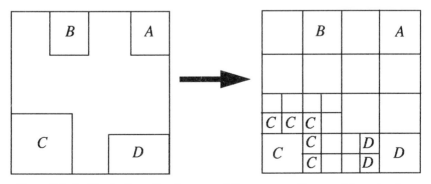

Figure 8.5 An illustration of an image and its corresponding quad-tree structure.

The second division step is

$$(u_1 \mid u_2) = ((u_{11} \mid u_{12}) \mid (u_{21} \mid u_{22})) = (((Ce) \mid (CeB \mid eB)) \mid ((e \mid De) \mid (DeA)))$$
$$(8.16)$$

$$(v_1 \mid v_2) = ((v_{11} \mid v_{12}) \mid (v_{21} \mid v_{22})) = (((CeD) \mid (Ce \mid e)) \mid ((e) \mid eBeA))) \qquad (8.17)$$

The third division step is

$$((u_{11} \mid u_{12}) \mid (u_{21} \mid u_{22})) = (((u_{111} \mid u_{112}) \mid (u_{121} \mid u_{122})) \mid ((u_{211} \mid u_{212}) \mid u_{22}))$$
$$= ((((Ce) \mid (Ce)) \mid ((CeB) \mid eB))) \mid (((e) \mid (De)) \mid (DeA)))$$
$$(8.18)$$

$$((v_{11} \mid v_{12}) \mid v_{21} \mid v_{22})) = ((v_{11} \mid v_{121} \mid v_{122}) \mid (v_{21} \mid v_{22}))$$
$$= (((CeD) \mid ((Ce) \mid (e))) \mid ((e) \mid (eBeA)))$$
$$(8.19)$$

Note that u_{22}, v_{11}, v_{21} and v_{22} need not be divided further after the second step.

8.2. GENERALIZED 2D STRINGS AND LOCAL OPERATORS

As stated earlier, local operators cannot be used to represent all the spatial relationships in one or two projection strings, although they are well suited for representing binary spatial relations between two picture objects. On the other hand, we can express the information represented by local operators using generalized 2D strings. This means that generalized 2D strings can be used to represent any kind of information that can be represented by local operators. In Table 8.1 all corresponding relations between local and generalized projection strings are listed, i.e. the v-string is either $a \mid b$ or $a < b$, because the two objects a and b may or may not be touching one another. Observe, also, the similarities between the local relations and the relations of Allen mentioned earlier.

Local relations can also be represented in another way, as demonstrated in Figure 8.6 where, for instance, pairs of objects can be regarded as subpictures from which substrings can be generated. The symbols "⟨" and "⟩" are used to define an area of interest, and describe the hierarchical structure of a picture. The object α composed from the objects a and b is illustrated in Figure 8.6 by the dashed rectangle. The sizes of the rectangles are of no importance, as long as they cover the intended objects. Thus from the image in Figure 8.6 the string pair

$$u: \quad \langle \alpha{:}\,ba \rangle \mid A \langle \alpha{:}\,b \rangle \mid \langle \alpha{:}\,b \rangle \qquad (8.20)$$

$$v: \quad A < \langle \alpha{:}\,b \mid a \rangle \qquad (8.21)$$

can be created.

Table 8.1

L_{op}	GPS	Pattern
$a \models b$	$U: a = b \, / \, b$ $V: a \, / \, b \lor a < b$	
$a \mathrel{\dashv\!\models} b$	$U: b = a \, / \, b$ $V: b \, / \, a \lor b < a$	
$a \dashv b$	$U: a \, / \, b = a$ $V: b \, / \, a \lor b < a$	
$a \dashv b$	$U: a \, / \, a = b$ $V: b \, / \, a \lor b < a$	
$a \backslash b$	$U: a \, / \, b = a \, / \, b$ $V: b \, / \, a \lor b < a$	
$a \, / \, b$	$U: a \, / \, a = b \, / \, b$ $V: a \, / \, b \lor a < b$	
$a \approx b$	$U: a \, / \, b = a \, / \, a$ $V: b \, / \, a \lor b < a$	
$b \approx a$	$U: a \, / \, a = b \, / \, a$ $V: a \, / \, b \lor a < b$	

This generalized projection string pair can be simplified to

$$u: \quad \alpha \, | \, A\alpha \, | \, \alpha \tag{8.22}$$

$$v: \quad A < \alpha \tag{8.23}$$

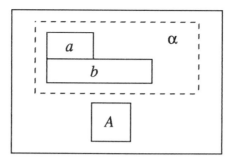

Figure 8.6 An example of local relations.

where the object α is defined at the next level by

$$\alpha.u: \quad ba \,|\, a \tag{8.24}$$

$$\alpha.u: \quad b \,|\, a \tag{8.25}$$

This string pair can be transformed and represented in terms of the local operators (L_{op}) given in Table 8.1, since they correspond to the general projection string type.

The generalized 2D string representation therefore supports hierarchical structuring of pictorial information. At the lowest level, or innermost bracket pairs, the local relations are stored. Such local relations can be used to support local spatial reasoning.

8.3. TRANSFORMATIONS OF ALGEBRAIC EXPRESSIONS INTO A TILE GRAPH

Holmes and Jungert (1992) showed that it is possible to generate a connectivity graph made up of nodes of empty space, where the pieces of empty space corresponding to the nodes are called *tiles*. The graph itself is part of a knowledge structure, which can be used for navigation or planning of the shortest path between two arbitrary points in a digitized map consisting of free space and obstacles, which is further discussed in Chapter 10. Figure 8.7 illustrates a simple image and the resulting tile structure. The objects A, B and C correspond to obstacles while the areas with the numbers 1 to 10 correspond to the tiles. As can be seen, the lines that correspond to the horizontal edges of the tiles are a subset of those lines in Figure 7.3, which creates $\mathbf{e}_{i,j}^{v}$ objects. It can be concluded that, since a tile graph can be generated from Figure 8.7, a tile graph can also be generated from the type of structure that was shown in Figure 7.3. However, the general structure of the tile graph is such that a node has at most four neighbors which are connected by arcs. The tile graph in Figure 8.8 corresponds to the image in Figure 8.7.

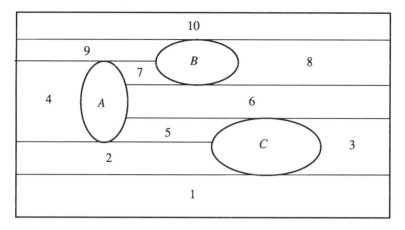

Figure 8.7 The tile structure of an image with three obstacles and free space.

A closer look at Figures 7.3 and 8.8 shows that, for instance, tile 4 corresponds to the expression:

$$\mathbf{e}^v_{3,1} \mid \mathbf{e}^v_{4,1} \mid \mathbf{e}^v_{5,1} \qquad (8.26)$$

tile 7 corresponds to $\mathbf{e}^v_{5,2}$, and so on.

It is easy to see that three categories of tiles exist, i.e. tiles between:

- the edge of the image and an object
- two objects
- the left and right edge of the image.

A tile corresponds generally to more than a single empty space object. This is true for the two first categories. This perspective must be considered when the rules for tile generation are defined.

For $k = 1, \dots, n$, the tiles t_k are generated according to one of the following four rules:

(i) $\mathbf{e}^s_{i,j} \mid$ or $\mid \mathbf{e}^s_{i,j} \mid$ or $\mid \mathbf{e}^s_{i,j} \Rightarrow t_k = \mathbf{e}^s_{i,j}$

(ii) $\mathbf{e}^s_{i,j} a \dots \mid \mathbf{e}^s_{i+1,j'} a \dots \mid \dots \Rightarrow t_k = \mathbf{e}^s_{i,j} \mid \mathbf{e}^s_{i+1,j'} \mid \dots$

(iii) $\dots a \, \mathbf{e}^s_{i,j} \mid \dots a \mathbf{e}^s_{i+1,j'} \mid \dots \Rightarrow t_k = \mathbf{e}^s_{i,j} \mid \mathbf{e}^s_{i+1,j'} \mid \dots$

(iv) $\dots \mid \dots a_1 \mathbf{e}^s_{i,j} a_2 \dots \mid \dots a_1 \mathbf{e}^s_{i+1,j} a_2 \dots \mid \dots \Rightarrow t_k = \mathbf{e}^s_{i,j} \mid \mathbf{e}^s_{i+1} \mid \dots$

where a, a_1, a_2 are obstacles. From rule (i) the generated tile will cross the entire image. Rules (ii) and (iii) correspond to tiles delimited by the edge of the image and an object, while finally rule (iv) results in tiles between two objects. Generation of the tiles takes place in a process that is similar to a scan of the images and goes on row by row. A row in this context occurs between two cutting lines or between the edge of the image and a cutting line when at the start or end of the image.

In rules (ii) and (iii), the 2D strings can be transformed in accordance with law

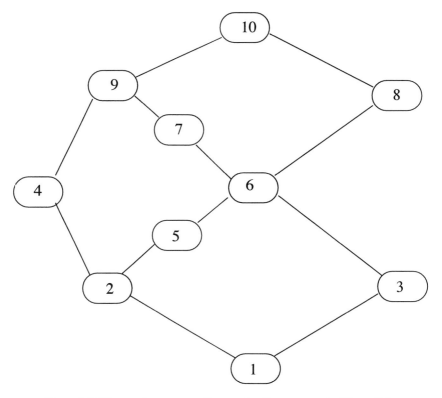

Figure 8.8 Tile graph corresponding to the tile structure in Figure 8.8.

(xii) of Chapter 7, i.e.

(ii′) $(\mathbf{e}^s_{i,j} \,|\, \mathbf{e}^s_{i+1,j'} |\, \ldots\,)a(\ldots)$

(iii′) $(\ldots)a(\mathbf{e}^s_{i,j} \,|\, \mathbf{e}^s_{i+1,j'} |\, \ldots\,)$

In (ii′) the first subexpression corresponds directly to the tile to be created and similarly the last subexpression in (iii′) to the tile.

8.4. ROTATIONS

8.4.1. Entity rotations (Hirakawa and Jungert)

For a single image pattern, which normally corresponds to an object, there exist several patterns which may have different orientation but are structurally the same. For example, consider the object in Figure 8.9(a). As can be imagined, seven different but relevant entities which exactly match the original object exist as shown in Figure 8.9(b). Applications exist where it is necessary to be able to retrieve images including any of these alternative orientations in order to identify the object somehow. This type of retrieval is normally part of a matching process. Several approaches to solving this problem can be identified. Here an object rotation approach that is to rotate the object "back" to the original pattern position will be discussed; this was originally proposed by Hirakawa and Jungert (1991).

Three basic operations can be defined to obtain the strings of all relevant images which can be associated with a given input image. The basic operations are called *u–v* transposition, unit-level reversion, and symbol-level reversion. The three operators are defined as follows.

***u–v* transposition.** This operation transforms the *u*-string expression of an image to its *v*-string expression and is denoted by "→".
 The *u*-string

$$u: \quad eee \mid e\#\# \mid ee\# \qquad (8.27)$$

corresponds to a simple image, shown in the example below, where the **e** elements correspond to the empty space while the **#** elements corresponds to any arbitrary object type. Thus a *u–v* transformation results in a new *u*-string:

$$u: \quad eee \mid e\#e \mid e\#\# \qquad (8.28)$$

This new *u*-string is equal to the *v*-string and vice versa. Hence, *u–v* transposition is analogous to a matrix transposition along the diagonal going from the lower left corner to the upper right corner.

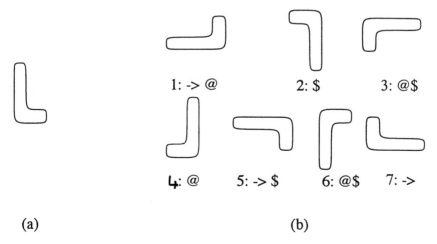

(a) (b)

Figure 8.9 Rotation of an entity, with the original orientation in (a), and the alternative orientations in (b) including their corresponding rotation operators.

$$v \uparrow \begin{array}{ccc} \mathbf{e} \# \# \\ \mathbf{e} \# \mathbf{e} \\ \mathbf{e} \ \mathbf{e} \ \mathbf{e} \end{array}$$
$$\xrightarrow{\quad} u$$

Unit-level reversion. This operation reverses the order of all substrings of unit type, where a unit represents a substring surrounded by edge-to-edge operators. This operation is denoted by "@".

An example of a unit-level reversion is

$$u: \quad \mathbf{eee} \,|\, \mathbf{e} \# \# \,|\, \mathbf{ee} \# \Rightarrow u: \quad \mathbf{ee} \# \,|\, \mathbf{e} \# \# \,|\, \mathbf{eee} \tag{8.29}$$

Symbol-level reversion. Symbol-level reversion reverses the order of all the symbols in a string, an operation that is denoted by "$".

Symbol-level reversion applied to the example from above gives

$$u: \quad \mathbf{eee} \,|\, \mathbf{e} \# \# \,|\, \mathbf{ee} \# \Rightarrow u: \quad \# \mathbf{ee} \,|\, \# \# \mathbf{e} \,|\, \mathbf{eee} \tag{8.30}$$

These three operations can be interpreted as actions of manipulating a sheet, as illustrated in Figure 8.10. It is assumed that the sheet is transparent and an image is drawn on it.

The u–v transposition operation is interpreted as follows. First rotate the sheet 90° counter-clockwise and then turn the front surface of the sheet back. This is equivalent to rotating the sheet half a revolution around the A–C diagonal. In practice, the u- and v-strings are obtained from the image and then interchanged. Furthermore, the view of the image after the transposition is seen from the back of the sheet. Applied to the object in Figure 8.9(a), this operation results in the orientation of the object in Figure 8.9(b)-7. Similarly, unit-level reversion, Figure 8.10(b), is described as turning the front of the sheet back, which is also illustrated in Figure 8.9(b)-4. Symbol-level inversion, finally, corresponds to rotating of the sheet 180° counter-clockwise as in Figure 8.10(c), or as illustrated in Figure 8.9(b)-2.

The object in Figure 8.9(b)-1 is obtained by rotating the original image 90° counter-clockwise. The string operation corresponding to this rotation can be obtained by applying u–v transposition ($\rightarrow$) and then unit-level reversion (@) to the original object string.

u–v transposition and unit-level reversion constitutes a minimum operator set because symbol-level reversion can be accomplished by the sequence "$\rightarrow$ @ $\rightarrow$ @". The order in which these operations are applied does not, however, affect the final result.

The illustrations above were only applied to a u-string, but clearly they can be applied to the v-strings as well.

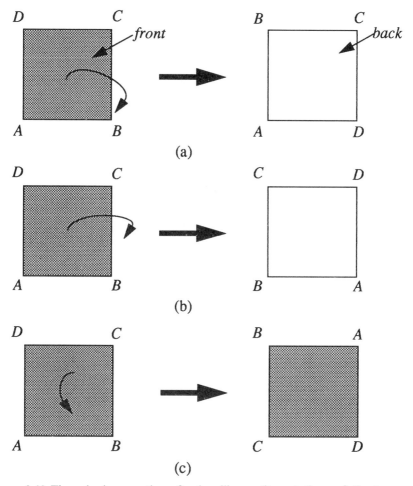

Figure 8.10 Three basic operations for handling entity rotations of the type $u-v$ transposition (a), unit level reversion (b) and symbol level reversion (c).

8.4.2. Rotation of virtual objects (Lee)

Lee (1992) proposes a rotation method that mainly concerns rotation of what he calls virtual objects (VB) which are similar to MBRs but with the exception that a VB may contain a set of objects instead of a single object. The motivation for including a set of objects in the VBs is to reduce the number of cutting lines. In Lee's work rotations are performed on pairs of virtual objects which partially overlap each other and where existing cutting lines are mainly a consequence of this overlap. The main purpose of the rotation technique is thus to completely eliminate the partial overlap such that, in the x-direction, the spatial relationship between the virtual object pair becomes either of type "]" or of type "|", see Chapter 5. The rotation technique is illustrated in Figure 8.11 which contains two virtual objects, A and B. Thus through the projection of the southwest and northeast corners onto the rotated coordinate axis, x' and y', and the projections back to the y-axis of the original coordinate system four control points, $v_1, \ldots, v_4$, can be determined. By means of these four control points the y-directional spatial relation between the two VBs after the rotation can be determined. The spatial relation between the VBs in the x-direction after the rotation can be determined as well. Obviously, the shape of the virtual objects is affected by the rotation and, as we shall see, this method is not qualitative since some calculations are necessary in the process.

In Lee's method each virtual object is identified by using its lower left coordinates, e.g. (x_a, y_a) for A and then the other three are determined by means of the length (l_a) and the width (w_a) of the VB. Hence, the rotated coordinate points can mainly be determined through normal coordinate transforms, i.e.

$$x_a' = x_a \sin(\alpha) + y_a \cos(\alpha) \tag{8.31}$$

$$y_a' = x_a \cos(\alpha) + y_a \sin(\alpha) \tag{8.32}$$

$$l_a' = l_a \cos(\alpha) + w_a \sin(\alpha) \tag{8.33}$$

$$w_a' = l_a \sin(\alpha) + w_a \cos(\alpha) \tag{8.34}$$

where the rotation angle can be calculated by means of

$$\alpha = \arctan((x_a + l_a - x_b - l_b)/(y_a - y_b)) \tag{8.35}$$

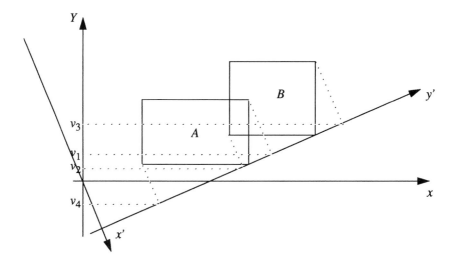

Figure 8.11 The original and the rotated coordinate system and the locations of the four control points.

The values of the four control points are

$$v_1 = y_b + w_b \cos(\alpha) \sin(\alpha) \qquad (8.36)$$

$$v_2 = y_a - l_a \cos(\alpha) \sin(\alpha) \qquad (8.37)$$

$$v_3 = y_a + w_a \cos(\alpha) \cos(\alpha) \qquad (8.38)$$

$$v_4 = y_b - l_b \cos(\alpha) \sin(\alpha) \qquad (8.39)$$

Lee has shown that there exist a number of rules that can be used to determine the spatial relations between the two VB after the rotation in both the x- and the y-directions. Thus the x' rules are:

$$l_a' > l_b' \Rightarrow A \mid B$$

$$l_a' < l_b' \Rightarrow B \mid A$$

$$l_a' = l_b' \Rightarrow A = B$$

Finally, in the y' rules the relations between the four control points are used in determination in the new relations between the virtual object pair. These rules are, however, more complicated than the x' rules. Thus, according to Lee, the following rules exist in the y-direction:

$$v_4 < v_2 < v_1 < v_3 \Rightarrow A \, [\, B \mid B$$

$$v_4 < v_1 = v_2 < v_3 \Rightarrow A \mid B$$

$$v_4 < v_1 < v_2 < v_3 \Rightarrow A < B$$

$$v_4 < v_2 < v_3 < v_1 \Rightarrow B \, [\, A \mid B$$

$$v_4 < v_2 < v_1 = v_3 \Rightarrow B \,] \, A$$

$$v_4 = v_2 < v_1 = v_3 \Rightarrow B = A$$

$$v_2 < v_4 < v_1 = v_3 \Rightarrow A \,] \, B$$

$$v_2 = v_4 < v_3 < v_1 \Rightarrow B \, [\, A$$

$$v_2 < v_4 < v_3 < v_1 \Rightarrow B \, [\, A \mid A$$

$$v_2 < v_3 = v_4 < v_1 \Rightarrow B \mid A$$

$$v_2 < v_3 < v_4 < v_1 \Rightarrow B < A$$

8.5. INTERVAL PROJECTION TO ACHIEVE ROTATION INVARIANCE

Symbolic Projection was developed as a means for iconic indexing. However, the application of the technique to a growing number of problems has resulted in the discovery of a number of problems. One is concerned with rotation invariance, i.e. a problem that occurs when, for instance, topological relations between objects must be identified. This happens when two pairs of objects may have the same topological relation although the objects have different orientation in space. This cannot, in a straightforward way, be described in terms of Symbolic Projection. An illustration to the problem is given in Figure 8.12 which shows one fourth of all existing object relations. It is quite simple to find all topological relations in the figure that are equal, since they have been separated with dotted lines.

It is not possible to use the original technique suggested by Chang *et al.* since it does not allow discrimination between all basic relations even when minimal bounding rectangles are used. The problem may be solved by using the generalized string type (GPS), but this is not sufficiently efficient. The reason is that in GPS the use of the "|" operator requires a large number of horizontal and vertical cuttings of the image. The result of these cuttings is that the image is split up into a relatively large number of small pieces. As a consequence, the projection strings become long and tedious to handle when the problem of rotation invariance has to be dealt with. To overcome this problem, a variation of Symbolic Projection can be used. The variation was originally suggested by Lee *et al.* (1992), who used the strings in a storage and access technique for signature files. They called this projection string type *2D B-strings*. Here, however, the 2D B-strings are subsequently called *interval projection strings* (IPS), since that name corresponds better to the general concept of this projection type. IPS was used by Jungert (1993) as a means of solving the

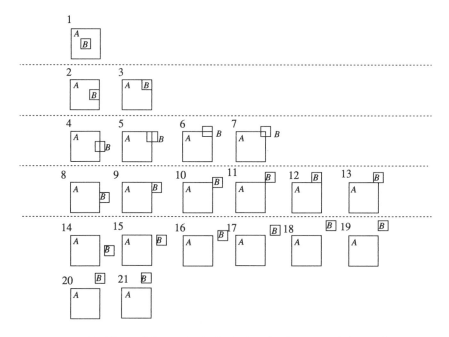

Figure 8.12 All possible topological relations in the upper right quadrant.

problem of rotation invariance when the objects are represented by their MBRs. It is obvious that, when applying IPS to objects approximated by their MBRs, the strings become exactly the same as if the objects themselves had been used for creation of the strings.

Figure 8.13 illustrates the principles of IPS, which are more or less trivial. In the example two objects i and j are given. Interval projection results, as usual, in two strings, each one corresponding to projections along a coordinate axis where the u-string corresponds to projections along the x-axis and the v-string to the y-axis. In interval projection the start and end points of the objects are projected. This projection type corresponds to the interval or extension of the objects in the projection directions. Subscript s is here used to indicate the beginning of the projection interval while e indicates the end.

The approach to solving the problem of rotation invariance is, in principle, to identify the basic object relations as they appear in the projection strings generated from the images. This is illustrated in Figure 8.14 for six topological relations. The shaded areas show the permitted positions of the smaller objects. Figure 8.14(a) illustrates the *inside* relation. Inverse relations can be identified as well, but here they are not considered since they do not change the basic principles. Hence, the main problem is to determine in which shaded area, corresponding to a certain topological relation, a specific object is situated. When this is done the corresponding topological relation is available. The alternatives in Figures 8.14(c)–(e) show the shaded areas that are open, i.e. without any outer limits.

The six different topological relations that can be identified are illustrated in Figure 8.14 and correspond to the following relations including their denotation:

(a) •%   inside
(b) •]   cover
(c) •/   overlap
(d) •| touch
(e) •<   outside
(f) •=   equal

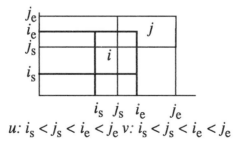

$$u: i_s < j_s < i_e < j_e \quad v: i_s < j_s < i_e < j_e$$

Figure 8.13 The interval projection strings for two objects.

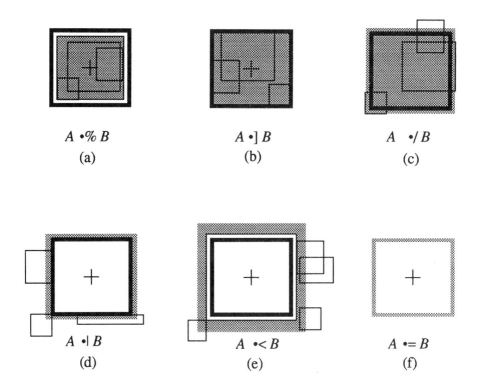

Figure 8.14 Illustrations of permitted (shaded) areas for objects in various topological relations and their symbolic notations (a) inside, (b) cover, (c) overlap, (d) touch, (e) outside and (f) equal.

The operators are equal to those defined and used by, for instance, Egenhofer, see Chapter 5.

To solve the rotation invariance problem for topological relations, all possible binary primitive relations in both x- and y-directions must be identified. There are 9 such possible primitive relations in each projection direction. This gives in all 81 different combinations. The 9 possible primitive relations are illustrated in terms of IPS for each direction in Table 8.2. The primitive relations are numbered from 1 through 9. However, the case for which the objects i and j are identical is a special case of 5 here called $5'$. The basic relations are also illustrated in Figure 8.15. The object relations are simple to identify by first matching the binary strings with the strings given in the table, hereby identify the actual u_k- and v_k-strings. Then from the pair of strings the topological relations can be determined from the definitions:

- $<$ $u_1 \lor u_9 \lor v_1 \lor v_9$
- $|$ $(u_2 \land (v_2 \lor v_3 \lor v_4 \lor v_5 \lor v_6 \lor v_7 \lor v_8)) \lor (u_8 \land (v_2 \lor v_3 \lor v_4 \lor v_5 \lor v_6 \lor v_7 \lor v_8))$
  $\lor (v_2) \land (u_3 \lor u_4 \lor u_5 \lor u_6 \lor u_7)) \lor (v_8 \land (u_3 \lor u_4 \lor u_5 \lor u_6 \lor u_7))$

Table 8.2 The definitions of the primitive binary relations of IPS

k	Interval projection strings	
	U_k^{IPS}	V_k^{IPS}
1	$j_s < j_e < i_s < i_e$	$j_s < j_e < i_s < i_e$
2	$j_s < j_e\, i_s < i_e$	$j_s < j_e\, i_e < i_e$
3	$j_s < i_s < j_e < i_e$	$j_s < i_s < j_e < i_e$
4	$j_s\, i_s < j_e < i_e$	$j_s\, i_s < j_e < i_e$
5	$i_s < j_s < j_e < i_e$	$i_s < j_s < j_e < i_e$
5'	$i_s\, j_s < j_e\, i_e$	$i_s\, j_s < j_e\, i_e$
6	$i_s < j_s < j_e\, i_e$	$i_s < j_s < j_e\, i_e$
7	$i_s < j_s < i_e < j_e$	$i_s < j_s < i_e < j_e$
8	$i_s < j_s\, i_e < j_e$	$i_s < j_s\, i_e < j_e$
9	$i_s < i_s < j_e < j_e$	$i_s < i_s < j_e < j_e$

- $/$ $(u_3 \wedge (v_3 \vee v_4 \vee v_5 \vee v_6 \vee v_7)) \vee (u_7 \wedge (v_3 \vee v_4 \vee v_5 \vee v_6 \vee v_7))$
  $\vee (v_3 \wedge (u_4 \vee u_5 \vee u_6)) \vee (v_7 \wedge (u_4 \vee u_5 \vee u_6))$
- $]$ $(u_4 \wedge (v_4 \vee v_5 \vee v_6)) \vee (u_6 \wedge (v_4 \vee v_5 \vee v_6)) \vee (v_4 \wedge u_5) \vee (v_6 \wedge u_5)$
- $=$ $u_5' \wedge v_5'$
- $\%$ $u_5 \wedge v_5$

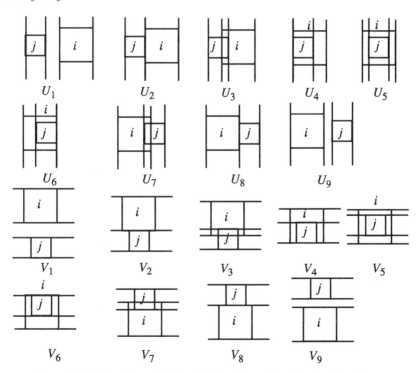

Figure 8.15 The nine primitive relations in each projection direction.

As an illustration to the technique, consider the image in Figure 8.16 from which the following interval projection strings are generated:

$$u_{\text{IPS}}: \quad A_s < B_s < A_e < D_s < B_e C_s < E_s < D_e E_e < C_e$$

$$v_{\text{IPS}}: \quad A_s D_s < E_s < B_s < E_e < D_e < A_e < C_s < B_e < C_e$$

If the problem is to identify the topological object relations between the objects A and B, B and C and finally between D and E, then the binary interval strings for these object relations must first be extracted from the complete interval strings. Secondly, the corresponding primitive string types, u_k and v_k are determined. That is:

in the u-direction

$$A_s < B_s < A_e < B_e \Rightarrow u_3$$

$$B_s < B_s C_s < C_e \Rightarrow u_8$$

$$D_s < E_s < D_e E_e \Rightarrow u_6$$

and in the v-direction

$$A_s < B_s < A_e < B_e \Rightarrow v_3$$

$$B_s < C_s < B_e < C_e \Rightarrow v_3$$

$$D_s < E_s < E_e < D_e \Rightarrow v_5$$

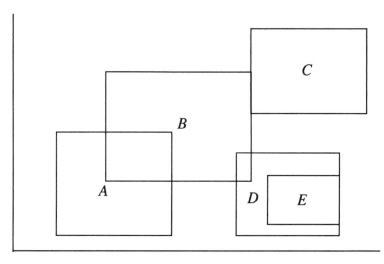

Figure 8.16 A structured image from which complete interval projection strings are generated.

Then for the given examples the relational operators become:

$$[u_3, v_3] \Rightarrow \bullet /$$

$$[u_8, v_3] \Rightarrow \bullet |$$

$$[u_6, v_5] \Rightarrow \bullet \,]$$

Hence, the following topological relations exist in the image for the given objects:

$$B \bullet / A$$

$$B \bullet | C$$

$$D \bullet \,] E$$

Various types of rotation-oriented rules can be applied to the $[u_k, v_k]$-string pairs that make it quite simple to identify all existing topological relations, not just those defined in the first quadrant. Two illustrations of this are given in Figure 8.17. In the first example the first step shows that when interchanging the u_k- and v_k-strings the result is a counter-clockwise rotation of the objects by 90°, which corresponds to a u–v transposition according to Hirakawa and Jungert (1991) as described in section 8.4.1. In the second step a further rotation of 90° is performed by another interchange followed by a substitution taken from the following set:

$$u_1 \Leftrightarrow u_9, \quad u_2 \Leftrightarrow u_8, \quad u_3 \Leftrightarrow u_7, \quad u_4 \Leftrightarrow u_6, \quad u_5 \Leftrightarrow u_5$$

This rotation step actually corresponds to a u–v transposition followed by a unit level reversion of the u_k-string, where again the latter operation is in accordance with the operations defined by Hirakawa and Jungert.

In the last step, finally an interchange is made between the u_4- and the v_5-strings which gives the final relation, i.e. a u–v transposition. By finally applying the same interchange rules as in step 2, a u–v transposition and a unit level reversion, the orientations of the two objects are back to the starting positions as shown on the left in Figure 8.17.

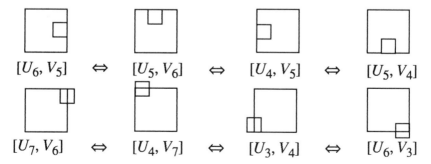

Figure 8.17 Transformations of u, v-strings for generation of related topological relations.

In the second example the rules are applied in a similar way. At each step there is a $u-v$ transposition and a unit level reversion. Consequently, it can be concluded that there must also be a unit level reversion in each step in the first example as well, but this is hidden by the fact that for u_5 the unit level reversion leaves the string unchanged. Clearly, all other relations corresponding to the same topological relationships can be rotated in similar ways.

Rotation of u, v-strings for identification of the topological relations can be used to simplify the handling of the object relations. Hence, just the relations in Figure 8.12 are required, excluding the first relation which always remains the same. In other words, there is no need to deal with all possible binary object string combinations. It is sufficient to rotate an arbitrary relation into the first quadrant, in other words into the primitive relations determined by $u_5, \ldots, u_9$ and $v_5, \ldots, v_9$. Clearly, all relations can be transformed into this set of relational strings from which the corresponding topological relations can be determined in a more simple way.

IPSs are here used as an indexing technique with the purpose of identifying topological object relations which as a first step are represented in lowest possible resolution, that is with their MBRs. The motivation for such an indexing technique is to allow as many false alternatives as possible to be pruned off at an early stage without the involvement of large volumes of data that would otherwise slow down the search process. An important observation

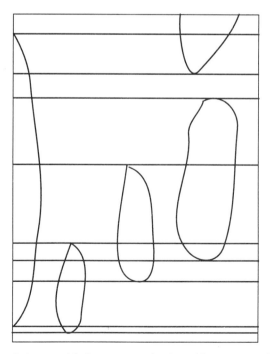

Figure 8.18 A simple image with free space and other object types and its cutting lines.

is, however, that the uncertainty of the described method increases as the objects get closer to each other, and there is no simple way to overcome this uncertainty. The interval string method can, however, delimit the problem somewhat, since extraction of the strings that include the information necessary for determination of the actual topological relations is trivial.

8.6. COMPACTING ALGEBRAIC EXPRESSIONS

One of the main reasons for the development of the symbolic image algebra was to find a structure suitable for spatial reasoning, i.e. the development of a knowledge structure for reasoning about spatial relationships particularly based on a syntactic formalism. Examples include reasoning about similarities among images and path planning. Of special interest is the use of the unified projection strings since for that type of projections only a single projection direction will suffice. A further aspect that must be taken into consideration is that the projection strings quite often become long and hence troublesome to handle. For this reason, we need transformations of strings into more compact structures (Jungert, 1989; Jungert and Chang, 1993) which are simpler to handle and can more easily be restored into their original structure. Figure 8.18 will be the starting point for this discussion, and from that image the following string can be created:

u: $(\mathbf{e} \backslash\backslash\sim)(\mathbf{e}\backslash\backslash u1|\#|\mathbf{e})(\#|\mathbf{e}|\#|\mathbf{e}\backslash\backslash\sim)(\#|\mathbf{e}|\#|\mathbf{e}|\#|\mathbf{e}\backslash\backslash\sim)(\#|\mathbf{e}|\#|\mathbf{e}|\#|\mathbf{e}|\#|\mathbf{e})$

$(\#|\mathbf{e}\backslash\backslash\wedge|\#|\mathbf{e}|\#|\mathbf{e}|\#|\mathbf{e}|)(\#|\mathbf{e}\backslash\backslash\wedge|\#|\mathbf{e})(\#|\mathbf{e}\backslash\backslash\wedge\backslash\backslash\sim)(\#|\mathbf{e}|\#|\mathbf{e})(\mathbf{e}\backslash\backslash1l|\#|\mathbf{e})$

$$(8.40)$$

The symbol "#" corresponds to subobjects of nonempty-space type. These subobjects are generalized objects, just like the empty space objects **e**. Hence, this particular image includes two different object types, both generalized. The image contains horizontal cutting lines which are numbered. The cutting strategy can be described as follows. At each point where a cutting line touches a #-object, a split point is obtained. By definition, a split point is situated either in the row below the cutting line which is a tangent of the object (lower split point) or in the row above (upper split point). Consequently, an upper split point is upper with respect to the tile to which it belongs and not to the touched object. The situation for the lower split point is equivalent. A tile is defined as the area in a row that is between two #-objects or an edge and an #-object or two facing edges of the image. The topic of split point is also discussed in Section 10.3. In expression (8.40), the split points are indicated with **e** followed by a pair of blackslashes and the type of split point. The various existing split point types and their notations are found in expressions (8.41)–(8.46).

$$\mathbf{e}\backslash\backslash\sim \qquad \text{upper split point} \qquad (8.41)$$

$$\mathbf{e}\backslash\backslash\wedge \qquad \text{lower split point} \qquad (8.42)$$

$$\mathbf{e}\backslash\backslash ll \qquad \text{lower split point at left edge} \qquad (8.43)$$

$$\mathbf{e}\backslash\backslash ul \qquad \text{upper split point at left edge} \qquad (8.44)$$

$$\mathbf{e}\backslash\backslash ur \qquad \text{upper split point at right} \qquad (8.45)$$

$$\mathbf{e}\backslash\backslash lr \qquad \text{lower split point at right} \qquad (8.46)$$

Examples of split points of type $\mathbf{e}\backslash\backslash ul$ and $\mathbf{e}\backslash\backslash ll$, which are situated at the edge of the image, are found in Figure 8.18 along lines 2 and 9, respectively.

When more than one split point of the same type is present along a line, this is indicated by a constant between the backslash and the character that indicates the number of split points along that line:

$$\mathbf{e}\backslash\backslash 2\wedge \qquad (8.47)$$

If more than one split point is present along the same line and if they are of different types then this is indicated as

$$\mathbf{e}\backslash\backslash ul\backslash\backslash 2\wedge \qquad (8.48)$$

Both (8.47) and (8.48) can be cut so as to contain one $\mathbf{e}$-object for each split point. Consequently, the expressions will become

$$\mathbf{e}\backslash\backslash\wedge\,|\,\mathbf{e}\backslash\backslash\wedge \qquad (8.49)$$

and

$$\mathbf{e}\backslash\backslash ul\,|\,\mathbf{e}\backslash\backslash\wedge\,|\,\mathbf{e}\backslash\backslash\wedge \qquad (8.50)$$

respectively.

Some of the strings are fairly long and have a pattern that is frequently repeated. For that reason, it is convenient to rewrite the strings in a more compact way:

$$(\#\,|\,\mathbf{e}\,|\,\#\,|\,\mathbf{e}\,|\,\#\,|\,\mathbf{e}\,|\,\#\,|\,\mathbf{e}\,|) \Rightarrow (4[\#\,|\,\mathbf{e}\,]) \qquad (8.51)$$

$$(\#\,|\,\mathbf{e}\,|\,\#\,|\,\mathbf{e}\,|\,\#\,|\,\mathbf{e}\,|\,\#) \Rightarrow (3[\#\,|\,\mathbf{e}\,]\,|\,\#) \text{ or } (\#\,|\,3[\mathbf{e}\,|\,\#]) \qquad (8.52)$$

This way of compacting the expressions can be described in more general form:

$$(\#\,|\,\mathbf{e}\,|\,\ldots\,|\,\#\,|\,\mathbf{e}) \Rightarrow n[\#\,|\,\mathbf{e}\,]] \qquad (8.53)$$

This way of compacting the strings works efficiently for patterns of homogeneous type, but is not as efficient for strings where, for instance, split points are included:

$$(\mathbf{e}\,|\,\#\,|\,\mathbf{e}\,|\,\#\,|\,\mathbf{e}\,|\,\#\,|\,\mathbf{e}\,|\,\#\,|\,\mathbf{e}\backslash\backslash\sim\,|\,\#\,|\,\mathbf{e}\,|\,\#\,|\,\mathbf{e}) \Rightarrow (4[\mathbf{e}\,|\,\#\,]\,|\,\mathbf{e}\backslash\backslash\sim\,|\,2[\#\,|\,\mathbf{e}\,]) \qquad (8.54)$$

The subexpressions in (8.54) can be compacted even further but this will not be discussed further here. Instead, for a more thorough study the reader is referred to the work by Jungert and Chang (1993).

REFERENCES

Chang, S. K. and Jungert, E. (1991) Pictorial data management based upon the theory of Symbolic Projection. *Journal of Visual Languages and Computing* **2**, 195–215.

Chang, S. K., Jungert, E. and Li, Y. (1990) The design of pictorial databases based upon the theory of symbolic projections. In *Design and Implementation of Large Spatial Databases*, ed. A. Buchmann, O. Gunther and T. R. Smith, pp. 303–23. Springer-Verlag, Berlin.

Hirakawa, M. and Jungert, E. (1991) An image database system facilitating icon-driven spatial information definition and retrieval. *Proceedings of IEEE Workshop on Visual Languages*, pp. 192–198, Kobe.

Holmes, P. D. and Jungert, E. (1992) Symbolic and geometric connectivity graph methods for route planning in digitized maps. *IEEE Transactions on Pattern Analysis and Machine Intelligence* **14**, 549–65.

Jungert, E. (1989) Symbolic expressions within a spatial algebra: unification and impact upon spatial reasoning. *Proceedings of the IEEE Workshop on Visual Languages*, Rome, pp. 157–62.

Jungert, E. (1993) Rotation invariance in symbolic projection as a means for determination of binary object relations. *Proceedings of the Workshop on Qualitative Reasoning and Decision Technologies (QUARDET'93), Barcelona*, ed. N. Piera Carreté and M. G. Singh, pp. 503–12. CIMNE, Barcelona.

Jungert, E. and Chang, S. K. (1989) An algebra for symbolic image manipulation and transformation. In *Visual Database Systems*, ed. T. L. Kunii, pp. 301–17. North-Holland, Amsterdam.

Jungert, E. and Chang, S. K. (1993) An image algebra for pictorial data manipulation. *Computer Vision, Graphics and Image Processing: Image Understanding* **58**, 147–60.

Lee, C. M. (1992) The unification of the 2D string and spatial query processing. PhD thesis, Graduate College of the Knowledge Systems Institute, Skokie, Illinois.

Lee, S. Y., Yang, M. C. and Chen, J. W., *2D B-string Knowledge Representation and Picture Retrieval for Image Database*, Second International Computer Science Conference – Data and Knowledge Engineering: Theory and Applications, Hong Kong, December 13–16, 1992, 609–615.

Lee, S.-Y., Yang, M.-C. and Chen, J.-W. (1992) Signature file as spatial filter for iconic image database. *Journal of Visual Languages and Computing* **3**, 373–97.

9

Generalized Symbolic Projection

Symbolic Projection is a qualitative reasoning technique which was originally deficient in certain respects. One of these deficiencies is the subject of this chapter. A consequence of the solution to that problem is an extension to Symbolic Projection that is in fact a generalization of the technique. The work presented here is also concerned with qualitative spatial reasoning for determination of, among other things, directions, distances and other object relations seen from the observer's perspective, i.e. the projections are in a majority of cases concerned with a singular point.

A generalization of Symbolic Projection, i.e. slope projection, is described in section 9.1. The generalization technique is extended and applied to various problems, line segments in section 9.2, directions in section 9.3 and distances in section 9.4. A further application to slope projection, in section 9.5, is concerned with path descriptions. Section 9.6 introduces polar projection while in section 9.7 it is demonstrated that qualitative transformations cannot be made between polar and cartesian projection string types.

9.1. GENERALIZED PROJECTION

In the original work on Symbolic Projection by Chang *et al.*, the projection strings described the relative positions of a set of objects along each coordinate axis by means of an operator set including just two relational operators, i.e. $\{<, =\}$. Later Jungert introduced the edge-to-edge operator. As a consequence, a minimal operator set was defined, i.e. $\{<, =, |\}$ which can be used for description of any image in terms of symbolic projections. However, since the introduction of Symbolic Projection some fundamental problems have been known. One of the most important of these problems is illustrated in Figure 9.1. The problem can be briefly stated as follows. When using Symbolic Projections there is no way to infer on which side of the line object l the two point objects p_1 and p_2 lie, if they are situated inside the rectangle spanned by l. This problem is due to the fact that lines that are neither horizontal nor vertical cannot be represented such that the knowledge of their slope is preserved in the strings.

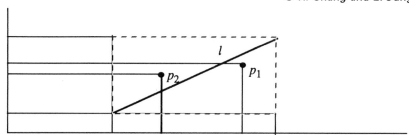

Figure 9.1 A fundamental problem that cannot be solved by the original approach to symbolic projection.

However, there is quite a simple solution to this problem, which is illustrated in Figure 9.2.

As can be seen in Figure 9.2, the point objects and the line object are all projected in the direction of the slope of the line object. These projections can be made either to the *x*-axis or the *y*-axis. It is not necessary to do both, since each of them contains all the necessary information. The result of these slope projections is the projection string u_l:

$$u_l: \quad x_{p_1} < x_l < x_{p_2} \tag{9.1}$$

where

$$x_{p_1} = x_1 - \frac{y_1}{k} \tag{9.2}$$

$$x_{p_2} = x_2 - \frac{y_2}{k} \tag{9.3}$$

$$x_l = x_3 - \frac{y_3}{k} \tag{9.4}$$

In (9.2), (9.3) and (9.4), k is the slope coefficient of the line object, that is:

$$k = \frac{y_3 - y_4}{x_3 - x_4} \tag{9.5}$$

Two extreme cases, $k = 0$ and $k = \infty$, need to be discussed further. Evidently, k can take any value beside the two extreme cases. The method also works for several lines at different orientations, in which case there must be separate slope projections for each line, see section 9.2.

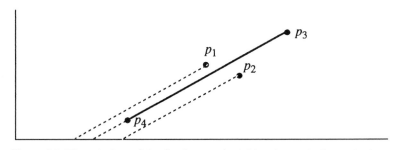

Figure 9.2 The solution of the fundamental problem in symbolic projection.

Basically, three further projection types exist beside the above. That is, a point can be projected along a certain slope to both the *x*- and *y*-axes, and to both axes perpendicular to the slope. Projections perpendicular to a line are simple to identify because of the well-known relationship between slope coefficients of two perpendicular lines:

$$k' = -\frac{1}{k} \tag{9.6}$$

Hence, the *y*-projection perpendicular to a given line, for instance point p_1 in Figure 9.2, can be determined from

$$y_{p_1} = y_1 + \frac{x_1}{k} \tag{9.7}$$

The two remaining projection points are

$$x'_{p_1} = x_1 + ky_1 \tag{9.8}$$

and

$$y'_{p_1} = y_1 - kx_1 \tag{9.9}$$

By looking at the first pair from a general view the projection points can be identified as

$$x_i - \frac{y_i}{k} \tag{9.10}$$

and

$$y_i + \frac{x_i}{k} \tag{9.11}$$

From these points the projection strings $\langle u, v \rangle$ can be generated. It is obvious that for $k = \infty$ the projection points become

$$x_i \tag{9.12}$$
$$y_i \tag{9.13}$$

which result in perpendicular projections points.

Generalization of the second projection pair gives the points

$$y_i - kx_i \tag{9.14}$$

and

$$x_i - ky_i \tag{9.15}$$

Hence, a second string pair, subsequently called $\langle u', v' \rangle$ can be created. For $k = 0$ the second pair is similar to the original perpendicular projection points. Furthermore, for $k = 0$ in the first pair and for $k = \infty$ in the second the projections become corrupted and are hence not useful.

When *k* is close to or equal to zero the $\langle u, v \rangle$ string pair must be substituted by the (u', v') pair, which can be expressed in terms of two rules:

if *k* is close to or equal to zero then use (u', v')
if *k* is very large then use (u, v)

Subsequently these rules are implicitly assumed even when this is not explicitly mentioned. However, if $\langle u, v \rangle$ is used for lines with moderately large coefficients and if suddenly a line appears with a slope coefficient that is equal to zero, then normally there must be a switch of focus from $\langle u, v \rangle$ to $\langle u', v' \rangle$. However, since the projection strings are of qualitative type it is not necessary to represent infinity directly in a quantitative way. For this reason they can, as an alternative, be represented as

$$u: \quad A_{k=0} < B_{k=1} < C_{k=\infty} \tag{9.16}$$

It is easy to see that the original approach to Symbolic Projection is just a special case of slope projection. Symbolic slope projection will thus have a number of consequences on the theory of Symbolic Projection which also have been discussed extensively by Jungert (1992, 1994) and which will be discussed further subsequently in this chapter. Consequently, *slope projection is a generalization of the original method with respect to the projection direction.* Slope projection includes a second pair of projection strings which can be combined with the first. Hence, further object information or knowledge can be inferred and explicitly be made available.

When applied to different problems it turns out that slope projection is used most frequently for a fairly limited set of angles, that is:

$$\pm \frac{\pi}{4}, \ \pm \frac{\pi}{8}, \ \pm \frac{3\pi}{8}$$

The price that has to be paid for the extra operations of this type of projection compared to the perpendicular is just one division and one subtraction, if the slope coefficient is preprocessed. This is reasonable when considering the information achieved.

9.2. LINE SEGMENT RELATIONS

Projection along the slope of a line can be expanded further, as can be seen in Figure 9.3, which illustrates how qualitative reasoning on line segments can be performed in order to identify relations like "crossing", "not crossing" and "parallel to". To identify valid relations between two line segments requires at most three projections each in two directions, a total of six projections. The two projection directions are determined by the directions of the two line segments. Thus projections in one direction are made along the corresponding segment and the end points of the other segment. Eventually the projection strings are created and a match between the strings of the image and the strings of the various relations can be performed. However, the segments can be projected to the y-axis as well, but it is sufficient just to apply the projections to just one coordinate axis.

Noncrossing is illustrated in Figure 9.3(a), from which the two projection

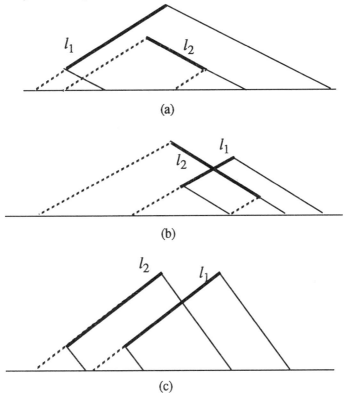

Figure 9.3 Slope projection applied line segments and corresponding to the spatial relations (a) not crossing, (b) crossing and (c) parallel.

strings corresponding to this spatial relationship can be determined:

$$u': \quad l_1 < l_{2s} < l_{2e} \tag{9.17}$$

$$u'': \quad l_{1s} < l_2 < l_{1e} \tag{9.18}$$

The start and end of a line segment are denoted by l_{is} and l_{ie} respectively, where i corresponds to the number of the segment. The u'-strings are determined from the projections corresponding to the dashed lines, and the u''-strings from the solid lines.

To determine that two segments are crossing each other one projection is not sufficient, as can be seen from the example in Figure 9.3(b) or from the following expressions:

$$u': \quad l_{1s} < l_2 < l_{1e} \tag{9.19}$$

$$u'': \quad l_{2s} < l_1 < l_{2e} \tag{9.20}$$

In both strings each start and end points of the segments must be on the opposite side of the other segment, otherwise the two segments are not crossing each other.

Determination of "parallel", Figure 9.3(c), is more trivial than the two other relations since in fact, no projections are required. To check parallelism, it is sufficient to prove that the slope coefficients are equal. On the other hand, if the order of the segments is needed as well then the projection strings are necessary. The projection string u' required to determine parallelism is

$$u': \quad l_{1s}l_{1e} < l_{2s}l_{2e} \tag{9.21}$$

Furthermore, if the relative position of the segments are of interest then the perpendicular projection (9.21) is needed as well:

$$u': \quad l_{1s} < l_{2s} < l_{1e} < l_{2e} \tag{9.22}$$

From (9.22) it can, depending on the direction, be concluded that l_1 starts a little bit before l_2 and that it also ends before l_2, and finally that l_1 is to the left of l_2. Clearly, several other alternatives may occur.

9.3. DIRECTIONS

When the direction of an object has to be determined, the object type can be either a point, a line or an extended object. Each requires its own technique. This is well known, and it is also true that not all methods are useful on all object types. Symbolic slope projection does, however, work for all three object categories, as we shall demonstrate subsequently.

Qualitative directions are of particular interest in reasoning about geographical directions such as north, west, northeast, etc. An example of such a directional system is: {N, NE, E, SE, S, SW, W, NW}. In the illustrations here only the three directions of the first quadrant are used, but the technique can easily be generalized for all possible directions. The first quadrant is thus split into three directions {E, NE, N} as illustrated in Figure 9.4. The angle between the two dividing lines is $\pi/4$ and the angles between the dividing lines and the coordinate axes are consequently $\pi/8$. The segment between the

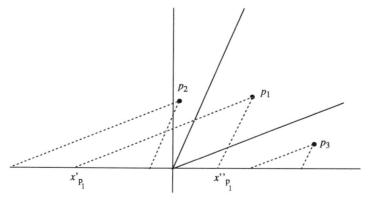

Figure 9.4 Determination of directions for point objects using symbolic slope projection.

dividing lines indicates the northeasterly while the area between a dividing line and a coordinate axis corresponds to half the segment of the northerly and easterly directions respectively.

9.3.1. Point Objects

The examples in Figure 9.4 illustrates how directions of point objects relative to a given origin are determined. The projection slopes of the point objects are parallel to the edges of the corresponding sector. For point p_2, both projections hit the x-axis to the left of the origin. The origin is subsequently called the *observer's position*, since it does not necessarily have to correspond to the origin of the current coordinate system. p_1, on the other hand, lies in the northeastern sector and the observer's position is here in between the projected points. The rule for deduction of the directions can hence be identified:

$$u_0: \quad x' < x'' < O \Rightarrow \text{north}(p, O) \tag{9.23}$$

That is, if two projection points on the x-axis of a point p, i.e. x' and x'', are less than the observer's position (O) in the u_0-string then the direction of p is north of O. The slopes correspond to lines separating north and northeast (northeast and east).

Similarly, the rule for "northeast" can be defined:

$$u_0: \quad x' < O < x'' \Rightarrow \text{northeast}(p, O) \tag{9.24}$$

The cost in execution time for generation of the projection strings is here two multiplications and two additions, which is acceptable. Conventional methods would require heavier computations including calculation of the arctan function. Observe that the slope coefficients can be preprocessed. In other words, since no extra calculations during run-time are required, then the method can be used to determine the direction of an object for angle intervals that are narrower. The extra time required for such calculations will still be acceptable. For instance, if the coefficients of all integer degrees are preprocessed then the required operations in the first quadrant are at most six multiplications and six additions if a binary search is applied. For intervals of $10°$ there will be at most four multiplications and four additions. Thus the general rule for determination of the object direction can be formulated as

$$\text{for all } p_i \text{ where } i = 1, 2, \ldots \quad \text{if } x_{p_i}^{(\alpha)} < O < x_{p_i}^{(\beta)} \Rightarrow p_i \in [\alpha, \beta] \tag{9.25}$$

Finally, it can also be shown that for $x_{p_i}^{(\alpha)} < O$ then the angle of p_i is greater than α and for $x_{p_i}^{(\alpha)} > O$ the angle is less than α.

9.3.2. Line Segments and their Orientations

Objects corresponding to line segments differ from point objects in that, as well as the direction from the observer, they can also have an orientation in the

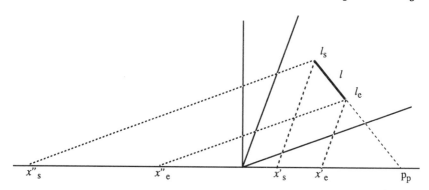

Figure 9.5 The determination of the direction of a line object.

plane. How the direction of a line segment can be determined by means slope projection is illustrated in Figure 9.5, where the segment is to the northeast of the observer. This can be inferred from the following rule:

$$u_s: \quad x_s'' < O < x_s' \text{ and } u_e: \quad x_e'' < O < x_e' \Rightarrow \text{northeast}(l, O) \qquad (9.26)$$

The direction seen from the observer is obtained from the types of slope projection introduced in section 9.3.1. For a line segment, both end points must be part of the projection rule. This may sometimes lead to directions that are not always as clear as in the example given, but that merely reflects the general problem of how to define certain directions and has nothing to do with the method as such. This problem will also be discussed further in section 9.3.3 for extended objects.

To determine the orientation of a line segment in the plane, a similar method can be applied. Consider the situation in Figure 9.5. If the segment is projected onto the x-axis along its own slope then a projection point p_p is created. The orientation of the line segment can now be obtained by using the same projection method as was used for determination of the direction of a point, although now the observer is not part of the process. Instead, one of the end points of the segment is used. From the example in Figure 9.5 it can be seen that when the correct directional is found, the projection point p_p will be inside the interval that is the result of the projections. If the two projection points are both on the same side of p_p then their direction does not correspond to the directional interval. In the example the following statement can be concluded:

$$\text{northwest}(l) \qquad (9.27)$$

This way of obtaining the orientation of the line segment is similar to a coordinate transformation or displacement of the projections along the current coordinate axis to which the points are projected.

9.3.3. Extended Objects

The method for determination of directions can be applied to extended objects as well. The minimal bounding rectangle (MBR) of the extended objects can be used for this purpose. Four points are needed in the worst case, i.e. all four corners of an MBR, which is illustrated in the examples in Figure 9.6(a)–(d). However, in most cases it is sufficient to use two points, the end points of one of the diagonals of the MBR. In the first quadrant the quadrant going from the upper left corner down to the lower right corner is used. For MBRs situated in any of the three other quadrants, it is easy to identify the appropriate diagonal. When an MBR crosses the border between two quadrants any of the diagonals can be used. Figure 9.6(a) illustrates a situation where all four corners of the rectangle are below the $\pi/8$ line, i.e. the object is east of the observer. In Figure 9.6(b) the object is northeast of the observer. Figures 9.6(c) and (d) are somewhat more complicated. Figure 9.6(c) ought to be northeast to east, since here the upper left point of the rectangle is northeast of the observer while the lower right point is to the east.

The situation in Figure 9.6(d), finally, causes some further problems. There are two points to the north, one point to the northeast and finally one to the east. If just the two points of the relevant diagonal are used then there is one point to the north and one to the east. Hence, in both cases, they most natural way to interpret the situation is simply to say that the direction is northeast. If, on the other hand, the two points at the north had been situated in the second quadrant then it would have been more correct to say that the direction is north. Clearly, a more or less exhaustive study of all possible alternatives must be

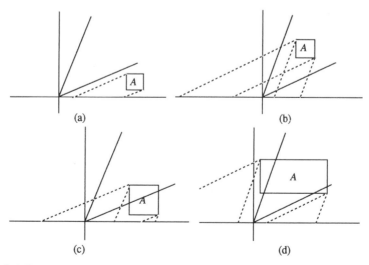

Figure 9.6 Four cases illustrating how the direction of extended objects can be determined.

performed. However, this will not change the method as such. As can be seen, it is quite easy to determine the directions of the two points but in some cases the directions of the objects may be subject to personal interpretation. The problem of determination of directional relationships has been addressed more deeply by Peuquet and Zhang (1987).

9.4. DISTANCES

Slope projection can also be used for determination of distances. Two different problems can be identified. The first is concerned with objects that may be far apart from each other but they will each have about the same distance from the observer. The solution to this problem requires approximations to absolute distances. The second problem is concerned with qualitative reasoning on distances between the observer and objects whose relative distances from the observer may vary widely. A recent work on qualitative distances is given by Hernández *et al.* (1995).

9.4.1. Approximation of Absolute Distances

It is important to differentiate objects with approximately the same distance from the observer, but situated in different directions with respect to the observer. The approach is an extension of symbolic slope projection. The method is accomplished by first determining the direction in qualitative terms and then find a distance approximation by means of slope projection towards one of the coordinate axes. Figure 9.7(a) and (b) illustrate this.

Illustrations of the approach to determination of distances given in Figure 9.7 will subsequently be called a *projection scheme*. The first scheme corresponds to a partition of the space into 8 sectors and the second into 16 sectors. In both

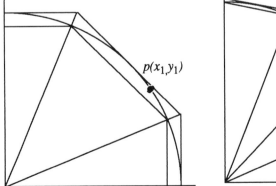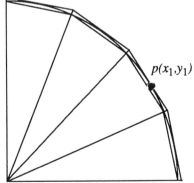

Figure 9.7 A distance projection scheme for points with approximately equal distances from the observer.

cases all points on the circle can directly or stepwise be projected down to either the x- or the y-axis so that the projected points fall inside the interval of the parallel lines, which is an estimation of the maximum error. In the scheme in Figure 9.7(a) points in the sector $0-\pi/8$ are approximated by $|x|$ while points in the sector $3\pi/8-\pi/2$ can be approximated by their corresponding y-value. Points inside the sector $\pi/8-3\pi/8$ may be projected to any of the coordinate axes in two steps: first to the $\pi/8$ line and then down to the x-axis, or in the other direction to the y-axis in a similar way. Here the former direction is demonstrated.

In the following $k^{(\alpha)}$ is used to denote a slope coefficient corresponding to the angle α. Thus $p(x_1, y_1)$ is a point in the $3\pi/8-\pi/2$ sector and on the line $y = -x + m$ where $m = y_1 + x_1$. This line crosses the line

$$y = k^{(\pi/8)}x \qquad (9.28)$$

at the x-coordinate determined by

$$x = \frac{x_1 + y_1}{1 + k^{(\pi/8)}} \qquad (9.29)$$

In (9.29), x is an approximation of the distance from the observer to the point object. It can be shown that

$$1 + k^{(\pi/8)} = 2^{1/2} \qquad (9.30)$$

Consequently, the distance from the observer can be approximated by

$$(x_1 + y_1)2^{-1/2} \qquad (9.31)$$

The maximum error, which occurs for the direction of $k^{(\pi/8)}$, is quite large:

$$\delta = \sqrt{x_1^2 + y_1^2} - x_1 = \left(1 - \frac{1}{\sqrt{1+k^2}}\right)\sqrt{x_1 + y_1} = 0.076\sqrt{x_1 + y_1} \qquad (9.32)$$

where δ is maximum for $k = \tan \pi/8$. This error is unacceptably large, and a better result can be obtained from the scheme in Figure 9.7(b).

The two projection schemes are essentially the same, except that for the 16-sector partition the point object in the $0-\pi/8$ sector is projected down to the x-axis using the approximation:

$$d = x - \frac{y}{k^{(9\pi/16)}} \qquad (9.33)$$

In the $\pi/8-\pi/4$ sector, the points are first projected to the $\pi/8$ line and then down to the x-axis. The two remaining sectors in the quadrant are analogously projected to the y-axis.

A point (x_1, y_1) in the $\pi/8-\pi/4$ sector is first projected down to the intersection between the following lines:

$$y = k^{(\pi/8)}x \qquad (9.34)$$

and

$$y = k^{(11\pi/16)}x + m \qquad (9.35)$$

where m becomes

$$m = y_1 - k^{(11\pi/16)}x_1 \tag{9.36}$$

Thus the point projected on to the line (9.34) is

$$x = \frac{y_1 - k^{(11\pi/16)}x_1}{k^{(\pi/8)} - k^{(11\pi/16)}} \tag{9.37}$$

$$y = k^{(\pi/8)}\frac{y_1 - k^{(11\pi/16)}x_1}{k^{(\pi/8)} - k^{(11\pi/16)}} \tag{9.38}$$

Finally, the point is projected down to the x-axis through the projection in (9.33). The final result is obtained from (9.33), (9.37) and (9.38):

$$d = (y_1 - k^{(11\pi/16)}x_1)C \tag{9.39}$$

where

$$C = \frac{k^{(9\pi/16)} + k^{(\pi/8)}}{k^{(9\pi/16)}(k^{(\pi/8)} - k^{(11\pi/16)})} \approx 0.48 \tag{9.40}$$

Now the maximum projection error, which occurs for $k^{(\pi/16)}$, can be determined from:

$$\delta = \sqrt{x_1 + y_1} - \left(x_1 - \frac{y_1}{k^{(9\pi/16)}}\right) \tag{9.41}$$

which gives:

$$\delta = 0.0196\sqrt{x_1 + y_1} \tag{9.42}$$

Compared to the 8-sector scheme the refinement corresponds to an improvement from about 8% to almost 2% in the 16-sector scheme. In distance transforms Borgefors (1986) has showed similar results, i.e. exactly the same result for the 16-sectors case while the 8-sectors case is somewhat larger here. However, distance transforms generally requires a number of operations that is less than for the distance transforms although their distances are generated globally in all directions.

In the worst case only two multiplications and one addition are required compared to the 8-direction scheme, which requires only one multiplication and one addition at the most. However, the operations for finding the directions must be included as well, which require two multiplications and two additions at the most. Hence, the total number of operations for getting both distance and direction for the two schemes are:

8-sector scheme   3 multiplications and 3 additions
16-sector scheme   4 multiplications and 3 additions

The extra calculations in the 16-sector scheme are the price for getting an error that is less than 2% compared to 7.6% in the 8-sector scheme. This is affordable when comparing the distances from the observer to two different points. In the 16-sector scheme the exact distance lies in the interval $[d, d + \delta]$. Because of δ, two distances can only be said to be approximately equal when compared. Hence, if two projected distances, d_1 and d_2, are compared, the following rules can be identified for the 16-sector scheme, which is however equivalent to the 8-sector scheme as well:

$$\text{if } d_1 < d_2 - \delta_2 \text{ where } \delta_2 \approx 0.02d_2 \text{ then } d_1 < d_2$$
$$\text{if } d_1 - \delta_1 > d_2 \text{ where } \delta_1 \approx 0.02d_1 \text{ then } d_1 > d_2$$

For all other cases $d_1 = d_2$. The conclusion that $d_1 \approx d_2$ is true when $d_1 > d_2 - 0.02d_2$ and $d_1 < d_2 + 0.02d_2$, since $\delta_1 \approx \delta_2$.

9.4.2. Qualitative Distances

Using qualitative methods for determination of distances is a technique similar to that introduced by Frank, see Chapter 10. Frank's technique is based on a method where the space is split into a set of intervals $\{c, m, f\}$ where c corresponds to "close to" a certain point which can be compared to the observer's position. Similarly f means "far from" the same point, while "m" is somewhere in the middle. Frank's method works for point objects only. This limitation can, however, be overcome by using symbolic slope projections. Determination of qualitative distances is entirely based on the technique introduced in section 9.2. The method is illustrated in Figure 9.8, where, however, the projection lines are only drawn for a few selected cases. The set

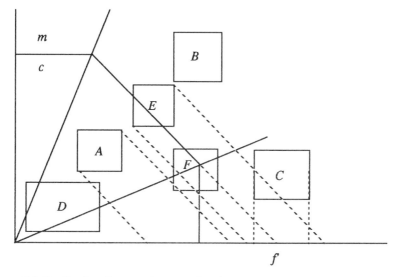

Figure 9.8 Qualitative distances between the observer and a set of extended objects.

of objects *A* through *F* in Figure 9.8 illustrates the occurrences of a number of distances between the observer and the corresponding extended objects.

The objects *A*, *B* and *C* have all distances that are unambiguous, since none of them crosses any of the lines that separates the various qualitative distance intervals, i.e. *A* is close to *O*, since all its projections are less than the projection of the line separating *c* and *m* when projected in the direction of *A*. Similarly, *B* and *C* are medium far away. *E* lies on the line separating *m* and *c*. This is indicated by the judgement that the object is "fairly close" to *O*. The same judgement is evidently valid for *F*. Finally, *D*, lies close to *O*. One aspect that is of concern here is that not just the objects themselves are of concern but the minimal enclosing rectangles. Consequently, a rectangle can be "fairly close" to the observer while the object itself can be "close" since all the points describing the object lies in the "close" area. However, because of the qualitative reasoning technique the difference between these two cases is of less importance. Table 9.1 shows all the object directions and the distances with respect to *O* in Figure 9.8.

Determination of qualitative distances is simply done by projecting the end points of the MBR diagonals that are perpendicular to the line separating the qualitative distances if the object is entirely or partly situated in the northeast sector. If the MBR is situated in the east sector then the projection is vertical and finally if in the north sector then the projection is horizontal, i.e. towards the *y*-axis. This is easily generalized for the whole space surrounding the observer. However, the distances cannot be determined alone; the direction must be determined at the same time. For this reason six slope projections for each MBR must be determined, which altogether requires six multiplications or divisions and the same number of additions or subtractions.

9.5. PATH PROJECTION

Path-finding is generally the problem of finding the shortest point-to-point description between a start point and an end point in an image that normally is a map. In many applications this is not sufficient. More information is required, including qualitative descriptions of the route and the various object relations along the route. Holmes and Jungert (see Chapter 10) showed how a primitive plan can be created by means of a set of rules in an inference system. That

Table 9.1

Object	Direction	Distance
A	northeast	close
B	northeast	medium
C	east	medium
D	northeast	close
E	northeast	fairly close
F	east-northeast	fairly close

technique was a step in the direction towards automatic generation of route descriptions. The solution to the problem proposed is based on symbolic slope projection and involves approximation of the route as a sequence of interconnected straight lines corresponding to road segments. For this a map represented at a relatively low resolution is required. A method for generation of route segments which can be used here is described in Persson and Jungert (1992). The objects along each route segment are projected down to the segments by means of slope projections which are perpendicular to the segments, i.e. the slope coefficient of the objects to be projected down to the route line is determined by the slope

$$k_0 = -\frac{1}{k_r}, \qquad k_r \neq 0 \tag{9.43}$$

In (9.43) k_r is the slope of the current route segment; this is illustrated in Figure 9.9.

The projection string (9.44) created from Figure 9.9 where the projections of the objects are made down to line segment l_i has the start point p_{is} and the end point p_{ie} is labelled OTR (object to route).

$$\text{OTR:} \quad A_L \,|\, A_L B_R \,|\, B_L < C_R \,|\, C_L C_R \,|\, D_L C_L C_R \,|\, C_L C_R \,|\, C_R \tag{9.44}$$

The subscripts L and R means to the left and to the right of the route segment respectively. As can be seen, the technique works for both extended and for point objects. Note also that the direction of the path must be defined otherwise left and right sides are not uniquely defined and may therefore be mixed up. Examples of simple relations that can be inferred from (9.44) are:

A lies to the left of the route segment (RS)
B lies to the right of RS
A and *B* lie on opposite sides of RS and partly face to face
The RS transects the object C
A comes before *D*, on the left-hand side.

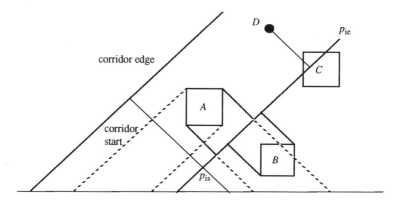

Figure 9.9 Object-to-route projection.

Obviously, some of the objects seen from l_i can be seen from the next route segment l_{i+1} as well. The technique can be used in navigation or just as a means for answering queries like "which is the next object" or "when will the next object of type X appear".

A few problems still remain to be solved. The first one concerns the size of the area to be searched for. The question is whether it is just those objects that are neighbors of a subroute that should be of concern. However, this problem is more or less application-dependent and for that reason it is assumed that the area has a width of $2d$ where the route segment divides the area in two equal halves, i.e. a road passing through a corridor with a width d on each side of the route segment. A second and more serious problem is how to find the objects in the corridor among all existing objects. There is a fairly elegant solution to this problem which at the same time solves the problem of generating the OTR string. The dashed projection lines from object A down to the x-axis and parallel to l_i in Figure 9.9 illustrate this. The corridor follows the slope of the route segment and can hence be projected down to the x-axis. All points can be projected down to the x-axis both along the direction of the corridor and perpendicular to it. As a consequence, two strings are generated along the x-axis. It is easy to see that the projections parallel to the route segment describes the relative distances of the objects seen from the route segment. Hence, of all objects projected down to the x-axis only those lying between the end points of the corridor are accepted in the OTR string. From the perpendicular projection direction the OTR string can be created. Hence, all information needed for the OTR string is generated in one step including the relative distances of the objects from the subroute. The later information may also be used in the inference process since it contains information about the relative distances between the objects with respect to the route segment. Hence, another string can be created that includes this information and which can be used to answer queries of the type: "which one of two objects is closest to the route segment?"

9.6. POLAR PROJECTION

The discussion so far has been concerned with projections in various directions primarily down to the x-axis. Since the projection direction at a point, from the observer's position, can be determined with respect to a line with a certain angle, it will also be possible to determine the relative angle between two points. Consequently, the relative orientation of an object with respect to O can be determined, i.e. to create strings corresponding to angular and radial projections. Figure 9.10 illustrates this where a radial projection is determined by the arc length between O and the point.

The technique for determination of polar projection can be described as: first create the supporting projection lines for each object point down to the x-axis

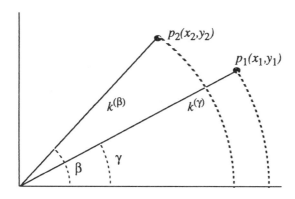

Figure 9.10 Determination of polar projection.

to determine the relative angles and distances. From this, two polar projection strings P_α and P_r can be generated. The example in Figure 9.10 thus gives the angular projection string

$$P_\alpha: \quad p_1 < p_2 \qquad (9.45)$$

while the radial projection string will be

$$P_r: \quad p_2 < p_1 \qquad (9.46)$$

In the angular projection the slope coefficients of the supporting lines can be used as a means for distinguishing all points in the space. Those coefficients are simple to calculate. For the points p_1 and p_2 in Figure 9.10 the angular projection string P_α (9.45) follows directly from

$$k^{(\gamma)} < k^{(\beta)} \qquad (9.47)$$

The supporting lines are used as cutting lines, see Chapter 5, and for that reason a modified edge-to-edge operation for polar projection, which here is denoted by "$|_\alpha$" where α stands for angular, can be defined. Applying polar projection to extended objects is not too complicated, as illustrated in the examples in Figures 9.11(a)–(d). From Figure 9.11(a) the angular projection string P_α can be determined:

$$P_\alpha: \quad B|_\alpha AB|_\alpha A \qquad (9.48)$$

In this particular case the following relation can, for instance, be inferred:

$$\text{partly-left-of}(B, A, O) \qquad (9.49)$$

which can be interpreted as: "B is partly to the left of A seen from O."
   In Figure 9.11(b) the angular projection string is

$$P_\alpha: \quad B < A \qquad (9.50)$$

which can be interpreted as "B is to the right of A seen from O":

$$\text{Right}(B, A, O) \qquad (9.51)$$

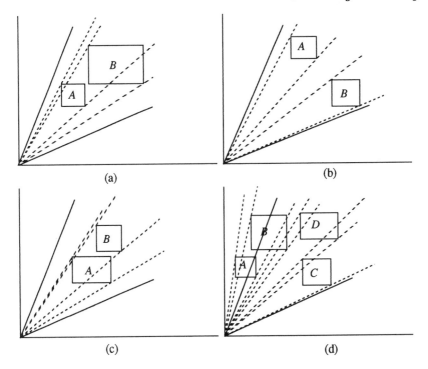

Figure 9.11 Angular projection from which it can be inferred that seen from O, (a) A is partly left of B, (b) B is right of A, (c) A and B have the same direction and (d) D is partly visible.

Figure 9.11(c) corresponds to:

$$P_\alpha: \quad A|_\alpha AB|_\alpha A \tag{9.52}$$

from which it can be inferred that:

$$\text{same\_direction}(B, A, O) \tag{9.53}$$

i.e. "B is in the same direction as A seen from O."

The situation in the image in Figure 9.11(d) is somewhat more complicated and so is the angular projection string:

$$P_\alpha: \quad C|_\alpha CD|_\alpha D|_\alpha BD|_\alpha B|_\alpha AB|_\alpha A \tag{9.54}$$

Several different conclusions can be inferred from the string in (9.54). One is that D is partly visible between C and B. This is determined from the substring $CD|_\alpha D|_\alpha BD$, from which it is easy to determine that D is between C and B but also that D can partly be seen from O since $\dots |_\alpha D|_\alpha \dots$ shows that there is no object in front of D in this sector.

The polar projection string technique has some similarities to Hough transforms, see for example Ballard and Brown (1982). However, Hough transforms are primarily concerned with the detection of boundaries of various types of objects such as straight lines and other types of curves. The method

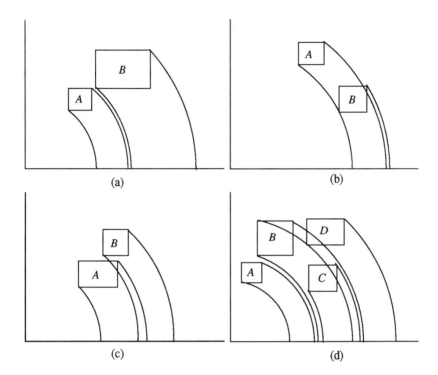

Figure 9.12 Illustrations to radial projection.

here, on the other hand, is concerned with identification of various types of relationships between objects. A similarity between the two methods is on a qualitative level since the accumulator that is generated as a result of the Hough transforms is a table with discrete values. Hence, that accumulator can be used as a means for qualitative reasoning about distances and directions in space.

Polar projection is not complete without distance projection. For calculation of the various distances either the method discussed in section 9.4.1. The four examples from Figure 9.12 can be used to illustrate this type of projection. As can be seen in Figure 9.12, the projected end points of the arcs corresponding to points of the MBRs are used in the creation of the radius projection strings. Hence, the following radial projections can be identified for the cases in Figure 9.12:

(a) $\qquad\qquad\qquad P_r: \quad A < B$

(b) $\qquad\qquad\qquad P_r: \quad AB|_\alpha B$

(c) $\qquad\qquad\quad P_r: \quad A|_\alpha AB|_\alpha B$

(d) $\qquad P_r: \quad A < B|_\alpha BC|_\alpha BDC|_\alpha BD|_\alpha D$

(9.55)

In this type of projection, as in the earlier cases, a variation of the edge-to-edge operator has been introduced, which is related to the general edge-to-edge operator. The operator is called *radius-edge-to-edge* and is denoted by "$|_r$".

The P_r strings can be used for deduction of various types of object relations, as in all earlier types of projection strings, although here the relations that are dealt with concern relative distances between the observer and the objects. An example from Figure 9.12(a) is:

$$\text{closer\_to}(A \ B \ O) \tag{9.56}$$

which is interpreted as: "*A* is closer to *O* than *B*". In Figure 9.12(b) it can be inferred that:

$$\text{equally-close-to}(A \ B \ O) \tag{9.57}$$

which means "*A* and *B* have the same distance from *O*." Finally, from Figure 9.12(c) the following relations can be identified:

$$\text{distance\_order}(A \ B \ C \ D \ O) \tag{9.58}$$

which is the order of distances of the full object set seen from *O*.

The main difference between radius projection and the original projection method, i.e. cartesian projection, is that the projected distances here correspond to absolute distances from the observer while in the cartesian case the projections are perpendicular to either the *x*- or *y*-coordinate axis and thus two distances must be dealt with. It is, however, quite clear that here also two strings have to be dealt with, i.e. the P_α and P_r strings, in order to be able to handle the full space and to be able to identify all possible relations. This is illustrated by the two strings created from Figures 9.11(c) and 9.12(a) which can be used to infer

$$\text{partly-behind}(B, A, O) \tag{9.59}$$

Another example that also requires knowledge from both strings is taken from the string given by Figures 9.11(c) and 9.12(c):

$$\text{behind}(B, A, O) \tag{9.60}$$

Clearly, for this qualitative method there is a relationship between polar and cartesian projection, just as in the traditional mathematical cases. This raises the question whether projection strings in one representation can be transformed into another, e.g. from polar to cartesian strings and vice versa. The problem can also be stated as whether the strings of one qualitative method can be translated into another method and back without using any extra information beside what is available in the expressions. This problem will be discussed further in the next section.

9.7. TRANSFORMATIONS BETWEEN THE QUALITATIVE SYSTEMS

It is sometimes of interest to transform projection strings in one representation into another. Hence, it could be of particular interest to transform strings in

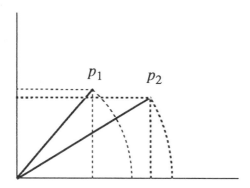

Figure 9.13 Illustration of the differences between cartesian and polar projections.

polar representation $\langle P_\alpha, P_r \rangle$ into strings in cartesian representation $\langle u, v \rangle$ corresponding to the original method of symbolic projections, and the other way around, i.e.

$$\langle P_\alpha, P_r \rangle \Leftrightarrow \langle u, v \rangle \tag{9.61}$$

It turns out that such transformations can only be made if quantitative information is used. To understand why this is so consider the situation in Figure 9.13 with the following strings:

$$\langle u, v \rangle: \quad \langle p_1 < p_2, p_2 < p_1 \rangle \tag{9.62}$$

$$\langle P_\alpha, P_r \rangle: \quad \langle p_2 < p_1, p_1 < p_2 \rangle \tag{9.63}$$

What matters here is that if polar strings are created directly from cartesian strings then they must be directly generated without use of any other information. This cannot be done by using the cartesian strings alone: coordinate information of the objects must be available as well. No such information is directly available in any of the strings.

The same type of argument can be used when trying to transform cartesian strings into polar strings. Thus for the cartesian strings the exact angles and distances are needed for the coordinates that must be used for generation of the polar strings. The conclusion is that no transformations can be made without the use of quantitative information.

A further question that can be raised finally is whether it is always impossible to transform one qualitative representation into another without the use of quantitative information. This problem is not solved here. It is probably not the case, but so far no proof exists.

REFERENCES

Ballard, D. H. and Brown, C. M. (1982) *Computer Vision*, pp. 123–31. Prentice-Hall, Englewood Cliffs, NJ.

Borgefors, G. (1986) Distance transforms in digital images. *Computer Vision, Graphics and Image Processing* **34**, 344–71.

Hernández, D., Clementini, E. and Di Felice, P. (1995) Qualitative distances, *Spatial Information Theory – A Theoretical Basis for GIS*, pp. 45–58, Springer Verlag, Heidelberg.

Jungert, E. (1992) The observer's point of view: An extension of symbolic projections. In *Theories and Methods of Spatio-Temporal Reasoning in Geographic Space*, pp. 179–95. Springer-Verlag, Heidelberg.

Jungert, E. (1994) Qualitative Spatial Reasoning from the Observer's Point of View – Towards a Generalization of Symbolic Projection, in *Pattern Recognition* **27**(6), 801–813.

Peuquet, D. J. and Zhang, C. X. (1987) An algorithm to determine the directional relationship between arbitrarily-shaped polygons in the plane. *Pattern Recognition* **20**, 65–74.

Persson, J. and Jungert, E. (1992) Generation of multi-resolution maps from run-length-encoded data. *International Journal of Geographical Information Systems* **6**, 497–510.

10

Spatial Reasoning and Applications

Symbolic Projection was originally developed as a technique for iconic indexing to image databases, and this still is an important application area. However, at an early point of the evolution of the theory it turned out that Symbolic Projection also includes other characteristics that made it suitable for various forms of spatial reasoning and in particular for qualitative spatial reasoning. The latter has become a very important research topic, not only because of the need for methods that allow deduction of complex object relations, but also because qualitative spatial reasoning techniques can be used as a means for cognitive modelling. This makes spatial reasoning particularly important, and for that reason it is not only methods based on Symbolic Projection that are of interest. Therefore, in this chapter, qualitative spatial reasoning in a wider perspective is considered and some other approaches besides Symbolic Projection are also discussed.

Section 10.1 gives an introduction to qualitative spatial reasoning; in section 10.2 a classification scheme of the subject is given together with some well-known approaches. The chapter ends with section 10.3, which contains an application of Symbolic Projection applied to route planning.

10.1. INTRODUCTION TO QUALITATIVE SPATIAL REASONING

Efficient methods for spatial reasoning are becoming increasingly necessary in a large variety of applications in which quite often temporal data must be combined and manipulated as well. The demand for such methods, which at the same time is a challenge, is fuelled by increasing volumes of spatial data represented in finer resolution than has hitherto been possible. Methods that can handle and analyze many terabytes of data will be required in the near future. Such methods must become a reality, although at present it is hard to see how the problem can be solved. The quantities of data that must be handled will doubtless increase for a long time to come. The reason for this is that various types of sensors that select information are being developed and are coming into use for a large number of applications. Such sensors can generate images with different characteristics. All these sensors are being installed on various types of platforms, and they can produce extremely large quantities of data in both high and low resolution. Furthermore, it is no longer sufficient to analyze data from just a single sensor. Methods that can support fusion and filtering of data from many different sensors as well as methods for data reduction, where loss of information is minimized, are also required. Needless to say, all these methods that need to be developed must be efficient because otherwise they will not be useful. At present however, there is a shortage of methods that can contribute to the solution of these problems. Symbolic Projection is one approach to the steps that must be taken to solve these problems.

All reasoning methods must be based on some kind of model or theory, and spatial reasoning is not different in this respect. Spatial reasoning must be "influenced by the concept of the underlying space" (Frank, 1992). To carry this further, in spatial reasoning the underlying space generally relies on euclidean geometry. However, the drawback with this type of geometry is that it cannot or does not serve as a theory for reasoning. Taking the increasing data quantities from the discussion above into account, the problem is that euclidean geometry does not consider the data reduction problem, and cannot handle the massive data quantities that will need to be analyzed in the future; not in any computer system and certainly not within any reasonable time. Hence, other methods must be considered. Clearly, such methods cannot go beyond euclidean geometry; on the contrary, they must be well rooted in that geometry, which for many reasons can be considered universal.

The main question is thus which direction should be chosen to solve these problems. The classical solution is to build up a search strategy that can be used as an indexing technique which can eventually make it possible to prune all the incorrect branches in the search space as early as possible. In image processing and related areas such methods are called *iconic indexing* and Symbolic Projection was originally developed as a method for this. Iconic indexing methods may be of different types but common to them all is that they are founded in techniques allowing both pruning and data reduction. Pruning means here that paths leading to incorrect solutions should be cut off early in the search process. Consequently, when we are closer to the final solution more acceptable quantities of high-resolution data can be comprehended which, in the end, will lead efficiently to correct or at least optimal solutions. Symbolic Projection, originally introduced for this purpose, displays metric information exactly in qualitative form. In other words, methods which are symbolic can be used for reasoning in qualitative terms and at the same time allow reduction of data without loss of information.

A good theory should always guarantee that it leads to correct inferences established by soundness and completeness of the method (Smith and Park, 1992), and spatial reasoning methods should not be permitted to diverge from this criterion. Hence, inferred relationships must always respond to these two requirements. Other requirements for identification of certain object attributes and relations exist and cannot be neglected either, although they are of less importance. Obviously, there is a need for various conceptualizations, corresponding to a formal language, that will eventually allow for a set of inference techniques to carry out the reasoning process and can finally infer whatever is required. Throughout a reasonably large set of applications, this has been demonstrated by Symbolic Projection. Thus it has been shown that the method is founded on a formal language, i.e. the projection strings. These projection strings make it possible to identify most types of spatial relations and properties in a way that demonstrates both completeness and soundness.

It has already been mentioned that Symbolic Projection is a method that performs spatial reasoning qualitatively. Another cornerstone, also pointed out earlier, is that in qualitative reasoning data can be reduced without loss of information. This is important, since correct conclusions can otherwise not be drawn. However, most qualitative spatial reasoning techniques generally contain other important and useful characteristics. For instance, Freksa and Röhrig (1993) also argue that "higher cognitive mechanisms employ qualitative rather than quantitative mechanisms even if the knowledge originally is available in quantitative forms through perception." Freksa and Röhrig argue furthermore that, "qualitative knowledge can be viewed as that aspect of knowledge which critically influences decisions." That is to say, methods based on qualitative knowledge can helpfully support the development of various types of decision support tools, which is true for symbolic reasoning as well.

The work by Allen that was discussed earlier is concerned with qualitative temporal reasoning, and has had an impact on much of the later work in the area of qualitative spatial reasoning. The reason for this is that the basic set of temporal relations that was introduced by Allen can directly be mapped into the spatial area, hence many similarities between spatial and temporal reasoning can be pointed out, as discussed in Chapter 5. However, temporal reasoning is simpler to deal with in that it includes only a single dimension and therefore can be more or less considered as a "linear" problem. The nature of spatial reasoning problems is much more complex. Despite this, Allen's work on temporal reasoning work must clearly not be underestimated, and it has certainly had strong impact on spatial reasoning. Several approaches exist that evidently have a clear relationship to his work, among them Symbolic Projection as illustrated in the work by Holmes and Jungert (1992) described in Section 10.3 be mentioned.

It may not be clear from this discussion on qualitative spatial reasoning that creating methods for solving problems on a high level of abstraction is not sufficient. This is due to the fact that sooner or later any method, no matter which level of abstraction it works on, must use basic data in full resolution. In other words, the "bits and pieces" must never be forgotten. Hence, when the methods are developed not only the qualitative aspects should be of concern but the quantitative ones as well. A lot of effort must also be spent on handling basic data so that both levels will work "hand in hand", allowing for efficiency from top to bottom and vice versa. In Symbolic Projection run-length code (Chapter 2) has been used as a basic data structure which has made it possible to include the low-level aspects efficiently in the reasoning process.

This discussion has mainly focused on the qualitative aspects of symbolic reasoning, but other important aspects have been pointed out as well, such as soundness and completeness of the method. There is also a further aspect of great importance that has been mentioned several times: this is the enormous quantities of data that must be analyzed in the future. Basically, all qualitative methods must allow reduction of data without loss of information. Symbolic Projection is no exception.

10.2. OTHER REASONING APPROACHES

The importance of qualitative reasoning methods in reasoning about space will in the future become extremely important, which motivates discussion not just on Symbolic Projection but also on other methods. For this reason it is necessary to give a classification of some existing qualitative approaches. However, it is not claimed that the classification given here is complete in the sense that all existing methods are covered: on the contrary, only a few relevant examples will be discussed. Furthermore, these examples have been chosen to represent different aspects that are highly relevant in qualitative spatial reasoning and for this reason must be considered when the methods are classified. The classification method that is used here was originally suggested by Freksa and Röhrig (1993) and is illustrated in Figure 10.1. The hierarchical method is for

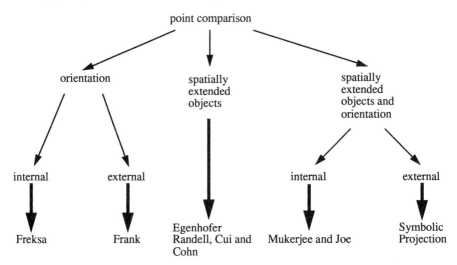

Figure 10.1. Classification scheme for various approaches to qualitative spatial reasoning according to Freksa, represented with the given approaches.

the most part self-descriptive, except for the terms "internal" and "external" which indicate whether the position of the reference axes is guided by *internal* properties of the object or by *external* properties. Symbolic Projection is an example of the latter, which is obvious when considering how the directions of the projections are determined – independently of whether the projections are made with a slope or perpendicular to a coordinate axis. The descriptions of the related qualitative approaches given in this section are not entirely complete: only the outlines of their basic theories are presented.

10.2.1. The Conceptual Neighbourhood Approach (Freksa)

The conceptual neighbourhood approach of Freksa (1992) and Freksa and Zimmerman (1994) is mainly concerned with spatial representation and reasoning through intuition. In other words, spatial reasoning should be concerned with cognitive aspects rather than with formalistic methods. Of interest is thus the use of spatial metaphors based on natural language including relations such as "after", "between", etc. Another main goal in Freksa's work has been to identify the concepts of a spatial inference engine which can deal with spatial knowledge similar to that of biological systems. As a consequence, the use of imprecise, partial or subjective knowledge must be dealt with.

The type of problems that have been dealt with so far can be described in terms of the following metaphoric example. Imagine someone walking through a forest, perhaps following a path. While doing so, landmarks of various types may be discovered. Such landmarks may not be visible to the walker at all times, since the trees of the forest may occlude them at certain positions. From experience, people know that even if a landmark cannot be seen at a certain time in space it is still there, and by observing it along the path it will be possible to estimate its position at any point along the path that is being followed. This type of problem can be dealt with by using the technique introduced by Freksa.

The basic so-called *orientation grid* that is used for representation of the qualitative spatial information originally introduced by Freksa (1992) is aligned to the orientation that can be determined from two points in a 2D space. These two points, normally denoted by **a** and **b**, together determine a vector with the start point at **a** and the end at **b**. Including a third addition point **c** then it is possible to determine the relative position of point **c** with respect to the vector **ab**. (The point **c** corresponds to the landmark in the example above.) The position of **c** relative to vector **ab** can thus, for instance, be to the right, left or even in front of it. The method is, however, not just limited to this: the vector can be viewed as a path between **a** and **b**. By taking these aspects into consideration it is possible to create a spatial model that can be used for qualitative spatial reasoning. This model is illustrated in Figures 10.2(a)–(c).

By taking the vector **ab** and its relative position in space it is possible to split the space surrounding **ab** into six different areas, thus creating a grid. From this grid structure it is possible to identify 15 qualitative positions relative to the vector **ab**. Among these, six appear in the grid areas, seven positions occur on the grid lines taking the vector itself into account, and two coincide with the points **a** and **b**.

A technique for spatial reasoning based on the 15 qualitative positions and the corresponding relations to **ab** can now be introduced. Given the position of **c** relative to the vector **ab** and the position **d** relative to the vector **bc**, a composition relation can be formed that describes the position of **d** relative to the vector **ab**. These relations are abbreviated to **ab:c** and **bc:d** respectively. The problem is thus to find **d** relative to the vector **ab**, i.e. **ab:d**. This relation cannot always be identified in terms of the 15 possible relationships defined by the basic grid structure; sometimes a multiple set will occur. Therefore, in addition, a set of primitive operations have also been defined. Three such primitive operations are inverse (INV), shortcut (SC) and homing (HM). Examples of the use of these operators are INV **ab:c** = **ba:c**, SC **ab:c** = **ac:b** and HM **ab:c** = **bc:a**. Combinations of these three operators occur frequently and can, for reason of convenience, be merged such that:

$$\text{HMI}(\mathbf{ab}:\mathbf{c}) = \text{INV}(\text{HM}(\mathbf{ab}:\mathbf{c})) \qquad (10.1)$$

$$\text{SCI}(\mathbf{ab}:\mathbf{c}) = \text{INV}(\text{SC}(\mathbf{ab}:\mathbf{c})) \qquad (10.2)$$

Finally, an identity operator (ID) has been identified as well.

The operations INV, SC and HM form a closed set of operations allowing for all possible inferences about qualitative static positions of objects with respect

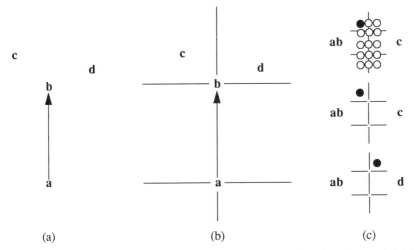

(a)   (b)   (c)

Figure 10.2 The vector path from **a** to **b** with two observed landmarks **c** and **d** (a). By introducing "horizontal" and "vertical" lines relative to the vector **ab** a grid for determining the position of **c** and **d** is created (b). From the 15 possible positions of the grid the positions of **c** and **d** can be determined (c).

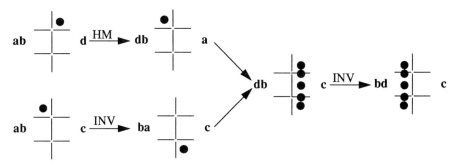

Figure 10.3 The composition of **bd : c** from **ab : c** and **ab : d**.

to their reference vectors. For example, consider the problem of inferring the relation of position **c** with respect to the vector **bd** from the two locations **c** and **d**. Both these two positions are known relative to the vector **ab**. This problem can be solved as follows. Apply the homing operator to the vector **ab** to get the relationship between **a** and the vector **db**, that is, HM **ab : d** = **db : a**. The next step is then to apply INV to **ab : c**, which corresponds to INV **ab : c** = **ba : c**. Now there is a direct connection between the two vectors **db** and **ba** which connects at **b**. Hence a composition can be made such that **c** can be obtained with respect to **db**. Finally, by applying INV to this, the desired result becomes available as illustrated in Figure 10.3. Observe, however, that the result may be obtained in a different way from other compositions as well.

10.2.2. Qualitative Determination of Distances and Directions (Frank)

Frank (1992) has introduced a qualitative reasoning technique for determination of cardinal distances and directions in which euclidean geometry is used as a source of intuition. This can be illustrated with examples of the following type:

> Given *Stockholm is north of Copenhagen*, and *Copenhagen is north of Berlin*, it can be deduced that *Stockholm is north of Berlin*.

The above illustrates deduction of a transitive relationship where data is imprecise and where the result becomes less precise than in quantitative problems. This is desirable when precise, quantitative information is not available and furthermore, when precision is not desirable.

   Frank's approach combines directional symbols in terms of an algebra where an identity symbol is introduced as well. The algebraic concept allows identification of rules for manipulation of both directional and distance symbols when combined with various operators. In the algebra vector addition ($+_v$) and subtraction ($-_v$), for example, are both qualified by a subscript, to indicate the type of algebra in which they occur. Hence typical axioms are expressed as:

$$a +_v 0_v = a \text{ where } 0_v \text{ is the identity element} \tag{10.3}$$

$$a +_v b = b +_v a \qquad\qquad\qquad \text{(commutative law)} \tag{10.4}$$

$$a +_v (b +_v c) = (a +_v b) +_v c = a +_v b +_v c \qquad \text{(associative law)} \tag{10.5}$$

$$-_v (-p_a) = a \tag{10.6}$$

$$a +_v (0 -_v a) = 0_v \qquad\qquad\qquad \text{(inverse of } +_v) \tag{10.7}$$

A rule is *euclidean exact* when the result is the same when compared to the result obtained from an analytical function with data in full resolution. This is also called a *homomorphism*. When the results differs for some data the quantitative rule is *euclidean approximate*. From this the following two definitions can be identified where $*_c$ corresponds to symbolic composition for directions and $+_d$ for distance.

Definition 10.1. A rule for quantitative reasoning on directions is called *euclidean exact* (or just exact) if dir (P_1, P_2) is a homomorphism:

$$\text{dir}(P_1, P_2) *_c \text{dir}(P_2, P_3) = \text{dir}((P_1, P_2) +_v \text{dir}(P_2, P_3)) \tag{10.8}$$

Definition 10.2. A rule for qualitative reasoning on distances is called *euclidean exact* if dist(P_1, P_2) is a homomorphism:

$$\text{dist}(P_1, P_2) +_d \text{dist}(P_2, P_3) = \text{dist}((P_1, P_2) +_v (P_2, P_3)) \tag{10.9}$$

In the above the function dir (P_i, P_j) maps the vector $\langle P_i, P_j \rangle$ into the direction symbol, while the function dist(P_i, P_j) maps the vector into the distance symbol. Thus, the vector addition $(+_v)$ is transformed into the composition $(*_c)$ for direction in (10.8) and into $(+_d)$ for distance in (10.9).

A direction path from a start point P_1 to an end point P_2 is called a *path*, denoted $\langle P_1, P_2 \rangle = s_1$. The inverse to this path, i.e. from P_2 to P_1, is denoted by "$-_p$", from which we get

$$-_p \langle P_i, P_j \rangle = \langle P_j, P_i \rangle \tag{10.10}$$

$$-_p(-_p s) = s \tag{10.11}$$

Composition is an operation that merges two contiguous paths, from P_1 to P_2 and P_2 to P_3, into the path P_1 to P_3, see Figure 10.4. The figure shows that composition is noncommutative, as in equation (10.14). The algebra for paths can be defined in the following way:

$$\langle P_1, P_2 \rangle *_p \langle P_2, P_3 \rangle = \langle P_1, P_3 \rangle \tag{10.12}$$

$$\langle P_1, P_2 \rangle *_p \langle P_2, P_3 \rangle = \langle P_1, P_4 \rangle *_p \langle P_4, P_3 \rangle = \langle P_1, P_3, P_4 \rangle \tag{10.13}$$

$$s_1 *_p s_2 \neq s_2 *_p s_1 \tag{10.14}$$

$$a *_p (b *_p c) = (a *_p b) *_p c = a *_p b *_p c \tag{10.15}$$

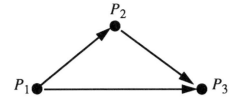

Figure 10.4 Composition of two contiguous paths.

Concatenation and composition of paths are not the same, since composition corresponds to a single path where the intermediate vertices are dropped while in concatenation the intermediate points are retained thus corresponding to a kind of compound path.

An identity path corresponds to a path from a point to itself and is denoted 0_p. The identity path follows directly from the composition definition in (10.12), and hence:

$$\langle P_1, P_1, P_2 \rangle = 0_p \tag{10.16}$$

$$p *_p 0_p = 0_p *_p p = p \tag{10.17}$$

$$-_p 0_p = 0_p \tag{10.18}$$

The inverse to a binary operation is defined such that a value combined with its inverse results in the identity value, that is:

$$p *_p (-_p p) = 0_p \quad \text{and} \quad (-_p p) *_p p = 0_p \tag{10.19}$$

$$\text{dir} \langle P_1, P_2 \rangle_c \, \text{dir} \langle P_2, P_1 \rangle = \text{dir} \langle P_1, P_1 \rangle \tag{10.20}$$

For any paths a and b a path x exists such that $x *_p a = b$ and $a *_p x = b$:

$$-_p (a_c b) = -_p b_c -_p a \tag{10.21}$$

We have now considered the various aspects of Frank's path algebra. Other aspects of interest in his work are the qualitative distance properties which will just be discussed briefly here. Frank defines distance as a function which maps a pair of points to some positive real numbers with the following properties:

$$\text{dist} (P_1, P_1) = 0 \tag{10.22}$$

$$\text{dist} (P_1, P_2) = \text{dist} (P_2, P_1) \tag{10.23}$$

$$\text{dist} (P_1, P_2) + \text{dist} (P_2, P_3) \geqslant \text{dist} (P_1, P_3) \tag{10.24}$$

In the qualitative distance cases the result of the mapping function may for example correspond to the qualitative value set $D = \{VC, C, M, F, VF\}$ which varies from "very close" to "very far". To fulfil the properties of distance addition $(+_d)$ and comparison $(\leqslant_d)$, operations are defined on the given qualitative distance symbols. The set must include a smallest element (0_d) which is the distance of the identity path, that is VC in D. From this, distance addition and monotonic increase can be identified which both inherit commutativity from addition of reals:

$$d_1 +_d d_2 = d_2 +_d d_1 \qquad (10.25)$$

$$d_2 \geqslant_d d_3 \Rightarrow d_1 +_d d_2 \geqslant_p d_1 +_d d_3 \qquad (10.26)$$

Summing up the distance values of qualitative type along a concatenated path now becomes possible. Adding one distance to another increases the value or leaves it unchanged, but does not decrease it.

$$\text{dist: point} \times \text{point} \to D \quad \text{or} \quad \text{dist: path} \to D \qquad (10.27)$$

$$\text{dist}\,(0_p) = 0_d \qquad (10.28)$$

$$d_1 +_d d_2 \geqslant_d d_1 \geqslant \max\,(d_1, d_2) \qquad (10.29)$$

Some further rules can also be identified that correspond to nearly homomorphic mappings. Thus composition of paths $*_p$ can be mapped onto additions of distances $+_d$, that is:

$$\text{dist}\,(p_1 *_p p_2) = \text{dist}\,(p_1) +_d \text{dist}\,(p_2) \qquad (10.30)$$

Considering again the three cities Stockholm, Copenhagen and Berlin as in the example in the beginning of this section, the distance between Stockholm and Berlin can be deduced given the following distances:

$$\text{dist}\,(\text{Stockholm, Copenhagen}) = F \qquad (10.31)$$

$$\text{dist}\,(\text{Copenhagen, Berlin}) = C \qquad (10.32)$$

Thus

$$\text{dist}\,(\text{Stockholm, Berlin}) = \text{dist}\,(\langle\text{Stockholm, Copenhagen}\rangle$$

$$*_p \langle\text{Copenhagen, Berlin}\rangle)$$

$$= \text{dist}\,(\langle\text{Stockholm, Copenhagen}\rangle)$$

$$+_d \text{dist}\,(\langle\text{Copenhagen, Berlin}\rangle)$$

$$= F +_d C = F \qquad (10.33)$$

This result is, of course, a qualitative or euclidean approximation.

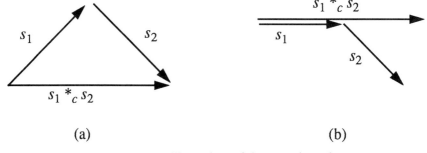

(a)   (b)

Figure 10.5 Two illustrations of the averaging rules.

Table 10.1

	N	NE	E	SE	S	SW	W	NW	0
N	N	n	ne	o	o	o	nw	nw	N
NE	n	NE	ne	e	o	o	o	n	NE
E	ne	ne	E	e	se	o	o	o	E
SE	o	e	e	SE	se	s	o	o	SE
S	o	o	se	se	S	s	sw	o	S
SW	o	o	o	s	s	SW	sw	w	SW
W	nw	o	o	o	sw	sw	W	w	W
NW	nw	n	o	o	o	w	w	NW	NW
0	N	NE	E	SE	S	SW	W	NW	0

Frank also discusses the properties of cardinal directions such as {N, NW, W, SW, S, SE, E, NE}, which corresponds to a structure of eight directional segments. For this structure, a table including a set of "averaging" rules where the entries correspond to two different directional segments and the actions are taken to conclude the final directions. This is illustrated in Figure 10.5. For instance, NE combined with SE results in E and E combined with SE gives E, while N combined with NW gives NW. To make the rules complete a 0 value is also required, thus resulting in 81 pairs of values. The lower-case instances in the table denote euclidean approximate instances. In Table 10.1 only 24 instances are inferred exactly, while 25 have a value that is equal to 0. Among the others, there are 32 with approximate results.

10.2.3. Composition of Binary Topological Relations (Egenhofer)

Egenhofer, whose information structure was introduced in Chapter 5, provides the basis for determination of transitive topological relations by using the topological relations demonstrated in section 5.6.2 and hence used for this purpose. Thus, given the topological relations represented by the 9-intersection, their composition can be determined by deriving the 9-intersection of the combined topological relation. Since spatial knowledge about topological relations is encoded into set intersections, standard transitive inference rules about point sets can be applied. Every object has exactly three parts (interior, boundary and exterior), which can be in a particular relationship to the parts of another object, and vice versa. Altogether, there are eight relevant combinations that form knowledge about the combined relation. They are statements of implications that can be proven to be theorems of set theory.

Theorem 10.1. A nonempty intersection between two parts A and B implies a nonempty intersection between the parts A and C if B is a subset of C, i.e.

$$A \cap B = \neg \emptyset \wedge B \subseteq C \Rightarrow A \cap C = \neg \emptyset \qquad (10.34)$$

Proof. Let x be a nonempty element such that $x \in A$ and $x \in B$. Since $B \subseteq C$, $x \in C$ as well. Thus, $x \in (A \cap C)$ and, therefore, $A \cap C = \neg \varnothing$. $\square$

Corollary 10.2

$$A \supseteq B \wedge B \cap C = \neg \varnothing \Rightarrow A \cap C = \neg \varnothing \qquad (10.35)$$

Proof. By replacing A and C in equation (10.35) with C and A, respectively, and reordering the terms on the left-hand side. $\square$

Theorem 10.3. An empty intersection between the parts A and B implies an empty intersection between the parts A and C if C is a subset of B, i.e.

$$A \cap B = \varnothing \wedge B \supseteq C \Rightarrow A \cap C = \varnothing \qquad (10.36)$$

Proof. Let $x \in C$. Since $B \supseteq C$, $x \in B$ as well. On the other hand, since $A \cap B = \varnothing$, $x \in B$ implies $x \notin A$. Therefore, $x \in C$ implies $x \notin A$, or $A \cap C = \varnothing$. $\square$

Corollary 10.4

$$A \subseteq B \wedge B \cap C = \varnothing \Rightarrow A \cap C = \varnothing \qquad (10.37)$$

Proof. By replacing A and C in equation (10.37) by C and A, respectively, and reordering the terms on the left-hand side. $\square$

Corollary 10.5. A nonempty intersection between the parts A and B implies a nonempty intersection with the union of the two parts C_0 and C_1 if B is a subset of the union of C_0 and C_1, i.e.

$$A \cap B = \neg \varnothing \wedge B \subseteq (C_0 \cup C_1) \Rightarrow A \cap (C_0 \cup C_1) = \neg \varnothing \qquad (10.38)$$

Proof. Immediately from equation (10.34) substituting C with $(C_0 \cup C_1)$. $\square$

Corollary 10.6

$$(A_0 \cup A_1) \supseteq B \wedge B \cap C = \neg \varnothing \Rightarrow (A_0 \cup A_1) \cap C = \neg \varnothing \qquad (10.39)$$

Proof. By replacing A and C_0 and C_1 in equation (10.38) with C, A_0 and A_1, respectively, and reorder the terms on the left-hand side. $\square$

Corollary 10.7. An empty intersection between A and the union of B_0 and B_1 implies an empty intersection between A and C if C is a subset of the union of B_0 and B_1, i.e.

$$A \cap (B_0 \cup B_1) = \varnothing \wedge (B_0 \cup B_1) \supseteq C \Rightarrow A \cap C = \varnothing \qquad (10.40)$$

Proof. Immediately from equation (10.36) substituting B with $(B_0 \cup B_1)$. $\square$

Corollary 10.8

$$A \subseteq (B_0 \cup B_1) \wedge (B_0 \cup B_1) \cap C = \emptyset \Rightarrow A \cap C = \emptyset \quad (10.41)$$

Proof. By replacing A and C in equation (10.40) with C and A, respectively, and reordering the terms on the left-hand side. □

This set of eight rules is sufficiently complete to describe the dependencies of the intersections. Further considerations about the union of three parts are unnecessary since these cases are trivial. For instance, the derived intersections of nonempty intersections over the union of three parts are impossible since every part must be included in the universe. On the other hand, the following constraint must hold true for every nonempty intersection, because it is impossible that all three intersections with another part are empty:

$$A \cap B = \neg\emptyset \wedge B \subseteq (C_0 \cup C_1 \cup C_2) \Rightarrow A \cap (C_0 \cup C_1 \cup C_2) = \neg\emptyset \quad (10.42)$$

The eight intersections are not orthogonal since Equations (10.34)–(10.37) are included in equations (10.38)–(10.41), respectively, if A is a subset of B_0 or B_1. This redundancy is eliminated if Equations (10.38)–(10.41) are modified so that they exclude the configurations covered by Equations (10.34)–(10.37). Let $\sqsubseteq$ be the relationship between a set A and the union of set B and C so that $A \subseteq (B \cup C)$ and $A \cap B = \neg\emptyset$ and $A \cap C = \neg\emptyset$:

$$A \cap B = \neg\emptyset \wedge B \sqsubseteq (C_0 \cup C_1) \Rightarrow A \cap (C_0 \cup C_1) = \neg\emptyset \quad (10.43)$$

$$(A_0 \cup A_1) \sqsupseteq B \wedge B \cap C = \neg\emptyset \Rightarrow (A_0 \cup A_1) \cap C = \neg\emptyset \quad (10.44)$$

$$A \cap (B_0 \cup B_1) = \emptyset \wedge (B_0 \cup B_1) \sqsupseteq C \Rightarrow A \cap C = \emptyset \quad (10.45)$$

$$A \sqsubseteq (B_0 \cup B_1) \wedge (B_0 \cup B_1) \cap C = \emptyset \Rightarrow A \cap C = \emptyset \quad (10.46)$$

The eight rules in Equations (10.34)–(10.37) and (10.43)–(10.46) can be applied to drive the 9-intersection of the combined topological relation if $A \subseteq B$, $A \sqsupseteq B$, $A \sqsubseteq (B_0 \cup B_1)$, $(A_0 \cup A_1) \sqsupseteq B$, $A \cap (B_0 \cup B_1) = \neg\emptyset$ and $A \cap (B_0 \cup B_1) = \emptyset$ can be represented in terms of the 9-intersection ($A \cap B = \emptyset$ and $A \cap B = \neg\emptyset$ are already in this canonical representation). The following transformations apply: $a_i \neq a_j \neq a_k$, $b_l \neq b_m \neq b_n$ and $c_0 \neq c_p \neq c_q$.

a_i is a subset of b_l if and only if $I[a_i, b_l]$ is nonempty, while the two intersections between a_i and the other two parts are empty (equation 10.47). This mapping of the subset relation onto the 9-intersection is obvious, because the nonempty intersection between a_i and b_l is immediately derived from the subset relation between nonempty sets. Since the three parts of B are pairwise disjoint, a nonempty intersection $a_i \cap b_m$ or $a_i \cap b_n$ would imply that there are some parts of a_i outside of b_l, which would contradict the subset relation:

$$I[a_i, b_l] = \neg\emptyset \wedge I[a_i, b_m] = \emptyset \wedge I[a_i, b_n] = \emptyset \quad (10.47)$$

Conversely, a_i is a superset of b_l if and only if:

$$I[a_i, b_l] = \neg\emptyset \wedge I[a_j, b_l] = \emptyset \wedge I[a_k, b_l] = \emptyset \quad (10.48)$$

$a \sqsubseteq (b_l \cup b_m)$ if the intersections $I[a_i, b_l]$ and $I[a_i, b_m]$ are nonempty, while the third intersection between a_i and b_n is empty (equation 10.49). By the definition of $\sqsubseteq$, $a_i \cap b_l = \neg\varnothing$ and $a_i \cap b_m = \neg\varnothing$. Since b_n is disjoint from both b_l and b_m, its intersection with a_i must be empty, otherwise a_i would have some parts outside of $b_l \cup b_m$, which would contradict the subset relation:

$$I[a_i, b_l] = \neg\varnothing \wedge I[a_i, b_m] = \neg\varnothing \wedge I[a_i, b_n] = \varnothing \qquad (10.49)$$

Conversely, $(a_i \cup a_j) \sqsupseteq b_l$ if:

$$I[a_i, b_l] = \neg\varnothing \wedge I[a_j, b_l] = \neg\varnothing \wedge I[a_k, b_l] = \varnothing \qquad (10.50)$$

The intersection of a_i with $b_l \cup b_m$ is nonempty if at least one of the two intersections $I[a_i, b_l]$ and $I[a_i, b_m]$ is nonempty:

$$\neg(I[a_i, b_l] = \varnothing \wedge I[a_i, b_m] = \varnothing) \qquad (10.51)$$

Complementary, $a_i \cap (b_l \cup b_m)$ is empty if and only if:

$$I[a_i, b_l] = \varnothing \wedge I[a_i, b_m] = \varnothing \qquad (10.52)$$

The eight inference rules about the intersections of the combined topological relations are derived by using equations (10.47)–(10.52) in (10.34)–(10.37) and (10.43)–(10.46). Thus eight new inference rules can be deducted for the 9-intersection and eventually the transitive relations.

Equation (10.47) in equation (10.34):

$$I_x[a_i, b_l] = \neg\varnothing \wedge I_y[b_l, c_0] = \varnothing \wedge I_y[b_l, c_p] = \varnothing \wedge I_y[b_l, c_q]$$
$$= \neg\varnothing \Rightarrow I_z[a_i, c_q] = \neg\varnothing \qquad (10.53)$$

Equation (10.48) in equation (10.35):

$$I_x[a_i, b_l] = \neg\varnothing \wedge I_x[a_j, b_l] = \varnothing \wedge I_x[a_k, b_l] = \varnothing \wedge I_y[b_l, c_0] = \neg\varnothing$$
$$\Rightarrow I_z[a_i, c_0] = \neg\varnothing \qquad (10.54)$$

Equation (10.48) in equation (10.36):

$$I_x[a_i, b_l] = \varnothing \wedge I_y[b_i, c_0] = \neg\varnothing \wedge I_y[b_m, c_0] = \varnothing \wedge I_y[b_n, c_0] = \varnothing$$
$$\Rightarrow I_z[a_i, c_0] = \varnothing \qquad (10.55)$$

Equation (10.47) in equation (10.37):

$$I_x[a_i, b_l] = \neg\varnothing \wedge I_x[a_i, b_m] = \varnothing \wedge I_x[a_i, b_n] = \varnothing \wedge I_y[b_l, c_0] = \varnothing$$
$$\Rightarrow I_z[a_i, c_0] = \varnothing \qquad (10.56)$$

Equation (10.49) and (10.51) in equation (43):

$$I_x[a_i, b_l] = \neg\varnothing \wedge I_y[b_l, c_0] = \neg\varnothing \wedge I_y[b_l, c_p] = \varnothing \wedge I_y[b_l, c_q] = \neg\varnothing$$
$$\Rightarrow \neg(I_z[a_i, c_0] = \varnothing \wedge I_z[a_i, c_q] = \varnothing) \qquad (10.57)$$

Equations (10.50) and (10.51) in equation (10.44):

$$I_x[a_i, b_l] = \neg\varnothing \wedge I_x[a_j, b_l] = \neg\varnothing \wedge I_x[a_k, b_l] = \varnothing \wedge I_y[b_l, c_0] = \neg\varnothing$$
$$\Rightarrow \neg(I_z[a_i, c_0] = \varnothing \wedge I_z[a_j, c_0] = \varnothing) \tag{10.58}$$

Equations (10.52) and (10.50) in equation (10.45):

$$I_x[a_i, b_l] = \varnothing \wedge I_x[a_i, b_m] = \varnothing \wedge I_y[b_l, c_0] = \neg\varnothing \wedge I_y[b_m, c_0]$$
$$= \neg\varnothing \wedge I_y[b_n, c_0] = \varnothing \Rightarrow I_z[a_i, c_0] = \varnothing \tag{10.59}$$

Equations (10.49) and (10.52) in equation (10.46):

$$I_x[a_i, b_l] = \neg\varnothing \wedge I_x[a_i, b_m] = \varnothing \wedge I_x[a_i, b_n] = \neg\varnothing \wedge I_y[b_l, c_0]$$
$$= \varnothing \wedge I_y[b_n, c_0] = \varnothing \Rightarrow I_z[a_i, c_0] = \varnothing \tag{10.60}$$

The intersections of the combined topological relations are described by the combination of the results of equations (10.56)–(10.60). Some intersections may be multiply derived, because the inference rules, Equations (10.56)–(10.60), determine each individual intersection over several paths; therefore, only those combined topological relations are valid which match with at least one of the eight intersections. A contradiction among two or more redundantly derived values exists if $\varnothing$ and $\neg\varnothing$ would be derived as the values of an intersection. This would indicate that the combined topological relation does not exist. Furthermore, the union of all compositions may be insufficient to identify a unique combined relation. In such ambiguous cases, the result is a set of possible relations which comprises all those relations whose intersections do not contradict the derived values. Since the set of topological relations is finite, the result can also be transformed so that it describes the complement, i.e. the set of impossible relations.

An example is determining the composition of a transitive topological relation from two given topological relations, where the resulting topological relation is unique. In the example the given topological relations are "meet" and "contains". The example is taken from Egenhofer (1994). The intersections of the composition of the topological relations "meet" and "contains" are determined in the following way with the 9-intersections of the two relations taken from Figure 5.23, i.e. "meet" corresponds to I_1 and "contains" to I_5.

$I_1[\partial, \partial], I_5[\partial, \_]$ in equation (10.53):

$$I_1[\partial, \partial] = \neg\varnothing \wedge I_5[\partial, °] = \varnothing \wedge I_5[\partial, \partial] = \varnothing \wedge I_5[\partial, ^-] = \neg\varnothing \Rightarrow I_{1,5}[\partial, ^-] = \neg\varnothing$$
$$\tag{10.61}$$

$I_1[\partial, ^-], I_5[^-, \_]$ in equation (10.53):

$$I_1[\partial, ^-] = \neg\varnothing \wedge I_5[^-, °] = \varnothing \wedge I_5[^-, \partial] = \varnothing \wedge I_5[^-, ^-] = \neg\varnothing \Rightarrow I_{1,5}[\partial, ^-] = \neg\varnothing$$
$$\tag{10.62}$$

$I_1[^\circ,^-], I_5[^-,\_]$ in equation (10.53):

$$I_1[^\circ,^-] = \neg\varnothing \wedge I_5[^-,^\circ] = \varnothing \wedge I_5[^-,\partial] = \varnothing \wedge I_5[^-,^-] = \neg\varnothing \Rightarrow I_{1,5}[^-,^-] = \neg\varnothing$$

(10.63)

$I_1[^-,\partial], I_5[\partial,\_]$ in equation (10.53):

$$I_1[^-,\partial] = \neg\varnothing \wedge I_5[\partial,^\circ] = \varnothing \wedge I_5[\partial,\partial] = \varnothing \wedge I_5[\partial,^-] = \neg\varnothing \Rightarrow I_{1,5}[^-,^-] = \neg\varnothing$$

(10.64)

$I_1[^-,^-], I_5[^-,\_]$ in equation (10.53):

$$I_1[^-,^-] = \neg\varnothing \wedge I_5[^-,^\circ] = \varnothing \wedge I_5[^-,\partial] = \varnothing \wedge I_5[^-,^-] = \neg\varnothing \Rightarrow I_{1,5}[^-,^-] = \neg\varnothing$$

(10.65)

$I_1[\_,^\circ], I_5[^\circ,\partial]$ in equation (10.54):

$$I_1[^\circ,^\circ] = \varnothing \wedge I_1[\partial,^\circ] = \varnothing \wedge I_1[^-,^\circ] = \neg\varnothing \wedge I_5[^\circ,\partial] = \neg\varnothing \Rightarrow I_{1,5}[^-,\partial] = \neg\varnothing$$

(10.66)

$I_1[\_,^\circ], I_5[^\circ,^\circ]$ in equation (10.54):

$$I_1[^\circ,^\circ] = \varnothing \wedge I_1[\partial,^\circ] = \varnothing \wedge I_1[^-,^\circ] = \neg\varnothing \wedge I_5[^\circ,^\circ] = \neg\varnothing \Rightarrow I_{1,5}[^-,^\circ] = \neg\varnothing$$

(10.67)

$I_1[\_,^\circ], I_5[^\circ,^-]$ in equation (10.54):

$$I_1[^\circ,^\circ] = \varnothing \wedge I_1[\partial,^\circ] = \varnothing \wedge I_1[^-,^\circ] = \neg\varnothing \wedge I_5[^\circ,^-] = \neg\varnothing \Rightarrow I_{1,5}[^-,^-] = \neg\varnothing$$

(10.68)

$I_1[^\circ,\_], I_5[^-,\partial]$ in equation (10.56):

$$I_1[^\circ,^\circ] = \varnothing \wedge I_1[^\circ,\partial] = \varnothing \wedge I_1[^\circ,^-] = \neg\varnothing \wedge I_5[^-,\partial] = \varnothing \Rightarrow I_{1,5}[^\circ,\partial] = \varnothing$$

(10.69)

$I_1[^\circ,\_], I_5[^-,^\circ]$ in equation (10.56):

$$I_1[^\circ,^\circ] = \varnothing \wedge I_1[^\circ,\partial] = \varnothing \wedge I_1[^\circ,^-] = \neg\varnothing \wedge I_5[^-,^\circ] = \varnothing \Rightarrow I_{1,5}[^\circ,^\circ] = \varnothing$$

(10.70)

$I_1[\partial,^\circ], I_5[\_,\partial]$ in equation (10.55):

$$I_1[\partial,^\circ] = \varnothing \wedge I_5[^\circ,\partial] = \neg\varnothing \wedge I_5[\partial,\partial] = \varnothing \wedge I_5[^-,\partial] = \varnothing \Rightarrow I_{1,5}[\partial,\partial] = \varnothing$$

(10.71)

$I_1[^\circ,^\circ], I_5[\_,\partial]$ in equation (10.55):

$$I_1[^\circ,^\circ] = \varnothing \wedge I_5[^\circ,\partial] = \neg\varnothing \wedge I_5[\partial,\partial] = \varnothing \wedge I_5[^-,\partial] = \varnothing \Rightarrow I_{1,5}[^\circ,\partial] = \varnothing$$

(10.72)

$I_1[\partial,^\circ], I_5[\_,^\circ]$ in equation (10.55):

$$I_1[\partial,^\circ] = \varnothing \wedge I_5[^\circ,^\circ] = \neg\varnothing \wedge I_5[\partial,^\circ] = \varnothing \wedge I_5[^-,^\circ] = \varnothing \Rightarrow I_{1,5}[\partial,^\circ] = \varnothing$$

(10.73)

$I_1[{}^{\circ},{}^{\circ}], I_5[\_,{}^{\circ}]$ in equation (10.55):

$$I_1[{}^{\circ},{}^{\circ}] = \varnothing \wedge I_5[{}^{\circ},{}^{\circ}] = \neg\varnothing \wedge I_5[\partial,{}^{\circ}] = \varnothing \wedge I_5[{}^{-},{}^{\circ}] = \varnothing \Rightarrow I_{1,5}[{}^{\circ},{}^{\circ}] = \varnothing$$

(10.74)

$I_1[\_,\partial], I_5[\partial,{}^{-}]$ in equation (10.58):

$$I_1[{}^{\circ},\partial] = \varnothing \wedge I_1[\partial,\partial] = \neg\varnothing \wedge I_1[{}^{-},\partial] = \neg\varnothing \wedge I_5[\partial,{}^{-}] = \neg\varnothing$$

$$\Rightarrow \neg(I_{1,5}[\partial,{}^{-}] = \varnothing \wedge I_{1,5}[{}^{-},{}^{-}] = \varnothing)$$

(10.75)

$I_1[\partial,\_], I_5[\_,\partial]$ in equation (10.60):

$$I_1[\partial,{}^{\circ}] = \varnothing \wedge I_1[\partial,\partial] = \neg\varnothing \wedge I_1[\partial,{}^{-}] = \neg\varnothing \wedge I_5[\partial,\partial] = \varnothing \wedge I_5[{}^{-},\partial] = \varnothing$$

$$\Rightarrow I_{1,5}[\partial,\partial] = \varnothing$$

(10.76)

$I_1[\partial,\_], I_5[\_,\partial]$ in equation (10.60):

$$I_1[\partial,{}^{\circ}] = \varnothing \wedge I_1[\partial,\partial] = \neg\varnothing \wedge I_1[\partial,{}^{-}] = \neg\varnothing \wedge I_5[\partial,{}^{\circ}] = \varnothing \wedge I_5[{}^{-},{}^{\circ}] = \varnothing$$

$$\Rightarrow I_{1,5}[\partial,{}^{\circ}] = \varnothing$$

(10.77)

The calculation of $I_{1,5}[a, c]$ yields redundant specifications. For instance, $I_{1,5}[{}^{-},{}^{-}]$ is determined three times. These redundancies are easily eliminated since they do not result in contradicting values for the intersections. The compilation of Equations (27)–(43) shows that all nine intersections $I_{1,5}[a, c]$ are determined, that is:

$$I_{1,5}(A, C) = \begin{bmatrix} \varnothing & \varnothing & \neg\varnothing \\ \varnothing & \varnothing & \neg\varnothing \\ \neg\varnothing & \neg\varnothing & \neg\varnothing \end{bmatrix}$$

(10.78)

The resulting 9-intersection is equal to I_0, in other words to *disjoint* and hence the combined and resulting topological relation is:

$$\text{meet}; \text{contains} \Rightarrow \text{disjoint}$$

The 64 combined binary topological relations between spatial regions without holes are given in Table 10.2, which reveals a number of observations. All compositions of topological relations are valid, because none of the 64 compositions produces a contradicting intersection and all derived intersections match with at least one of the eight topological relations. "equal" is the identity relation, because its composition with any other relation results in the original relation. Furthermore, excluding the compositions involving "equal" just 12 unique compositions can be made. The outcome of three compositions are fully undetermined, since the combination of the constraints does not exclude any of the eight relations. The three compositions are "disjoint; disjoint", "inside; contains" and "overlap; overlap". The table also shows that three relations are transitive ("equal", "inside" and "contains") and finally two pairs are commutative ("coveredBy; inside = inside; coveredBy" and "covers; contains = contains; covers").

Table 2 The 64 compositions of the binary topological relations $A r_i B$ and $B r_j C$ (d = Disjoint, m = Meet, e = Equal, i = Inside, cB = CoveredBy, ct = Contains, cv = Covers, and o = Overlap).

	Disjoint (B, C)	Meet (B, C)	Equal (B, C)	Inside (B, C)	CoveredBy (B, C)	Contains (B, C)	Covers (B, C)	Overlap (B, C)
Disjoint (A, B)	d∨m∨ e∨i∨ cB∨ct ∨cv∨o	d∨m∨ i∨cB∧o	d	d∨m∨ i∨cB∨o	d∨m∨i∨ cB∨o	d	d	d∨m∨ i∨cB∧o
Meet (A, B)	d∨m∨ ct∨cv∨o	d∨m∨ e∨cB∨ cv∨o	m	i∨cB∨o	m∨i∨ cB∨o	d	d∨m	d∨m∨ i∨cB∨o
Equal (A, B)	d	m	e	i	cB	ct	cv	o
Inside (A, B)	d	d	i	i	i	d∨m∨ e∨i∨ cB∨ct ∨cv∨o	d∨m∨ i∨cB∨o	d∨m∨ i∨cB∨o
CoveredBy (A, B)	d	d∨m	cB	i	i∨cB	d∨m∨ ct∨cv∨o	d∨m∨ e∨cB∨ cv∨o	d∨m∨ i∨cB∨o
Contains (A, B)	d∨m∨ ct∨cv∨o	ct∨cv∨o	ct	e∨i∨ cB∨ct ∨cv∨o	ct∨cv∨o	ct	ct	ct∨cv∨o
Covers (A, B)	d∨m∨ ct∨cv∨o	m∨ct∨ cv∨o	cv	i∨cB∨o	e∨cB∨ cv∨o	ct	ct∨cv	ct∨cv∨o
Overlap (A, B)	d∨m∨ ct∨cv∨o	d∨m∨ ct∨cv∨o	o	i∨cB∨o	i∨cB∨o	d∨m∨ ct∨cv∨o	d∨m∨ ct∨cv∨o	d∨m∨ e∨i∨ cB∨ct ∨cv∨o

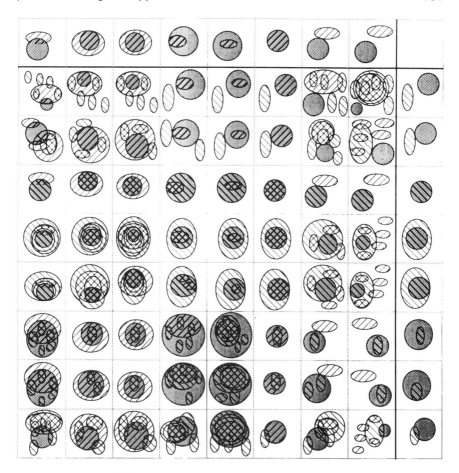

Figure 10.6 Geometric interpretations of the 64 combinations of binary topological relations.

Figure 10.6 shows geometric examples for the derived relations. All combined topological relations could be realized among three 2D objects in $\mathbb{R}^2$.

10.2.4. Regional Relations (Randell *et al.*)

Some approaches to qualitative spatial reasoning based on logic have been proposed. Of these two are closely related in that they both are based on a theory which was originally developed and presented by Clarke (1981) and which goes back as far as 1981. The two approaches are by Randell *et al.* (1992) and View (1992). The subsequent discussion will mainly be concerned with that of Randell *et al.*

Basic concepts in the theory of Randell *et al.* are the regions and the primitive dyadic relations which can be defined between the regions. The regions in the theory support either a spatial or a temporal interpretation. Informally these regions may be thought to be potentially infinite in number, and any degree of connection between them is allowed in the intended model, from external contact to identity in terms of mutually shared parts.

The basic part of the formalism, i.e. the primitive dyadic relation, is denoted by $C(x, y)$, which is read as "x connects with y". For the basic part of the theory, the individuals can be interpreted as either spatial or temporal regions. The relation $C(x, y)$ is reflexive and symmetric. A topology can be given to interpret the theory, namely that $C(x, y)$ holds when the topological closures of regions x and y share a common point. Two important axioms are thus introduced:

$$\forall x \quad C(x, x) \tag{10.79}$$

$$\forall x \quad y[C(x, y) \Rightarrow C(y, x)] \tag{10.80}$$

From $C(x, y)$, a basic set of dyadic relations have been defined:

$DC(x, y)$	(x is disconnected from y)	(10.81)
$P(x, y)$	(x is part of y)	(10.82)
$PP(x, y)$	(x is a proper part of y)	(10.83)
$x = y$	(x is identical with y)	(10.84)
$O(x, y)$	(x overlaps y)	(10.85)
$DR(x, y)$	(x is discrete from y)	(10.86)
$PO(x, y)$	(x partially overlaps y)	(10.87)
$EC(x, y)$	(x is externally connected with y)	(10.88)
$TPP(x, y)$	(x is a tangential proper part of y)	(10.89)
$NTPP(x, y)$	(x is a non-tangential proper part of y)	(10.90)

Among these relations P, PP, TPP and NTPP are nonsymmetrical and thus support inverses. For the inverses the notation Φ^{-1}, where $\Phi \in \{P, PP, TPP, NTPP\}$, is used. Of the defined relations DC, EC, PO, =, TPP, NTPP and the inverses for TPP and NTPP are provably mutually exhaustive and pairwise disjoint. Hence, a complete set of relations can be embedded in a relational lattice which can be found in Randell *et al.* (1992). In Figure 10.7 some examples of the relations are illustrated. The definition of the major part of the theory, which has been verified by Randell *et al.* thus becomes:

$$DC(x, y) \equiv_{\text{def}} \neg C(x, y) \tag{10.91}$$

$$P(x, y) \equiv_{\text{def}} \forall z[C(z, x) \Rightarrow C(z, y)] \tag{10.92}$$

$$PP(x, y) \equiv_{\text{def}} P(x, y) \wedge \neg P(x, y) \tag{10.93}$$

$$x = y \equiv_{\text{def}} P(x, y) \wedge P(y, x) \tag{10.94}$$

$$O(x, y) \equiv_{\text{def}} \exists z[P(z, x) \wedge P(z, y)] \tag{10.95}$$

$$PO(x, z) \equiv_{def} O(x, y) \wedge \neg P(x, y) \wedge \neg P(y, x) \tag{10.96}$$

$$DR(x, y) \equiv_{def} \neg O(x, y) \tag{10.97}$$

$$TPP(x, y) \equiv_{def} PP(x, y) \wedge \exists z[EC(z, x) \wedge EC(z, y)] \tag{10.98}$$

$$EC(x, y) \equiv_{def} C(x, y) \wedge \neg O(x, y) \tag{10.99}$$

$$NTPP(x, y) \equiv_{def} PP(x, y) \wedge \neg \exists z[EC(z, x) \wedge EC(z, y)] \tag{10.100}$$

$$P^{-1}(x, y) \equiv_{def} P(y, x) \tag{10.101}$$

$$TPP^{-1}(x, y) \equiv_{def} TPP(y, x) \tag{10.102}$$

$$NTPP^{-1}(x, y) \equiv_{def} NTPP(y, x) \tag{10.103}$$

In this theory, which is somewhat revised compared to the original theory by Clarke, there is no formal distinction between open, semi-open and closed regions used to interpret this part of the formalism.

Boolean functions following the theory have also been defined: "sum(x, y)" which is read as "the sum of x and y", "Us" as "the universal (spatial) region", "compl(x)" as "the complement of x", "prod(x, y)" as "the product of x and y" (that is the intersection of x and y) and diff(x, y) as "the difference of x and y". "compl(x)", "prod(x, y)" and "diff(x, y)" are partial but are made total in the sorted logic by simply specifying sort restrictions and by introducing a new sort called NULL. The sorts NULL and REGION are disjoint.

$$sum(x, y) =_{def} \iota y[\forall z[C(z, y) \Leftrightarrow [C(z, y) \vee C(z, y)]]] \tag{10.104}$$

$$compl(x) =_{def} \iota y[\forall z[[C(z, y) \Leftrightarrow \neg NTPP(z, x)] \wedge [O(z, y) \Leftrightarrow \neg P(z, x)]]] \tag{10.105}$$

$$Us =_{def} \iota y[\forall z[C(z, y)]] \tag{10.106}$$

$$prod(x, y) =_{def} \iota z[\forall u[C(u, z) \Leftrightarrow \exists v[P(v, x) \wedge P(v, y) \wedge C(u, v)]]] \tag{10.107}$$

$$diff(x, y) =_{def} \iota w[\forall z[C(z, w) \Leftrightarrow C(z, prod(x, compl(y)))]] \tag{10.108}$$

$$\forall xy \quad [NULL(prod(x, y)) \Leftrightarrow DR(x, y)] \tag{10.109}$$

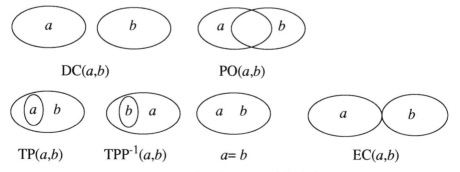

Figure 10.7 Examples of some spatial relations.

Table 10.3

Rel. Rep.	1	2	3	4	5	6	7	8	9	10	11	12	13	14	15	16	17	18	19	20	21	22	23
1 NTPP																							
2 TPP																							
3 PO																							
4 EQ																							
5 NTPP^{-1}																							
6 TPP^{-1}																							
7 [IIE]																							
8 [IPE]																							
9 [IPD]																							
10 [IOE]																							
11 [IOD]																							

| 12 [PIE] | 13 [PID] | 14 [PPE] | 15 [PPD] | 16 [POE] | 17 [POD] | 18 [OIE] | 19 [OID] | 20 [OPE] | 21 [OPD] | 22 [OOE] | 23 [OOD] |

Here $\alpha(\bar{x}) =_{\mathrm{def}} \iota y[\Phi[\alpha(\bar{y})]]$ means $\forall x[\Phi[\alpha(\bar{x})]]$; thus, for instance, the definition of prod (x, y) in the object language is translated into $\forall xyz[C(z, \mathrm{prod}\,(x, y)) \leftrightarrow \exists w[P(w, x) \wedge P(w, y) \wedge C(u, w)]]$.

A primitive function $\mathrm{conv}(x)$ corresponding to the convex hull of x is also provided and axiomatized. It is assumed that conv is only well sorted when defined on one-piece regions:

$$\forall x \quad P(x, \mathrm{conv}(x)) \tag{10.110}$$

$$\forall x \quad P(\mathrm{conv}\,(\mathrm{conv}(x)), \mathrm{conv}(x)) \tag{10.111}$$

$$\forall x \forall y \forall z \quad [[P(x, \mathrm{conv}(y)) \wedge P(y, \mathrm{conv}(z))] \Rightarrow P(x, \mathrm{conv}(z))] \tag{10.112}$$

$$\forall x \forall y \quad [[P(x, \mathrm{conv}(y)) \wedge P(y, \mathrm{conv}(x))] \Rightarrow O(x, y)] \tag{10.113}$$

$$\forall x \forall y \quad [[DR(x, \mathrm{conv}(y)) \wedge DR(y, \mathrm{conv}(x))] \Leftrightarrow DR(\mathrm{conv}(x), \mathrm{conv}(y))] \tag{10.114}$$

These functions can now be used to define a set of relations describing regions being inside, partly inside, and outside (INSIDE, P-INSIDE and OUTSIDE) and their inverses as well, and where for instance OUTSIDE (x, y) means "x is outside y". Furthermore, it is also possible to distinguish between geometrically inside and typologically inside. This means that an object that is topologically inside another object requires a pass through the surrounding object, from any direction, to reach the first object, while this is not necessary for the geometrically inside relation.

In this approach a composition table has been built which makes transitivity inferences possible when the relations between regions x and y, and between y and z, are known. Such a table can be used to infer the possible relations between x and z, in other words, the objective was to infer the transitive relations. The transitivity table according to Bennett (1994) is found in Table 10.3. The variation of the table presented here is constructed for 22×22 cases. The table is relatively sparse; note also that there is a total of 142 different types of cells. Each entry in the table is in the form of an array indicating which relations are possible. The presence of a relation is indicated by a ○ or ∗ where the coding is given just to the right of each relation name at the start of each row of the table. The top row of each array relates to the relations TPP, NTPP, PO, =, TPP^{-1} and NTPP^{-1}. The bottom three rows relate to the INSIDE, P-INSIDE and OUTSIDE relations, with each of the three columns relating to the inverse of these three relations. In this case a ○ indicates a DC variant and a ∗ the EC variant. Notice also that the first column, second row is absent since INSIDE_INSIDE^{-1} is impossible. In all cases a ● indicates the presence of both the ○ and ∗ relations. If all 11 positions are occupied by a ●, then all the 22 relations are present and all relations are consequently possible.

10.2.5. A Qualitative Model for Space (Mukerjee and Joe)

This approach to qualitative spatial reasoning by Mukerjee and Joe (1990) is concerned with extended objects and their interrelationships when the angles between the objects are arbitrary. The approach is mainly a generalization in space of that of Allen. As an introduction to the technique just a single dimension is necessary. Objects in one dimension can be either points or intervals. It can also be assumed that such points can be ordered in some spatial or temporal direction. Thus three possible cases between two objects, *A* and *B*, can be identified:

- *A* and *B* are both points. Three relations can be identified, i.e. *A* is behind *B* $(-)$, *A* is equal to *B* $(=)$ and *A* is ahead of *B* $(+)$.
- *A* is a point and *B* is an interval. Five different qualitative relations are then possible, i.e. *A* is behind *B* $(-)$, *A* is equal to the back of *B* (b), *A* is inside *B* (i), *A* is at the front of *B* (f) and finally *A* is ahead of *B* $(+)$.
- *A* and *B* are both intervals. The set of possible relations is equal to Allen's 13 temporal events.

Of the above three cases the third is the most important. If an end point of an interval *C* is considered then, as in the second case above, this end point can have five different relations to *B*, i.e. the set $\{+, f, i, b, -\}$ or "ahead", "front", "interior", "back" and "posterior". Now the two end points of *C* may be at any of these five positions relative to the interval *B*, thus the complete relations between *C* and *B* can be expressed as for example as $C++B$ which is interpreted as, the interval *C* is after the interval *B*. In the conventional way this is interpreted as "the interval *B* is less than the interval *C*", thus when looking from left to right *B* comes ahead of *C*. Again, it is clear that the

Table 10.4

C/B	A/B C/B	+	f	i	b	−
C after B	++	?	−	−	−	−
C met-by B	f+	>	b	−	−	−
C overlap-by B	i+	>	i	<	−	−
C finish-as B	if	+	f	<	−	−
C contained-in B	ii	+	+	?	−	−
C starts B	bi	+	+	>	b	−
C equals B	bf	+	f	i	b	−
C started-by B	b+	>	i	i	b	−
C contains B	−+	>	i	i	i	<
C finished-by B	−f	+	f	i	i	<
C overlaps B	−i	+	+	>	i	<
C meets B	−b	+	+	+	b	<
C before B	−−	+	+	+	+	?

complete number of such relations is 13, which can be seen in the left-most column in Table 10.4.

Table 10.4 shows the relationship of an end point of A with respect to the interval C if the relations of both A and C are known with respect to an interval B. The symbol "?" denotes that the relation A/C may correspond to any of the five possible relations, while ">" corresponds to a member of the subset $\{i, f, +\}$ and finally "<" corresponds to one of $\{-b, i\}$. Hence, given the local relations between "neighbouring" intervals (A/B) and (B/C), the transitive relation (A/C) is often disjunctive. The inference relations shown in Table 10.4 are more compact than Allen's, in other words 5×13 instead of 13×13. This is because they exploit independence between the two ends of an interval. This can simply be illustrated by the following example: if A is an interval and it is known that A is *overlap-by* B, that is $A/B = i+$, and that C *starts* B, that is $C/B = bi$, then it can be concluded from the transitivity Table 10.4 that with respect to C the rear end of A is ">", or $\{i, f, +\}$. Consequently, A/C is either i+, f+ or "++", which is a constraint in the possible positions of A with respect to C.

The above illustrates how Allen's relations can be modified and applied to a one-dimensional spatial environment. However, Mukerjee and Joe have generalized this method such that it can be applied to a two-dimensional space where the angles between the object intervals are arbitrary. Thus, a comprehensive mapping of the relations between two objects at an arbitrary angle, in the qualitative contact sense, was necessary. The angular relations of B with respect to A are dependent on A's direction, which in generally not related to the direction of B. The operators in such a formalism will not be commutative and neither will it have well-defined inverses, in other words given the position A/B, the position of B/A cannot be determined. Furthermore, consider the relations with respect to a single object A with a designated front, then four angular quadrants with respect to A and that front can be defined. Thus the front of another object may be oriented in any

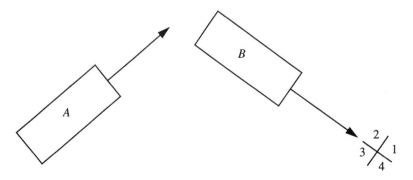

Figure 10.8 Qualitative directions. Object A is following object B in the direction of the first quadrant of the latter.

of these four quadrants, which is illustrated in Figure 10.8. Mukherjee and Joe use just four cardinal directions while for instance Frank has shown how eight can be used, see section 10.2.3.

Two spatial attributes can be identified for a relationship between two objects at an angle, the *relative direction* and the *relative position*. The representation of the angular information in this approach is based on a predefined direction called the *front*, as for the direction of object *B* in Figure 10.8. Clearly, most objects have a front and a direction and even for those cases when there is no direction there may, at least, be a front. In two-dimensional space the assignment of a front to an object makes it possible to identify several different qualitative zones, besides the front, such as left, right and back. Including also the four quadrants from above gives us in all a total of eight qualitative angular relationships for the two-dimensional case. In the three-dimensional case the number of qualitative relations becomes 26. The directional information in this approach is represented as dir(A/B) = "actual quadrant" = 1, which is interpreted as "object *A* is pointing at quadrant 1 of object *B*". Furthermore, given dir(A/B) then dir(B/A) is uniquely defined. By including parallel lines of the objects in forward and backward direction the lines of travel can be identified. These lines are illustrated in Figure 10.9.

When the pair of lines of travel for two separate objects meet, a parallelogram, called the *collision parallelogram* (CP), is created. The CP defines an area that under certain circumstances is common to the two objects. Considering the CP in Figure 10.10(b) the same set of position relations as was discussed for the one-dimensional case can now be identified for these objects moving in arbitrary directions in two dimensions. These relations are relative to the fronts of each object. Thus an end line of an object goes from "behind" the CP, passes through the CP and finally ends up after the CP. From this observation it is simple to identify the positions of the various relations in the set $\{-, b, i, f, +\}$ for each end point of the object. Furthermore, in Figure 10.10(b) the four-directional quadrants for two objects are presented.

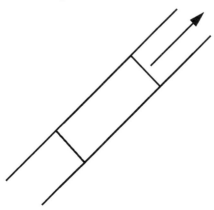

Figure 10.9 Illustration of the lines of travel for a rectangular object.

Now, three different object relations must be present in order to completely describe all relevant object relations between a pair of objects:

- the quadrant information of dir(A/B)
- where is the object located relative to the other object, pos where (A/B)
- where is the other object located relative to the first, pos(B/A).

The quadrant information, dir(B/A), can be derived once the corresponding information for A/B is known. The relationship between two objects is determined by considering each end line at a time. This is done similarly for the one-dimensional case. In other words the position of the front of an object is determined with respect to CP as in Figure 10.10(a).

Some further observations can be made concerning the interrelationships between the relations dir $(\cdot)$ and the positional relation pos $(\cdot)$. Here the front directions of the objects are of particular importance, e.g. changing the front direction of one of the objects will affect the position as well as the direction relations. However, the position of the other object will not be affected since the CP will remain the same.

Mukerjee and Joe have shown how their approach can be applied to identification of transitive relations for the 2D case, that is if the relations (A, B) and (B, C) are known, what can be inferred about (A, C). An illustration to this problem is given in Figure 10.11. To solve this problem a solution is applied that is analogous to the 1D case. In other words, the same type of transitivity table can be used here as well. Clearly, the 2D approach is somewhat more complicated than the 1D one. Basically, the approach is to use the end line information and again achieve a 7/13 savings. A shorthand illustration of this can be found in Table 10.5. Observe that each quadrant group contains 13×13 such tables, for instance for dir$(B/A) = 1$ and dir$(B/C) = 1$.

The formalism contains an uncertainty of $90°$ in each angular relation, hence the uncertainty in the output is two quadrants or $180°$. This is also clear from Table 10.5 which includes two rows, one for each quadrant, for each of the relations A/C and C/A.

In order to use Table 10.5 five parameters must be known: pos(B/A),

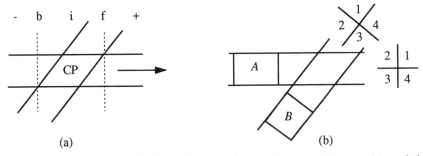

(a)   (b)

Figure 10.10 The position relations relative to the collision parallelogram (a), and the directional quadrants for two objects (b).

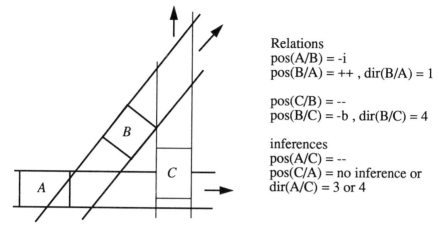

Relations
pos(A/B) = -i
pos(B/A) = ++ , dir(B/A) = 1

pos(C/B) = --
pos(B/C) = -b , dir(B/C) = 4

inferences
pos(A/C) = --
pos(C/A) = no inference or
dir(A/C) = 3 or 4

Figure 10.11 An illustration of transitivity inference according to Mukerjee and Joe.

$\mathrm{dir}(B/A)$, $\mathrm{pos}(B/C)$, $\mathrm{dir}(B/C)$ and the relationship of one end line of either A or B relative to B. The resulting relations inferred from the transitive table may be disjunctive.

Table 10.5 represents the case for which, for instance, the following input tuples are valid:

$$B/A = \langle 1, ++, \mathrm{i}+ \rangle$$
$$B/C = \langle 1, --, \mathrm{bf} \rangle$$

Then from the table the following relations can be inferred:

$$A/C = 1; >+, \qquad \text{i.e. } \langle \mathrm{i}+, \mathrm{f}+, ++ \rangle$$
$$4; -?, \qquad \text{i.e. } \langle --, -\mathrm{b}, -\mathrm{i}, -\mathrm{f}, -+ \rangle$$
$$C/A = 1; -<, \qquad \text{i.e. } \langle --, -\mathrm{b}, -\mathrm{i} \rangle$$
$$4; ++$$

Each relation between two objects has $13 \times 13 \times 4 = 676$ possible results. Consequently, for two objects there are $676 \times 676 = 456\,976$ entries in the table. However, the number of entries in the table can be reduced since the quadrant and the position information are interrelated, for example, if the direction of an object is reserved then the position changes in a certain way.

Table 10.5

$B/C = \langle --, 1 \rangle$	Quadrant	+	f	i	b	−
$A/B \Rightarrow A/C$	1	+	+	>	>	?
$A/B \Rightarrow A/C$	4	?	−	−	−	−
$C/B \Rightarrow C/A$	1	?	<	<	−	−
$C/B \Rightarrow C/A$	4	+	+	+	+	?

Similarly, certain properties are preserved when considering the reflexive configurations. Observe also that $pos(A/C)$ is independent of $pos(C/A)$ and the other way around. Specifically, $pos(A/B)$ only affects $pos(C/A)$ while $pos(C/B)$ only affects $pos(C/A)$. The transitive table thus allows C/A and A/C to be partitioned into two spaces.

The set of quadrants in the transitive tables form two groups under the operation of reflection and direction inversion. The arrangement of the groupings illustrates the structure of the relationship that exists between the groupings. For example, the group $dir(B/A) = 1$ and $dir(B/C) = 1$, when the direction of B is reversed this results, for instance, in $dir(B/A) = 3$ and $dir(B/C) = 3$. Thus by determining the effects of the reflection operator, it becomes sufficient to maintain only one of these two quadrant groups.

Considering the reflection operation then, for example, Table 10.6 shows that the configuration is the same as in Table 10.3. The position relations are essentially the same, because reflection does not affect $pos(\cdot)$ relations. Instead the rows are interchanged, that is, the quadrants are not affected.

A further operation that reduces the table size is directional inversion. When the direction of an object is reversed, the object will remain at the same physical position. In a transitive relationship any one of the objects, A, B or C, may change direction. A directional change of two objects, say A and B, will result in a quadrant group obtained by changing the direction of the third object alone, in other words just C in this particular case. Thus, the directional changes form a cycle of operations in the group.

Consider a transitive inference involving A, B and C. If the direction of B is changed then the only things that change are the quadrant relations of B/A and B/C and B's positional relations with respect to A and C. Since there has been a quadrant change for B/A and B/C, the result is a movement from one quadrant group to another.

In Figure 10.12 each node represents the transitivity tables for two quadrants for B/A and B/C. The arcs between the nodes represent the operations that map one quadrant group into another. In the figure the following notation is used: R = reflection; A, B, C = direction change on either of the corresponding objects, i.e. A, B or C. As a result of these interrelations it is sufficient to maintain the table for only two quadrants, say (1 1) and (2 1), that is, one node of each group set.

Table 10.6

$B/C = \langle --, 1 \rangle$	Q	$+$	f	i	b	$-$
$A/B \Rightarrow A/C$	1	?	$-$	$-$	$-$	$-$
$A/B \Rightarrow A/C$	4	$+$	$+$	$>$	$>$	?
$C/B \Rightarrow C/A$	1	$+$	$+$	$+$	$+$	?
$C/B \Rightarrow C/A$	4	?	$<$	$<$	$-$	$-$

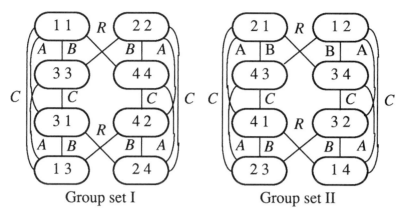

Figure 10.12 Quadrant interrelations representing the transitivity tables.

10.3. SYMBOLIC PROJECTION APPLIED TO ROUTE PLANNING (HOLMES AND JUNGERT)

The qualitative approach to route planning presented here was originally proposed by Holmes and Jungert (1992) and concerns reasoning on digitized maps. It focuses on techniques for 2D route planning in the presence of obstacles where the basic data structure is run-length coded. Eventually a symbolic description of the planned route is generated. Originally two alternative approaches to route planning were included: one involved heuristic, symbolic processing based on Symbolic Projection, while the other employed geometric calculations. Here only the former will be presented, since methods based on computational geometry are outside the scope of this book.

10.3.1. The Tile Graph

Given obstacles which may not be traversed, it is possible to treat the free space surrounding the obstacles as a single object, with a homogeneous

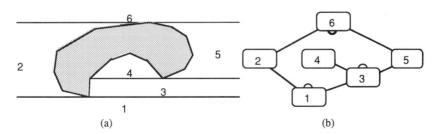

Figure 10.13 Partitioning into tiles (a) and the corresponding tile graph (b).

background. When addressing the route planning problem then, the obstacles can be ignored; only computations upon the run-length coded free space need be considered.

During preprocessing, a data structure called a *tile graph* or *free space graph* is created. This data structure is the connectivity graph resulting from a systematic planar subdivision of the free space. Each distinct region within the planar subdivision is called a tile. An example which illustrates free space partitioned into tiles is given in Figure 10.13. As shown in the figure, the free space is partitioned using "slices" parallel to the x-axis. This choice is natural, given the structure of the RLC database. Splitting of the background occurs at those points which, with respect to the y-axis, are local maximum and minimum vertices of obstacles. *Split-points* are points where one tile splits into two or, alternatively, where two tiles join to become one. Split-points are defined as belonging to the tile which is being split. This example contains three split-points: upper split-points in tile 1 and tile 3, and a lower split-point in tile 6. In the tile graph, the split-points are denoted by small half-circles.

In Plate 10.1 a realistic illustration of a tile graph is given. The map that is used for this purpose has also, as will be seen subsequently, been used for test purposes. The map has 660×860 pixels and the number of obstacles is approximately 300, which gave rise to about 2250 tiles. However, for such a large map it is not possible to generate and display the complete graph. For this reason, the tiles in a part of the map were painted in different colours to illustrate the final structure of the free space.

At this point, it is important to mention certain topological properties that this method enforces when partitioning the free space. To be explicit, when any tile splits, it splits into exactly two new tiles. This creates a situation where each tile has at most four neighbouring tiles. In this way, a constant branching factor is assured within the tile graph. This property is induced by using "empty tiles" (i.e. tiles containing no lines) when necessary, as shown in Figure 10.14. Introduction of empty tiles becomes necessary when two or more split-points have equal y-coordinates.

A free space graph structure has the following properties:

- nodes in the graph correspond to regions where an object is allowed to move without hindrance
- an arc between two nodes in the graph reflects connectivity between the regions represented by those nodes.

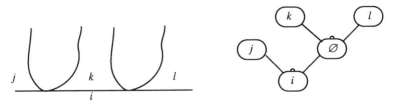

Figure 10.14 Illustration of an empty tile.

10.3.2. Routes and Paths

Within this system, descriptions of traversal from a start point to a goal point may be given at two different levels of abstraction. For this reason, it is necessary to introduce two working definitions:

- a *route* is a sequence of tiles (i.e. a sequence of regions in free space)
- a *path* is a sequence of coordinate points.

In addition, these definitions imply that:

- given any path, there exists exactly one non-cyclic route which contains that path
- within a route, there exists an infinite number of paths
- for each route, there is exactly one optimal path.

10.3.3. System Overview

As seen in Figure 10.15, the symbolic system contains not only a heuristic module (shown enclosed by the dashed lines), but an algorithmic module as well. Within the heuristic module lies an A^* graph search/construction process (Nilsson 1979), along with an inference engine; details concerning their tandem operation are covered in section 10.3.5. In this system's organization, the purpose of the algorithmic module is to optimize and refine symbolic plans produced by the heuristic module. Of utmost significance here is that given such symbolic plans, we are free to choose among any number of potential optimization schemes.

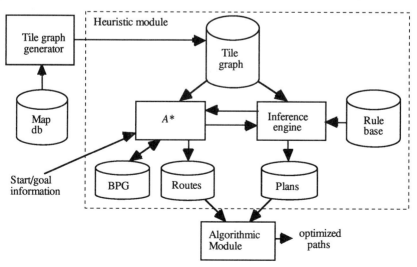

Figure 10.15 Block diagram describing the symbolic route planning process.

As shown in the figure, the heuristic module receives a start point and goal point as input. At this stage, query processing relies upon the spatial information encoded in the free space graph, as well as a set of inference rules. When given a connected sequence of three tiles, the inference engine and rules can inspect the free space graph and classify the local, spatial topology inherent in that sequence. For such a three-tile sequence, the output of the inference process provides a symbolic description of the topological "events" or features encountered within the second tile of that sequence; for this reason, this tile is called the *tile-in-focus*. Using a special graph search/construction process, the heuristic module can produce alternative sequences of tiles (i.e. routes extending from the start to the goal), where each route has a corresponding symbolic plan.

The result from the heuristic module is a sorted list of alternative routes, along with their respective symbolic plans. Each route and plan represents a different order of traversal amongst the obstacles; that is, the order and side upon which the obstacles are traversed is different for each alternative. Each symbolic plan contains a description of the stepwise, tile-by-tile events along its respective route.

At present, the symbolic plans produced by the heuristic module contain only a limited amount of information. That is, the information produced by the inference rules has been specialized in order to accommodate the implementation of a particular, geometric optimization module: an algorithmic module which optimizes in terms of euclidean length. Here, it is important to note that since the inference rules work with *objects*, the output from those rules (i.e. the resulting symbolic plans) can be made extraordinarily rich – without changing the underlying machinery.

As an alternative to the symbolic system for route planning, a purely geometric system for path derivation has been developed see Holmes and Jungert (1992). The organization of the geometric system is such that the start and goal points are fed directly into an algorithmic module and, as output, an optimal euclidean path between those points is produced. As mentioned above, paths are represented as simple sequences of coordinate points and, as such, essentially devoid of symbolic content. At the same time, however, the geometric system has been useful for providing benchmarks regarding the quality of results produced by the symbolic system.

10.3.4. The Inference Rules

The extended symbolic projection operators are used within a set of 64 inference rules. These rules are evaluated by an inference engine located within the heuristic module. As mentioned earlier, the inference rules classify local topologies found within connected regions of free space. Within a rule, a *local topology* is associated with a sequence of three connected tiles. It can be shown that since a tile or node in the tile graph is restricted to having at

Plate 10.1 Illustration of the tile graph structure when applied to a real map.

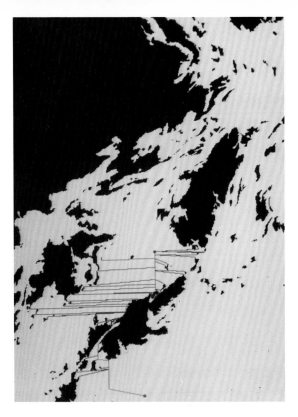

Plate 10.2 An example of the bearing point graph in a real map.

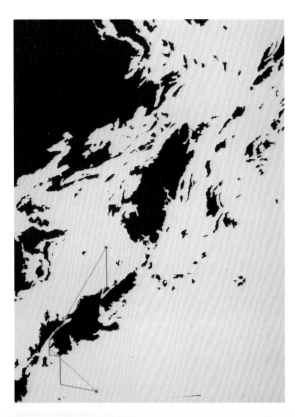

Plate 10.3 The optimized path for the route found in Figure 10.22.

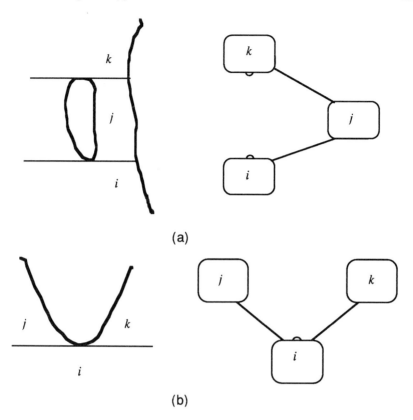

(a)

(b)

Figure 10.16 Illustration of map and graph corresponding to (a) a strait with a small obstacle to the left and (b) a left turn.

most four neighbors, the number of such local, topological patterns is limited to 34.

Within the inference rules, the tiles are regarded as rectangles. This sort of approximation (qualitative view) never causes any ambiguity, as long as the positions of the split-points are considered as well. The conditional part of the inference rules contains three subconditions. The first is a subsequence of three tiles given in the order in which they are to be traversed. This subsequence is labelled "*r*:". The other two subconditions concern the relative positions of the three tiles, expressed in terms of Symbolic Projection. Again, the string labelled "*u*:" corresponds to the projections along the *x*-axis, while the string labelled "*v*:" corresponds to the projections along the *y*-axis. A typical rule thus looks like:

$$\text{if } r: \quad i, j, k$$
$$u: \quad i \dashv j \text{ and } k \dashv j$$
$$v: \quad i \,|\, j \text{ and } j \,|\, k$$

then

> Direction_in j north   Context_in j closed_sp_left

This rule is satisfied when the spatial relationship between the subsequence of tiles $\langle i, j, k \rangle$ corresponds to the symbolic projections given in the u- and v-strings. The relations between the tiles correspond to the operator set described in Chapter 5. The output from the rule contains both direction and context information. The direction information "north" roughly summarizes the direction of planned movement through tile j, the tile-in-focus. The context information, "closed_sp_left", provides a symbolic description of the local topological features to be found in tile j, while moving from tile i to tile k. Here, that context corresponds to a strait with a small obstacle to the left, as depicted in Figure 10.16(a). Further, this context indicates that in traversing from tile i to tile k, two split-points are passed: one at the southern end of the obstacle and one at the northern end. There might also be an obstacle to the right, but there is no split-point. The subgraph corresponding to this context is also illustrated in Figure 10.17(a). Among the inference rules, there is also a rule corresponding to a movement within the same local context, but in the opposite direction (i.e. south).

Turns in which direction changes from north to south, or vice versa, can also be defined by means of these rules. For example, consider the rule:

> if r: j, i, k
>
> u: $j < k$
>
> v: $i \mid j$ and $i \mid k$
>
> then   Direction_in i south-north   Context_in i left_turn

This rule, whose context is illustrated in Figure 10.16(b), states that the subroute has a direction which is first south, then north. The context indicates a left turn. Implicitly the rule also indicates that during the turn in tile i, the path passes a split-point on the left, i.e. an upper split-point in tile i.

Before moving on, we give an illustration of how a symbolic plan is constructed. Temporarily assume that a route from a start point to a goal point is given. In this case, the inference engine selects the first three tiles of the route and searches the rulebase to find the matching rule. When the correct rule is found, the system begins building a plan – a plan which, when completed, describes the tile-by-tile events along that route. After successfully locating the first matching rule, the inference engine selects the second, third and fourth tiles of the route and again, searches for the rule matching that local situation. Such

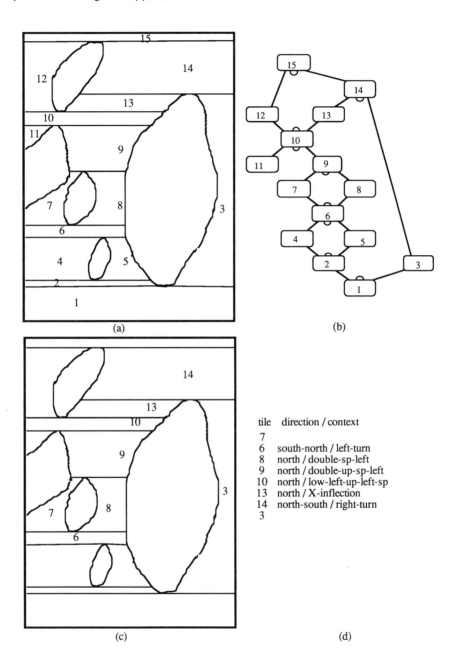

Figure 10.17 Illustration of (a) map, (b) tile graph, (c) route, and (d) plan.

processing may be conceived to continue until all tiles in the route have been examined. An example of the inference engine's operation is given in Figure 10.17.

10.3.5. The Bearing Point Graph

In the previous section it was described how the inference engine can produce a symbolic plan which describes events and features along that route. The method for generating alternative routes from the start to the goal is the subject of this section. As various routes are generated, they should be sorted according to some relevant cost criterion. With respect to various applications within route planning, it is nearly always the case that the cost associated with some planned motion takes into account the length of the final path. For this reason, a technique was developed which generates and sorts routes according to the length of an approximate path through each route.

Within the heuristic module, the A^* module and the inference engine operate in tandem. One output from the inference engine leads into the A^* module. This output represents the symbolic selection of *symbolic* reference points within tiles; here, such points are called points of bearing (or bearing points, BPs). The factors influencing their selection are discussed below in section 10.3.6. The task of the A^* module is to construct a special graph using the BPs generated by the inference engine. This new graph is called the *bearing point graph* (BPG).

The A^* module constructs the BPG dynamically, choosing to expand branches depending upon which appears most promising at any time. Having selected some branch for expansion, the A^* module then consults the free space graph for information regarding connectivity amongst the tiles. The A^* module passes information to the inference engine about the different possible "moves" out of the current tile. Receiving this information, the inference engine generates BPs for each of the possible next moves. This cycle continues until a desired number of alternative routes reach the goal or until there exist no further alternatives. It should also be noted here that although the general form of the BPG is constrained to resemble that of the free space graph, the BPG's actual shape is dependent upon the positions of the start and goal points.

At this point, it is necessary to define the concept of "branch", as used here. A branch is a sequence of one or more points of bearing within the BPG. A branch is bound to a sequence of tiles, but it is not necessarily uniquely defined by those tiles; that is, it is possible for two or more branches to "compete" within the same route. Such competition helps cope with the lack of knowledge about future contexts, when selecting BPs for the tile-in-focus.

Figure 10.18 illustrates a completely expanded BPG for a map having five obstacles. Note the start and goal points given in tiles 1 and 12, respectively. In this illustration, branches normally pruned during BPG construction are included as well. Notice that redundant branches exist within certain routes;

this is more clearly depicted in Figure 10.19. In this figure, the portion of the BPG reflecting the tile sequence ⟨1, 2, 4, 6, 8, 9, 10, 12⟩ is shown. Redundant branches arise as a result of branch competition; that is, it often happens that more than one branch begins growing within the same route. When the BPs for two or more competing branches meet, the redundant, less optimal branch(es) can be inhibited from further growth. In Figure 10.19, the final segments of the redundant (i.e. pruned) branches are shown by dashed lines.

10.3.6. Heuristics for Bearing Point Selection

In order to construct branches in the BPG, heuristics must be defined which select point(s) of bearing for the current tile in focus. In this system, some heuristics used to develop the bearing point graph are found in the inference rules, while others are enforced within the A^* module. These heuristics are listed below, while the selection method is described later in the section.

   When expanding a branch:

1. Only one point of bearing is generated or selected within the tile-in-focus, unless that tile contains a split-point.

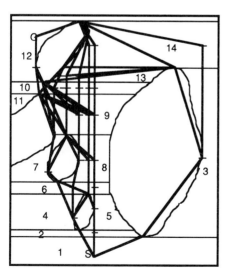

Figure 10.18 A complete bearing point graph.

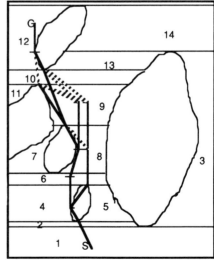

Figure 10.19 An example of a single route and its redundant branches in the bearing point graph in Figure 10.18.

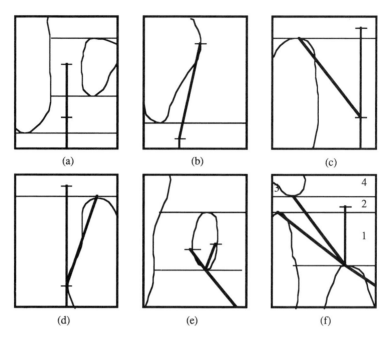

Figure 10.20 Illustrations of the various rules for generation of branches in a bearing
point graph.

2. The next point of bearing is located by extending the selected branch north or
   south along the same latitude, until reaching the "longitudinal center" of the
   tile-in-focus. Should such extension be obstructed, the BP is instead selected
   to be
   (a) the easternmost point along the west side of the tile-in-focus (i.e. the
       obstacle appears to the west), or
   (b) the westernmost point along the east side of the tile-in-focus (i.e. the
       obstacle appears to the east).
3. Should the tile-in-focus contain exactly one split-point, a second, alternative
   BP is generated at that split-point; this second BP defines a new branch
   within the BPG.
4. If the tile in focus contains two split-points, either one or both split-points
   are selected as BPs, depending upon their geometric relationships.
5. If the BPs of two branches merge and those branches compete within the
   same route, the redundant (more expensive) branch should be pruned.
6. If the branch leads into a cycle, it should be pruned.

Figure 10.20 illustrates the BP selection heuristics; for all of these examples,
expansion proceeds in a northerly direction. The first four figures depict the
primary type of BP selection, as described in the first two heuristics above. In
Figure 10.20(b), observe how the BP is selected to be the easternmost point
along the west side of the tile-in-focus. Figures 10.20(c) and 10.20(d) depict the

case described by heuristic 3 above. Note in those figures that two alternative BPs are generated for the tile-in-focus (i.e. the upper tile), since it contains a single split-point.

Figure 10.20(e) also depicts use of the second heuristic. In this case, the split-point at the southern end of the obstacle functions as the end point of two alternative branches: one passing the obstacle on the left and the other, on the right. When both alternatives have been expanded, the BPG contains the arcs shown in the figure.

In Figure 10.20(f), the branch currently ending at the split-point is being expanded; this split-point is a member of tile 1. At this expansion step, tile 2 is the tile-in-focus; note that it contains two split-points. Given tile 1 and tile 2 as the first and second members of a three-tile sequence, there exist two sequences which must be considered: $\langle 1, 2, 3 \rangle$ and $\langle 1, 2, 4 \rangle$. For the $\langle 1, 2, 3 \rangle$ sequence, two alternative BPs are selected for the focused tile; these BPs are the two split-points within that tile. Two alternative BPs are selected for the sequence $\langle 1, 2, 4 \rangle$ as well; here, one of those BPs is the easternmost of those two split-points while the other is derived by extending the selected branch upwards along its current latitude. Observe here that two or more branches within the BPG might temporarily overlap, i.e. within some sequence of tiles, they follow the same path. This is permitted as long as unique routes are created. When branches compete within the same route, the less optimal branch is pruned when the BPs for those branches merge.

The heuristics for bearing point selection are found within the inference rules. Here, the format of the rules is only slightly more complex than the simplified version presented earlier. The new format includes an extra subcondition, along with extra output information. Given a branch to be expanded, the extra subcondition checks the position of the last BP of that branch; the extra output information provides a symbolic description of the new BP(s) to which the branch should be extended. As an example, consider the first rule presented in

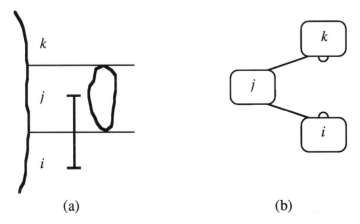

(a)   (b)

Figure 10.21 A rule context and the corresponding tile graph.

section 10.3.4. Note that Figure 10.16(a) reflects a symmetric situation. The rule creating the extension of the branch shown in Figure 10.16(a) looks like:

> if r: i, j, k
>
> u: $i \vDash j$ and $k \vDash j$
>
> v: $i \,|\, j$ and $j \,|\, k$
>
> p: $\mathrm{BP}(i) > \mathrm{max\_start\_pt}(j)$ and $\mathrm{BP}(i) < \mathrm{min\_end\_pt}(j)$
>
> then Direction_in j north   Context_in j closed_sp_right
>
> Bearing-Point_in j midpoint

In this rule, which is also illustrated in Figure 10.21, the subconditions r, u and v are interpreted as before. Subcondition p checks to see that the current point of bearing in tile i lies between the easternmost point on the west side of tile j and the westernmost point on the east side of tile j. When these subconditions are fulfilled, the new BP is symbolically output as the longitudinal "midpoint" within tile j. The symbolic BP information is returned to the A^* module and there transformed into the (x, y)-coordinates used for extension of the current branch. So far, 172 rules have been identified.

10.3.6. Dynamic BPG Construction

The purpose of this section is to provide a concrete illustration of the tandem processing between A^* and the inference engine. This example employs the map shown in Figure 10.22. Assume a start point and a goal point are given in tiles 1 and 12 respectively. The possible three-tile sequences leading out of tile 1 are $\langle 1, 2, 4 \rangle, \langle 1, 2, 5 \rangle$ and $\langle 1, 3, 14 \rangle$; for two of these sequences, tile 2 is in focus. For $\langle 1, 2, 4 \rangle$, the BP selected is point a; point b is selected for $\langle 1, 2, 5 \rangle$; and point d for $\langle 1, 3, 14 \rangle$.

Given the three BPG branches (*S-a*, *S-b* and *S-c-d*), A^* judges point a to be the most promising and selects *S-a* for expansion; this branch corresponds to the tile sequence $\langle 1, 2, 4 \rangle$. Within the tile graph, all three-tile sequences beginning with $\langle 2, 4, \ldots \rangle$ are considered. The only such sequence is $\langle 2, 4, 6 \rangle$ and, as specified by the second heuristic in section 10.3.4, point e is selected within tile 4.

At this stage of BPG construction, the possible branches for expansion are *S-a-e*, *S-b* and *S-c-d*. Here, point b returns the best evaluation within A^* and *S-b* is thereby selected for expansion. Analogous to the case above, $\langle 2, 5, 6 \rangle$ is the only three-tile sequence beginning with $\langle 2, 5, \ldots \rangle$; for this sequence, point f is chosen to be the BP within the focused tile. This expansion is depicted along the middle branch in Figure 10.22.

Among the possible BPG branches for expansion, *S-b-f* is selected next, due to point f's favorable A^* evaluation. Here the corresponding route $\langle 1, 2, 5, 6 \rangle$

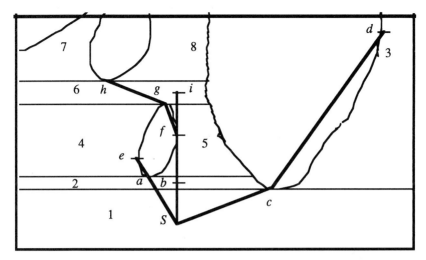

Figure 10.22 An illustration of the construction of the BPG.

may be extended in three different ways; the sequences $\langle \ldots, 5, 6, 4 \rangle$, $\langle \ldots, 5, 6, 7 \rangle$ and $\langle \ldots, 5, 6, 8 \rangle$ are all possible extensions of this route. Although tile 6 is the focused tile for each of these three-tile sequences, each possible extension is treated independently. Given $\langle \ldots, 5, 6, 4 \rangle$, point g is chosen as the BP in tile 6. For the sequence $\langle \ldots, 5, 6, 7 \rangle$, the f-h extension is instantiated as f-g-h due to the geometric relationships among the three points. Considering the $\langle \ldots, 5, 6, 8 \rangle$ extension, two alternative BPs are generated within tile 6: point g and point i. These alternatives result from interactions among the heuristics depicted in Figure 10.20, and they also illustrate the first example wherein two different branches begin competing within the same route. At this point in processing, the BPG branches and their corresponding routes are:

S-a-e	$\langle 1, 2, 4, 6 \rangle$
S-b-f-g	$\langle 1, 2, 5, 6, 4 \rangle$
S-b-f-g-h	$\langle 1, 2, 5, 6, 7 \rangle$
S-b-f-g	$\langle 1, 2, 5, 6, 8 \rangle$
S-b-f-i	$\langle 1, 2, 5, 6, 8 \rangle$
S-c-d	$\langle 1, 3, 14 \rangle$.

The tandem operation between A^* and the inference engine continues according to the principles described above, until some user-selected number of routes reach the goal or until there exist no further alternatives. The map used was illustrated in Plate 10.2. Plate 10.3, which includes the BPG for a realistic case, shows only about one-fourth of that map.

　　The euclidean optimizer includes two processing phases: plan segmentation and path construction. Briefly stated, *plan segmentation* is a phase wherein plans, along with their corresponding routes, are divided up into a number of independent segments. Each of the resulting plan segments represents a highly structured subproblem. Each of these subproblems can thereafter be solved in linear time during the path construction phase.

10.3.7. Optimization – From Routes to Paths

When processing reaches this stage, the inference engine and A^* have produced a number of alternative routes from the start to the goal, along with the corresponding plan for each route. These plans and routes are then fed into an algorithmic module for euclidean optimization i.e. the path construction phase.

The *path construction* phase employs a specialized version of an algorithm originally developed by Lee and Preparata (1984). Lee and Preparata's algorithm is designed to find the euclidean shortest path within simple (hole-free) polygons in linear time, given triangulations of those polygons. The specialization of their algorithm is restricted to the class of simple monotone polygons, yet it requires no triangulation and still performs in linear time. For convenience here, this specialized version of Lee and Preparata's algorithm is called the shortest path, monotone (SPM) algorithm. The value of the SPM algorithm lies in the fact that it can be exploited within individual tiles, as well as certain sequences of tiles, since every tile has sides which are monotone with respect to the y-axis.

Since each plan segment may be treated as a monotone polygon, the task of finding the optimal Euclidean path for a given route is greatly simplified. It is only necessary to invoke the SPM algorithm upon each plan segment, then join the various results at their unique, respective points of intersection. An example from the system is given in Plate 10.3.

Acknowledgement

The major contributions of A. Frank and M. Egenhofer to Sections 10.2, 10.2.2 and 10.2.3 are acknowledged.

REFERENCES

Bennett, B. (1994) Spatial reasoning with propositional logics. In *Principles of Knowledge Representation and Reasoning*, pp. 51–62. Morgan Kaufmann, San Francisco, CA.

Clarke, B. L. (1981) A calculus of individual based on connection. *Notre Dame Journal of Formal Logic* 2(3).

Egenhofer, M. (1994) Deriving the combination of binary topological relations. *Journal of Visual Languages and Computing* 5, 133–49.

Frank, A. (1992) Qualitative spatial reasoning about distances and directions in geographical information systems. *Journal of Visual Languages and Computing* **3**, 343–71.

Freksa, Ch. (1992) Using orientation information for qualitative spatial reasoning. In *Theories and Methods of Spatio-Temporal Reasoning in Geographic Space*, pp. 162–78. Springer-Verlag, Heidelberg.

Freksa, Ch. and Röhrig, R. (1993) Dimensions of qualitative reasoning. *Proceedings of the Workshop on Qualitative Reasoning and Decision Technologies (QUARDET'93), Barcelona*, ed. N. Piera Carreté and M. G. Singh, pp. 483–92. CIMNE, Barcelona.

Freksa, Ch. (1992) Enhancing spatial reasoning by the concept of motion. *Proceedings of AISB'93*, Birmingham, March 29–April 2.

Holmes, P. D. and Jungert, E. (1992) Symbolic and geometric connectivity graph methods for route planning in digitized maps. *IEEE Transactions on Pattern Analysis and Machine Intelligence* **14**, 549–65.

Lee, D. T. and Preparata, F. P. (1984) Euclidean shortest paths in the presence of rectilinear barriers. *Networks* **14**, 393–410.

Mukerjee, A. and Joe, G. (1990) A qualitative model for space. *Proceedings of the AAAI*, pp. 721–7.

Nilsson, N. J. (1979) *Principles of Artificial Intelligence*. Tioga Publishing, Palo Alto, California.

Randell, D. A., Cui, Z. and Cohn, A. G. (1992) A spatial logic based on regions and connection. *3rd International Conference on Knowledge Representation and Reasoning*. Morgan Kaufmann, San Francisco, CA.

Smith, T. R. and Park, K. K. (1992) Algebraic approach to spatial reasoning. *International Journal of Geographical Information Systems* **6**, 177–92.

View, L. (1992) A logical framework for reasoning about space. *Proceedings of the Conference on Spatial Information Theories, Elba*, pp. 25–53.

PART III

ACTIVE IMAGE INFORMATION SYSTEMS

11

Visual Queries

In Part I of the book we covered the basic theory of Symbolic Projection and illustrated the applications to image information retrieval. Part II of the book presented the advanced theory of Symbolic Projection with applications to spatial reasoning. In Part III, we will extend the theory and deal with various issues in active image information systems design. The emphasis of Part III is on the systems aspects, and on how active database concepts can be incorporated into image information systems.

To access information from image databases effectively and efficiently, various query mechanisms need to be combined. The user interface should be visual highly and should also enable visual relevance feedback and user-guided navigation. This chapter explores the design of multiparadigmatic visual interfaces to image databases. In section 11.1, we discuss the visual representation of the information space. Strategies for visual reasoning are surveyed in section 11.2. Visual querying systems for image databases are introduced in section 11.3. Section 11.4 gives a taxonomy of visual querying paradigms, and section 11.5 deals with database interaction techniques. Finally, in section 11.6 we describe an experimental prototype to illustrate the concept of multiparadigmatic visual interface. This chapter is adapted from Chang and Costabile (1996).

11.1. REPRESENTATION OF THE INFORMATION SPACE

In a visual interface for image databases, the information stored in the image database needs to be visualized in an information space. This visualization can either be carried out by the user in his or her mind, in which case it is essentially the user's conceptualization of the database; or the visualization may be accomplished by the system, in which case the visualization is generated on the display screen. In this section, we describe the different representations of the information space.

Database objects, in general, are abstracted from real-life objects in the real world. Therefore, we can distinguish the *logical information space* and the

Typical Visual Reasoning Problem

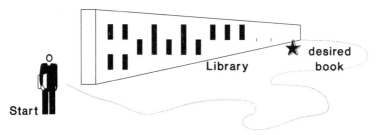

Figure 11.1 The logical information space for visual reasoning.

physical information space. In the logical space, abstract database objects are represented. In the physical space, abstract database objects are materialized and represented as physical objects such as images, animation, video, voice, etc. The physical objects either *mimic* real-life objects such as objects in a virtual reality, or *reflect* real-life objects such as diagrams, icons and sketches.

The real world, from which the database objects are abstracted, is the environment that the database objects must relate to. The real world also is often abstracted in the information space. Only in the virtual reality information space will the real world be represented in a direct way (see later).

An example of a logical information space for a library is illustrated in Figure 11.1. The logical objects are the books, and the user must apply visual reasoning and navigate in this logical information space to find a desired book. Figure 11.2, on the other hand, illustrates a physical information space, where the person must apply spatial reasoning and navigate in this physical information space, avoiding the obstacles to reach the goal.

The *logical information space* is a multidimensional space, where each point represents an object (a record, a tuple, etc.) from the database. A database object, e_j, or an *example*, is a point in this space. Conceptually, the entire information space then corresponds to all the database objects in a database.

Typical Spatial Reasoning Problem

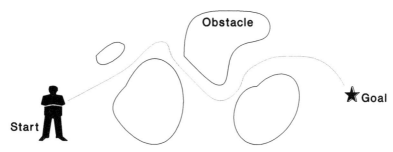

Figure 11.2 The physical information space for spatial reasoning.

The logical information space is thus a unified view of the database, i.e. a universal relation.

Each attribute of a database object represents one dimension in this multidimensional space. Therefore, in the logical information space, different dimensions actually have different characteristics: continuous, numerical, discrete, or logical.

A *query* q_i is an arbitrary region in this information space. A *clue* x_k is also an arbitrary region in the logical information space, but it may contain additional directional information to indicate *visual momentum*. Therefore, an example e_j is a clue. A visual query q_i is also a clue.

The information retrieval problem is to construct the "most desirable" query q_i with respect to the examples e_j and the clues x_k presented by the user. The "most desirable" query is one which will retrieve the largest number of relevant database objects and whose "size" in the formation space is relatively small. The process of visual reasoning, which is discussed in section 11.2, may help the user find the most desirable query from examples and clues.

The logical information space may be further structured into a *logical information hyperspace*, where the *clues* becomes hyperlinks that provides directional information, and the information space can be navigated by the user following the directional clues. Information is "chunked", and each chunk is illustrated by an example (the hypernode).

The *physical information space* consists of the materialized database objects. The simplest example is as follows. Each object is materialized as an icon, and the physical information space consists of a collection of icons. These icons can be arranged spatially, so that the spatial locations approximately reflect the relations among database objects. More recently, *intelligent visualization systems* have been developed, for task-specific visualization assistance (Ignatius *et al.*, 1994). Such systems can offer assistance in deriving perceptually effective materialization of database objects.

In the physical information space, the objects reflect real-world objects, but the world is still an abstract world. One further step is to present information in a *virtual reality information space*. Virtual reality (VR) allows the users to be placed in a 3D environment they can directly manipulate. What the users see on the screen will be the same as what can be experienced in the real world. 3D features can be used to present the results in a VR setting. For example, the physical location of medical records can be indicated in a (simplified) 3D presentation of a virtual medical laboratory by blinking icons. If the database refers to the books of a library, we can represent a virtual library in which the physical locations of books are indicated by blinking icons in a 3D presentation of the book stacks of the library. What the user sees on the screen will be the same (after simplifications) as what can be experienced in the real world. VR such as the virtual library or the virtual medical laboratory can become a new query paradigm. For example, the user can select a book by picking it from the shelf, as in the real world.

It is worth noting that we are talking about *nonimmersive* VR (Roberts *et al.*, 1993). In nonimmersive VR the user is placed in a 3D environment he or she can directly manipulate without wearing head-mounted stereo displays or special gloves, acting only with the mouse, keyboard and monitor of a conventional graphics workstation. This is an alternative form of VR that is being explored in several research laboratories. The use of 3D modeling and rendering is the same as in immersive VR, because the scene is displayed with the same depth cues: perspective view, hidden-surface elimination, color, texture, lighting, shading. Researchers working in nonimmersive VR report that the user is drawn into the 3D world, since mental and emotional immersion takes place in spite of the lack of visual or perceptual immersion. Moreover, mouse/keyboard controlled interaction techniques are easy to learn and use and are often faster than Dataglove interaction techniques. Therefore significant advantages come from using such familiar and inexpensive tools, as well as lower startup costs. Indeed, immersive VR technology has still many limits and problems (producing and synchronizing stereo images, handling of immersive input devices, etc.), so that researchers spend much more time focusing on the devices rather than on applications and interaction techniques. As a further advantage, non-immersive VR does not force office workers to wear special equipment that isolates them from their usual environment, and thus minimizes psycho-logical and physical stress that most users will not tolerate. On the other hand, new approaches to the use of immersive VR, such as interactive Worlds-in-Miniature (WIM) (Stoakley *et al.*, 1995) may pave the way for more applications of this technique.

Nonimmersive VR is a valuable interaction paradigm that will be fruitful in image database applications, as well as in general business applications. When displays and input devices that are easily manageable and not intrusive become available, immersive VR will become acceptable.

The above categorization is summarized in Table 11.1.

Table 11.1 Summary of information spaces

Space	Objects	World
Logical information space	Abstract	Abstract
Logical information hyperspace	Abstract clustered	Abstract
Physical information space	Real*	Abstract
VR information space	Real	Real

* Real objects in the physical information space reflect real-life objects, rather than mimic real-life objects.

11.2. STRATEGIES FOR VISUAL REASONING

Visual reasoning is the process of reasoning and making inferences, based upon visually presented clues. As mentioned in section 11.1, visual reasoning may help the user find the most desirable query from examples and clues. In this section, we survey strategies for visual reasoning.

Visual reasoning is widely used in human-to-human communication. For example, a teacher draws a diagram on the blackboard. Although the diagram is incomplete and imprecise, the students are able to make inferences to fill in the details, and gain an understanding of the concepts presented. Such *diagram understanding* relies on visual reasoning so that concepts can be communicated. Humans also use *gestures* (Hanne and Bullinger, 1992) to communicate. Again, gestures are imprecise visual clues for the receiving person to interpret.

In human-to-computer communication, a recent trend is for the human to communicate with the computer using visual expressions. Typically, the human draws a picture, a structured diagram, or a visual example, and the computer interprets the visual expression to understand the user's intention. This has been called variously *visual coaching*, *programming by example* (Myers, 1988) or *programming by rehearsal* (Gould and Finzer, 1984 and Huang, 1990).

Visual reasoning is related to spatial reasoning, example-based programming, and approximate/vague retrieval. As illustrated in Figure 11.2, *spatial reasoning* is the process of reasoning and making inferences about problems dealing with objects occupying space (Dutta, 1989). These objects can be either *physical objects* (e.g., books, chairs, cars, etc.) or *abstract objects* visualized in space (e.g. database objects). Physical objects are tangible and occupy physical space in some measurable sense. Abstract objects are intangible but nevertheless can be associated with a certain space in some coordinate system. Therefore, visual reasoning can be defined as spatial reasoning on abstract objects visualized in space. An example is illustrated in Figure 11.1.

Example-based programming refers to systems that allow the programmer to use examples of input and output data during the programming process (Myers, 1986). There are two types of example-based programming: *programming-by-example* and *programming-with-example*. Programming-by-example refers to systems that try to guess or *infer* the program from examples of input and output or sample traces of execution. This is often called *automated programming* and has been an area of AI research. Programming-with-example requires the user to specify everything about the program (there is no inferencing involved), but the programmer can work out the program on a specific example. The system executes the programmer's commands normally, but remembers them for later re-use. Halbert (1984) characterizes programming-with-example as "do what I did" whereas inferential programming-by-example might be called "do what I mean". Many recently developed visual programming systems have utilized the example-based programming approach (Gould and Finzer, 1984; Myers, 1988; Sirovich and Smith, 1977). The approach described in Section 11.6 combines presentation of visual clues (programming-by-example) with query augmentation techniques (programming-with-example).

We now discuss visual reasoning approaches for databases. Most research in database systems is based on the assumptions of precision and specificity of both the data stored in the database, and the requests to retrieve data. In reality, however, both may be imprecise or vague. Motro characterizes three categories of imprecision and/or vagueness:

- the data stored in the database is imprecise
- the retrieval request is imprecise
- the user does not have a precise notion of the contents of the database (Motro, 1988).

Imprecision in the user's model may be classified as follows (Motro, 1986): incomplete knowledge of the data model, imprecise information on the database schema and/or its instance, vagueness of user goals, and incomplete knowledge about the interaction tools.

To deal with imprecision in user's model, several approaches have been investigated:

- *browsing* techniques to provide different views of the database (Motro, 1986)
- *heuristic interpretation of user's query* to transform the user's query by a connective approach (Chang and Ke, 1979; D'Atri and Tarantino, 1989; Wald and Sorenson, 1984)
- *example-based techniques* to generalize from selected examples (Zloof, 1977) or to modify the original query if the answer is not considered satisfactory (Motro, 1988; Williams, 1984). The modification is done either interactively or automatically.

Limitations of each of these approaches are discussed in D'Atri and Tarantino (1989). Two main points are worth mentioning here:

- the browsing environment and the querying environment are usually distinct, thus separating the learning and the querying activities
- knowledge about the user must be gathered to build the user profile (user model).

The approach described in section 11.6 integrates the querying environment (using the visual query) and the browsing environment.

11.3. VISUAL QUERY SYSTEMS FOR IMAGE DATABASES

Image and multimedia databases, when compared to traditional databases, have the following special requirements (Faloutsous *et al.*, 1994; Fox, 1991):

- The size of the data items may be very large. The management of image information therefore requires the accessing and manipulation of very large data items.
- Storage and delivery of video data requires guaranteed and synchronized delivery of data.

- Various query mechanisms need to be combined, and the user interface should be highly visual and should also enable visual relevance feedback and user-guided navigation.
- On-line, real-time processing of large volumes of data may be required for some types of image databases.

For image databases, there are different ways to query the databases. The query mechanisms may include free text search, SQL-like querying, icon-based techniques, querying based upon the entity–relationship diagram, content-based querying, as well as virtual reality (VR) techniques. Some of these query mechanisms are based upon traditional approaches, such as free text search and SQL query language (see Section 6.1). Some are developed in response to the special needs of image databases such as content-based querying for image or video databases (Faloutsous *et al.*, 1994). Some are dictated by new software/hardware technologies, such as icon-based queries and VR queries. Except for the traditional approaches, the other techniques share the common characteristic of being highly visual.

A visual interface to image databases, in general, must support some type of visual querying language. *Visual query languages* (VQLs) are query languages based on the use of visual representations to depict the domain of interest and express the related requests. Systems implementing a visual query language are called *visual query systems* (VQSs) (Battini *et al.*, 1991/93; Catarci and Costabile, 1995). They include both a language to express the queries in a pictorial form and a variety of functionalities to facilitate human–computer interaction. As such, they are suitable for a wide spectrum of users, ranging from people with limited technical skills to highly sophisticated specialists. In recent years, many VQSs have been proposed, adopting a range of different visual representations and interaction strategies, but most existing VQSs restrict human–computer interaction to only one kind of interaction paradigm. However, the presence of several paradigms, each one with different characteristics and advantages, will help both naive and experienced users in interacting with the system. For instance, icons may well evoke the objects present in the database, while relationships among them may be better expressed through the edges of a graph, and collections of instances may be easily clustered into a form. Moreover, the user is not required to adapt his or her perception of the reality of interest to the different views presented by the various data models and interfaces.

The way in which the query is expressed depends on the visual representations as well. In fact, in the existing VQSs, icons are typically combined following a spatial syntax (Chang *et al.*, 1995), while queries on diagrammatic representation are mainly expressed by following links and forms are often filled with prototypical values. Moreover, the same interface can offer the user different interaction mechanisms for expressing a query, depending on both the experience of the user and the kind of the query itself (Catarci *et al.*, 1995).

11.4. TAXONOMY OF VISUAL QUERYING PARADIGMS

As discussed in section 11.1, the information stored in an image database is organized in a logical information space. Such logical information needs to be materialized in the physical information space in order to allow the user to view it. We are particularly interested in materializations performed by using visual techniques. Therefore, visual query systems, as defined in section 11.3, are needed. A survey of VQSs proposed in recent years is presented in Batini *et al.* (1991/93) where the VQSs are also compared with respect to three taxonomy criteria:

- the visual representation that is adopted to present the reality of interest and the applicable language operators
- the expressive power, that indicates what can be done by using the query language
- the interaction strategies that are available for performing the queries.

The query paradigm, which settles the way the query is performed and represented, is very much dependent on the way the data in the database (the query operands) are visualized. The basic types of visual representations analyzed in Batini *et al.* (1991/93) are form-based, diagrammatic, and iconic, according to the visual formalism primarily employed, namely forms, diagrams, and icons. A fourth type is the hybrid representation, which uses two or more visual formalisms.

A form can be seen as a rectangular grid having components that may be any combination of cells of groups of cells (subform). A form is intended to be a generalization of a table. It helps the users by exploiting the usual tendency of people to use regular structures for information processing. Moreover, computer forms are abstracted from conventional paper forms familiar to people in their daily activities. Form-based representations have been the first attempt to provide users with friendly interfaces for data manipulation, taking advantage of the bidimensionality of the computer screen. QBE has been a pioneer form-based query language (Zloof, 1977). The queries are formulated by filling appropriate fields of prototypical tables that are visualized on the screen.

Representations based on diagrams are widely adopted in existing VQSs. We use the word "diagram" with a very broad meaning, referring to any graphics that encode information using position and magnitude of geometrical objects and/or show the relationships among components. Referring to the different types of visual representations analyzed in Lohse *et al.* (1994) our broad definition of diagram include graphs (such as bar chart, pie chart, histogram, scatterplot, etc.), graphic tables, network charts, structure diagrams, process diagrams. An important and useful characteristics of a diagram is that, if we modify its expression by following certain rules, its content can show new relationships (Eco, 1975). Often, a diagram uses visual elements that are one-to-one associated with specific concept types. Diagrammatic representations adopt as typical query operators the selection of elements, the traversal on adjacent elements and the creation of a bridge among disconnected elements.

The iconic representation uses sets of icons to denote both the objects of the database and the operations to be performed on them. In an icon we distinguish the *pictorial part*, i.e. the image shown on the screen, and the *semantic part*, i.e. the meaning that such an image conveys. The simplest way to associate a meaning to an icon is by exploiting the similarity with the referred object. If we have to represent an abstract concept, or an action, that does not have a natural visual counterpart, we have to take into account different correlation modalities between the pictorial and the semantic part, like analogy, metonymy, convention, etc. (see Batini *et al.*, 1991/93, for more details). In iconic VQSs, a query is expressed primarily by combining icons. For example, icons may be vertically combined to denote conjunction (logical AND) and horizontally combined to denote disjunction (logical OR) (Chang, 1990).

All the above representations present complementary advantages and disadvantages. In existing systems, only one type of representation is usually available. This significantly restricts the database users that can benefit from the system. An effective database interface should supply multiple representations, in order to provide different interaction paradigms, each one with different characteristics. Therefore, each user, whether novice or expert, can choose the most appropriate paradigm to interact with the system. Such a multiparadigmatic interface for databases has been proposed in Catarci *et al.* (1995), where the selection of the appropriate interaction paradigm is made with reference to a user model that describes the user's interest and skills. Another interesting query paradigm is introduced in Chang *et al.* (1994). It is based on the idea that a VR representation of the database application domain is available. An example is presented in section 11.6.

The research on multiparadigmatic visual interfaces is conceptually similar to the research on multimodal interfaces for image databases (Blattner and Dannenberg, 1992). The rationale for providing different input and output mechanisms is to accumulate user diversity. Humans, by their very nature, have unpredictable behavior, different skills and a wide range of interests. Since we cannot obtain *a priori* information on how each user wishes to interact with the computer system, we need to create customizable human–computer interfaces, so that the users themselves will choose the best way to interact with the system, possibly by exploiting multiple input and output media.

Effective user interfaces are difficult to build. Multimodal and multimedia user interfaces are even more difficult to build. They have further requirements that need to be fully satisfied. The qualities for multimodal and multimedia interfaces have been studied in Hill *et al.* (1992). To give an example, "blended modalities" is one such quality. Blending of modes means that at any point a user can continue input in a new, more pragmatically appropriate mode. The requirement "at any point" is not easy to achieve. In the multiparadigmatic interfaces described in (Catarci *et al.*, 1995), conditions for allowing a paradigm switch during query formulation are carefully demonstrated. The problem needs to be investigated both from the system's viewpoint and from the cognitive viewpoint. Besides any model that can help predict user behavior, extensive experimentation is needed with users in order to make sure that the presence of several modes does not generate mental overload.

The expressive power achievable in the different modes, i.e. the kind of database operations that can be performed, may not be the same. For the different visual paradigms analyzed above, form-based and diagrammatic paradigms often provide the same expressive power as the relational algebra (Batini *et al.*, 1991/93), but VR only allows selection of objects and retrieval of objects for which similarity functions have been specified. This is even more evident when we consider interaction through different media. A database expert will be very comfortable when performing queries with SQL, which is even more powerful than relational algebra. The same expressive power cannot be achieved, with the current technology, if we use either speech or stylus-drawn gesture; such modes have the further disadvantage of providing ambiguous or probabilistic input. So far the design of interfaces has avoided the use of such inputs, because their ambiguity is unmanageable. Next-generation interfaces should include such input modes, if appropriate to the task the interface is for, and provide means to resolve specific ambiguities. One possibility is changing the interaction mode, so that in the new mode a certain operation is no longer ambiguous.

11.5. DATABASE INTERACTION TECHNIQUES

Computer technology is providing everyone with the possibility of exploring information resources directly. On one hand, this is extremely useful and exciting. On the other hand, the ever-growing amount of information at our disposal generates cognitive overload and even anxiety, especially in novice or occasional users. The current user interfaces are usually too difficult for novice users but inadequate for experts, who need tools with many options, and so limit the actual power of the computer.

We recognize three different needs of people exploring information:

- to understand the content of the database
- to extract the information of interest
- to browse the retrieved information in order to verify that it matches what they wanted.

To satisfy such needs, the user-interface designers are challenged to invent more powerful search techniques, simpler query facilities, and more effective presentation methods. When creating new techniques, we have to keep in mind the variability of the user population, ranging from first-time or occasional users to frequent users, from task-domain novices to experts, from naive users (requesting very basic information) to sophisticated users (interested in very detailed and specific information). Since not one technique is capable of satisfying the needs of all such classes of users, the proposed techniques should be conceived as having a basic set of features, while additional features can be requested as users gain experience with the system.

A user interacting the first time with an information system should be allowed to navigate easily into the system in order to get a better idea of the kind of data that can be accessed. As information systems become larger and larger, while each user is generally interested in only a small portion of data, one of the primary goals of a designer is to develop some kind of filter to reduce the set of data that need to be taken into account. At Xerox in recent years a group of researchers has developed several information visualizations techniques with the aim of helping users to understand and process the information stored in the system (Robertson *et al.*, 1993). They have created *information workspaces*, i.e. computer environments in which the information is moved from the original source, such as networked databases, and where several tools are at the disposal of users for browsing and manipulating the information. One of the main characteristics of such workspaces is that they offer graphical representations of information that facilitate rapid perception of the overall patterns. Moreover, they use 3D and/or distortion techniques to show some portion of the information at a greater level of detail, but keeping it within a larger context. These are usually called *fisheye techniques*, but it is clearer to call them *focus + context*, which better expresses the idea of showing an area of interest (the focus) quite large and with detail, while the other areas are shown successively smaller and in less detail. Such an approach is very effective when applied to documents, and also to graphs. It achieves a smooth integration of local detail and global context. It has advantages over other approaches to filtering information, such as zooming or the use of two or more views, one of the entire structure and the other of a zoomed portion. The former approach shows local details but loses the overall structure, the latter requires extra screen space and forces the viewer to mentally integrate the views. In the focus + context approach, it is effective to provide animated transitions when changing the focus, so that the user remains oriented across dynamic changes of the display, avoiding unnecessary cognitive load.

Search techniques applicable to image data may include other media. For instance, sound is included among data types of image databases, and it could constitute both an output (as a response of the system) or an input (as a query). Some existing electronic dictionaries already provide not only the meaning of words but also their pronunciation, so offering full information on every requested word. Madhyastha and Reed (1995) present *sonification*, i.e. the mapping of data to sound parameters, as a rich but still unexplored technique for understanding complex data. The current technology has favored the development of the graphical dimension of user interfaces while limiting the use of the auditory dimension. This is also because the properties of the aural cues are not yet as well understood as those of visual signals. Moreover, sound alone cannot convey accurate information about a visual context. The tool described in Madhyastha and Reed (1995) uses sound to complement visualization, thus enhancing the presentation of complex data. Sound can be useful in some situations, for instance we prefer to set up an alarm when working with the computer to remember to do something at a certain time. The opposite is also true, i.e. visualization can help in analyzing sound. For example, it is useful for an expert performing a detailed analysis of a certain sound to look at the graphics of its amplitude in a given time interval.

A system called Hyperbook uses sounds as an imitation of bird calls (either in the melody or in the tone) to retrieve specific bird families within an electronic book on birds (Tabuchi *et al.*, 1991). The user can also retrieve a bird by drawing a silhouette of the bird. The descriptions provided by both techniques are incomplete since it is difficult for the user to give an exact specification. Hyperbook solves such queries on the basis of a data model, called *metric spatial object data*, which represents objects in the real world as points in a metric space. In order to select the candidate objects, distances are evaluated by the system, enabling the user to choose those objects (birds) which have a minimal distance from the query in the metric space.

Techniques can be exploited for searching images in a database on the basis of their pictorial contents. Given a sketch of a house, the user may want to find all pictures that contain that house. With the visual query system called *pictorial query-by-example* (PQBE), Papadias and Sellis proposes an approach to the problem of content-based querying geographic and image databases (Papadis and Sellis, 1995). PQBE makes use of the spatial nature of spatial relations in the formulation of a query. This should allow users to formulate queries in a language close to their thinking. As in the case of the well-known query-by-example, PQBE generalizes from the example given by the user, but, instead of having skeleton relational tables, there are skeleton images.

Traditional languages such as SQL allow the user to specify exact queries that indicate matches on specific field values. Nonexpert and/or occasional uses of the database are generally not able to formulate directly a query whose result fully satisfies their needs, at least at their first attempts. Therefore, the users may prefer to formulate a complex query by a succession of progressive simple queries, i.e. step by step, by first asking general questions, obtaining preliminary results, and then revisiting such outcomes to further refine the query in order to extract the result they are interested in. Since the results obtained up to a certain point may not converge to the expected data, a nonmonotone query progression should be allowed. During this process of progressive querying, an appropriate visualization of the preliminary results could give significant feedback to the user. Moreover, it will provide hints about the right way to proceed towards the most appropriate final query. Otherwise, the user will immediately backtrack and try a different alternative path. Often, even if the user is satisfied with the result, he or she is also challenged to investigate the database further, and as a result may acquire more information from it.

The above-described advantages of performing a progressive query through visual interaction, also displaying the obtained partial results in a suitable representation, has led to the *visual querying and result hypercube* (VQRH), which is a tool that provides a multiparadigmatic approach for progressive querying and result visualization in database interaction (Chang *et al.*, 1994). Using the VQRH tool, the user interacts with the database by means of a sequence of partial queries, each displayed, together with the corresponding result, as one slice of the VQRH. Successive slices on the VQRH store partial queries performed at successive times. Therefore, the query history is presented in a 3D perspective, and a particular partial query on a slice may be brought to the front of the VQRH for further refinement by means of a simple mouse click.

Another powerful technique for querying a database is *dynamic query*, which allows range search on multikey data sets. The query is formulated through direct manipulation of graphical widgets such as buttons and sliders, one widget being used for every key. The result of the query is displayed graphically and quickly on the screen. It is important that the results fit on a single screen and that they are displayed quickly, since the users should be able to perform tens of queries in a few seconds and immediately see the results. Giving a query, a new query is easily formulated by moving with the mouse the position of a slider. This gives a sense of power but also fun to the user, who is challenged to try other queries and see how the result is modified. As in the case of the progressive query, the user can ask general queries and see what the results are, then he or she can better refine the query. An application of dyanamic queries is shown in Shneiderman (1992), and refers to a real-estate database. There are sliders for location, number of bedrooms, and price of homes in the Washington DC area. The user moves these sliders to find appropriate homes. Selected homes are indicated by bright points of light on a map of Washington shown on the screen.

11.6. AN EXPERIMENTAL MULTIPARADIGMATIC VISUAL INTERFACE

In this section we describe an experimental multiparadigmatic visual interface that allows the user to perform progressive queries, called the *visual query and result hypercube* (VQRH) (Chang *et al.*, 1994). The main features of VQRH are:

- the screen is divided into two main windows, in the left one the user formulates its query, and the results are shown in the right window
- both query and results can be visualized in any of the available paradigms for query and data representation
- the queries formulated during an interaction section are stored, with the

corresponding results, as successive slides of the VQRH, so each slice can be easily recalled.

Information retrieval experiments using VQRH in two application domains have been investigated: medical databases and library databases. The subjects are students with no previous experience in using VQRH. The preliminary experiments indicate that the users have little difficulty in learning VQRH, and they can formulate queries after half an hour of interaction. They generally like the idea of progressive querying, and find it useful to be able to recall any past query-and-result slice. From such experiments, it is already clear that the visualization of the retrieved result is very important for the success of this approach. While in the initial design of VQRH only physical information spaces were used for presenting the data, in a second version of the prototype a VR information space was added. VR serves therefore as a query paradigm, that is, the user selects with the mouse the items of this 3D space he or she is interested in.

When performing a query, the admissibility conditions to switch between a *logical paradigm* and a *VR paradigm* (such as the virtual library) can be defined as follows. For a logical paradigm, a *VR-admissible query* is an admissible query whose retrieval target object is also an object in VR. For example, the VR for the virtual library contains stacks of books, and a VR-admissible query could be

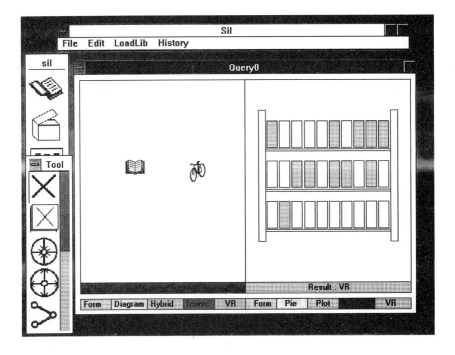

Figure 11.3 A logical query with result in virtual reality.

any admissible query about books, because the result of that query can be indicated by blinking book icons in the virtual library. Conversely, for a VR paradigm, an *LQ-admissible query* is a VR where there is a single marked VR object that is also a database object, and the marking is achieved by an operation icon such as "similar_to" (find objects similar to this object), "near" (find objects near this object), "above" (find objects above this object), "below" (find objects below this object), and other spatial operators. For example, in the VR for the virtual library, a book marked by the operation icon "similar_to" is LQ-admissible and can be translated into the following query: "find all books similar to this book." These spatial operators, and more complex ones, can be formulated based upon the theory of Symbolic Projection. For example, in Figure 11.4, the operator "dot above a horizontal bar" indicates "above", and the operator "dot below a horizontal bar" indicates "below". The first (top) operator in that menu of four operators indicates "similar", and the second operator indicates "close to".

An example of a VR-admissible logical query is illustrated in Figure 11.3. The query is to find books about bicycles. It is performed with the iconic paradigm. The result is presented as marked objects in a virtual library. The user can then navigate in this virtual library, and switch to the VR query paradigm. Figure 11.4 illustrates an LQ-admissible query. The query is to find similar books to the right and below a specific book about bicycles that has been marked by the user.

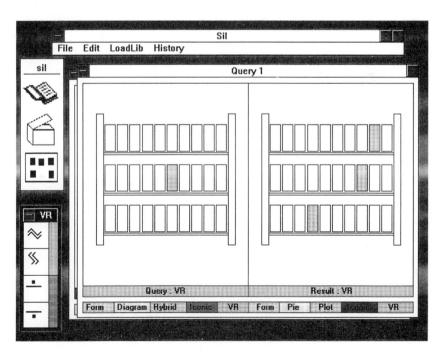

Figure 11.4 A VR query with result also in virtual reality.

The result is again rendered as marked objects in a virtual library. If we switch to a form-based representation, the result could also be rendered as items in a form. This example illustrates that progressive querying can be accomplished with greater flexibility by combining the logical paradigms and the VR paradigms. The experimental VQRH system supports VR paradigms, but the similarity function must be supplied for the problem domain.

The experimental was very useful for understanding the limitations of the screen design and their impact on the system's usability. Some interesting characteristics of the VR paradigm emerged, which led to a revised screen design. Indeed, the distinction between query space and result space does not make sense in VR, since a query is performed by acting, with either the mouse or another pointer device, in the environment the user is in. The result of the query usually determines some modification of such an environment, and this new situation is the one on which a successive request has to be performed. As a consequence, when working with the VR paradigm, the user gets confused by the separation of the query window and the result window. This is easily understandable by looking at Figures 11.3 and 11.4. The situation depicted in Figure 11.3 does not create any problem for the user, who is formulating a query in the iconic paradigm but wants to see the result in VR, since VR actually

Figure 11.5 The virtual library.

provides visual indication of where the requested books can be found. Moreover, showing the results in separate windows gives the user the possibility of viewing them in a different representation, providing the user with the full advantage of viewing the data in different ways (Batini *et al.*, 1991/93). For example, while VR gives immediate indication about the physical location of a book, in a form-based representation more details can be provided at once, such as title of the book, authors, exact number of pages, etc. Therefore, in a situation involving a change of paradigm between query and result representations, the user is perfectly comfortable with the two windows shown on the screen.

However, the user gets confused when working in VR, in the situation depicted in Figure 11.4. The two windows both show the bookshelf, but the left-hand window shows the VR query and the right-hand window shows the VR result. When the user is visualizing the VR result, it is unnatural to go to a different window to modify the query. There should be only one window, showing both the VR result and the VR query. Therefore, in the new version of the VQRH prototype, the computer screen appears as shown in Figures 11.5 and 11.6. The user first navigates in the virtual library (Figure 11.5), and clicks on a bookshelf. The bookshelf is shown in Figure 11.6. The user then proceeds to click on individual books, and uses operators such as "near", "above", "similar_to", etc. to retrieve other books.

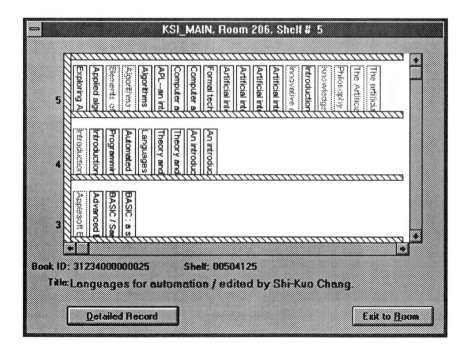

Fig. 11.6 A VR query in the VR information space.

11.7. CONCLUSION

The use of visual interfaces combining different query mechanisms represents a step toward a truly effective utilization of image information systems by many classes of users. The user interface should be highly visual to enable the user to gradually grasp the database contents, the navigation technique and the visual reasoning strategy, as well as the querying process. Since the success of a complex application is largely due to the way it matches the users' expectations, as well as their skills and learning ability, a significant part of the designer's efforts should be devoted to testing and validating the proposed interfaces, in order to provide an accurate evaluation of their usability.

Acknowledgement

The major contributions of M. F. Costabile are acknowledged.

REFERENCES

Batini, C., Catarci, T., Cosrabile, M. F. and Levialdi, S. (1991/93) Visual query systems. Technical report 04.91, Dipartimento di Informatica e Sistemistica, Universita de Roma "La Sapienza" (revised version 1993).

Blottner, M. M. and Dannenberg, R. B. (Eds.) (1991) *Multimedia Interface Design.* Addison-Wesley, Reading, MA.

Catarci, T. and Costabile, M. F. (eds.) (1995) *Journal of Visual Languages and Computing* **6**, Special issue on visual query systems.

Catarci, T., Chang, S. K., Costabile, M. F. *et al.* (1995) A graph-based framework for multiparadigmatic visual access to databases. *IEEE Transactions on Knowledge and Data Engineering.*

Chang, S. K. (1990) Visual reasoning for information retrieval from very large databases. *Journal of Visual Languages and Computing* **1**, 41–58.

Chang, S. K. and Costabile, M. F. (1996) Visual interface to multimedia databases. In *Handbook of Multimedia Information Management*, ed. W. I. Grosky, R. Jain and R. Mehrotra. Prentice-Hall, Englewood Cliffs, NJ.

Chang, S. K. and Ke, J. S. (1979) Translation of fuzzy queries for relational database system. *IEEE Transactions on Pattern Analysis and Machine Intelligence* **1**, 281–94.

Chang, S. K., Costabile, M. F. and Levialdi, S. (1994) Reality bites – progressive querying and result visualization in logical and VR spaces. Proceedings of the *IEEE Symposium on Visual Languages, St Louis*, pp. 100–9.

Chang, S. K., Costogliola, G., Pacini, G. *et al.* (1995) Visual language system for user interfaces. *IEEE Software*, 33–44.

D'Atri, A. and Tarantino, L. (1989) From browsing to querying. *Data Engineering* **12**, 46–53.

Dutta, S. (1989) Qualitative spatial reasoning: A semi-quantitative approach using fuzzy logic. *Conference Proceedings on Very Large Spatial Databases*, pp. 345–364. Santa Barbara.

Eco, U. (1975) *A Theory of Semotics.* Indiana University Press, Bloomington.

Faloutsous, C., Barber, R., Flickner, M. *et al.* (1994) Efficient and effective querying by image content. *Journal of Intelligent Information Systems* **3**, 231–62.

Fox, E. A. (1991) Advances in interactive digital multimedia systems. *IEEE Computer Journal* **24**(10), 9–21.

Gould, L. and Finzer, W. (1984) Programming by rehearsal. *Byte*, June issue, 187–210.

Halbert, D. C. (1984) Programming by example. Xerox Office Systems Division, TR OSD-T8402.

Hanne, K. and Bullinger, H. (1992) Multimodal communication: integrating text and gestures. *Multimedia Interface Design*, pp. 127–38. Addison-Wesley, Reading, MA.

Hill, W., Wroblewski, T., McCandless, T. and Cohen, R. (1992) Architectural qualities and principles for multimodal and multimedia interfaces. *Multimedia Interface Design*, pp. 311–18. Addison-Wesley, Reading, MA.

Huang, K. T. (1990) Visual interface design systems. In *Principles of Visual Programming Systems*. Prentice-Hall, Englewood Cliffs, NJ.

Ignatius, E., Senay, H. and Favre, J. (1994) An intelligent system for task-specific visualization assistance. *Journal of Visual Languages and Computing* **5**, 321–8.

Lohse, G. L., Biolsi, K. A., Walker, N. and Rueter, H. H. (1994) A classification of visual representations. *Communications of the ACM* **37**(12), 36–49.

Madhyastha, T. M. and Reed, D. A. (1995) Data sonification: Do you see what I hear? *IEEE Software* **12**(2), 45–56.

Motro, A. (1986) BAROQUE: An exploratory interface to relational databases. *ACM Transactions on Office Information Systems* **4**, 164–81.

Motro, A. (1988) VAGUE: A user interface to relational database that permits vague queries. *ACM Transactions on Office Information Systems* **6**, 187–214.

Myres, B. A. (1986) Visual programming, programming by example and program visualization: A taxonomy. *Proceedings of SIGCHI'86, Boston, MA*, pp. 59–66.

Myers, B. A. (1988) *Creating User Interfaces by Demonstration*. Academic Press, Boston.

Papadias, D. and Sellis, T. (1995) Pictorial query-by-example. *Journal of Visual Languages and Computing* **6**, 53–72.

Robertson, G. G., Card, S. K. and Mackinley, J. D. (1993a) Nonimmersive visual reality. *IEEE Computer* **26**(2), 81–3.

Robertson, G. G., Card, S. K. and Mackinley, J. D. (1993b) Information visualization using 3D interactive animation. *Communications of the ACM* **36**(4), 57–71.

Shneiderman, B. (1992) *Designing the User Interface*. Addison-Wesley, Reading, MA.

Sirovich, L. and Smith, D. C. (1977) *Pygmalion: A Computer Program to Model and Stimulate Creative Thought*. Birkhauser, Stuttgart.

Stoakley, R., Conway, M. J. and Pausch, R. (1995) Virtual reality on a WIM: Interactive worlds in miniature. *Proceedings of CHI-95, Denver, CO*, pp. 265–72.

Tabuchi, M., Yagawa, Y., Fujisawa, M. *et al.* (1991) Hyperbook: A multimedia information system that permits incomplete queries. *Proceedings of the International Conference on Multimedia Information Systems, Singapore*, pp. 3–16.

Wald, J. A. and Sorensen, P. G. (1984) Resolving the query inference problem using Steiner trees. *ACM Transactions on Database Systems* **9**, 348–68.

Williams, M. D. (1984) What makes RUBBIT run? *International Journal on Man–Machine Studies* **21**, 333–52.

Zloof, M. M. (1977) Query by example. *IBM Systems Journal* **16**, 324–43.

12

Active Indexing

This chapter introduces the *active index* for content-based image information retrieval. The dynamic nature of the active index is its most important characteristic. The main applications of the active index are to prefetch image and multimedia data, and to facilitate similarity retrieval. The experimental active index system is described.

12.1. INTRODUCTION

In image information retrieval, an important issue is how to index image objects, so that the visual objects can be accessed quickly and certain actions can be performed automatically. In conventional database systems, keyword-based indexing techniques are adequate to support users' needs. In image information systems, there are many applications that cannot be properly supported by keyword-based techniques. In addition to keywords, users often want to access or manipulate image objects by shape, texture, spatial relationships, etc. (Chang *et al.*, 1992; Grosky, 1986; Grosky and Jiang, 1992; Mehrotra and Grosky, 1989; Tagare *et al.*, 1991). That is, certain features of the image objects are used as indexes, and, in many cases, they cannot be represented as keywords.

Faced with these problems, new indexing structures must be explored which should also support similarity retrieval (Faloutsous *et al.*, 1994; Jagadish, 1991; Klinger and Pizano, 1989). Moreover, the index structures should be highly flexible and dynamic, with the following characteristics:

- *active index* instead of *passive index*: the index can be used to perform actions
- *partial index* instead of *total index*: only a few image objects are indexed
- *dynamic index* instead of *static index*: the index can evolve, grow and shrink
- *visible index* instead of *transparent index*: the user is aware of the existence of the index, perhaps as part of the knowledge structure, so the index is not necessarily transparent
- *imprecise index* instead of *precise index*: the index can be used in processing imprecise or approximate queries.

This chapter introduces the active index as a new approach to index image data. The main applications are to prefetch image and multimedia data, and to facilitate similarity retrieval. The index cell is described in Section 12.2. Section 12.3 presents the active index system. To illustrate its application to the active image information system, in Section 12.4 we present a three-level active index. This three-level active index also demonstrates the concept of the index cell hierarchy. The reversible index presented in Section 12.5 facilitates similarity retrieval. Section 12.6 presents an example of applying indexing and spatial reasoning to disaster management, based upon a real scenario of an oil spill in the Pittsburgh area. Some concluding remarks are given in Section 12.7.

12.2. THE INDEX CELL

We first introduce the index cell, which is the fundamental building block of an active index.

An index cell (IC) accepts input messages and performs some actions. It then posts an output message to a group of output index cells. Depending on the internal state of the index cell and the input messages, the index cell can post different messages to different groups of output index cells. Therefore the connection between an index cell and its output cells is not static, but dynamic. This is the first characteristic of the index cell: *the interconnection among cells is dynamically changing.*

An index cell can be either *live* or *dead*. If the cell is in a special internal state called the *dead state*, it is considered dead. If the cell is in any other state, it is considered live. The entire collection of index cells, either live or dead, forms the *index cell base* (ICB). This index cell base ICB may consist of infinitely many cells, but the set of live cells is finite and forms the *active index* (IX). This is the second characteristic of the index cell: *only a finite number of cells are live at any time.*

When an index cell posts an output message to a group of output index cells, these output index cells are activated. If an output index cell is in a dead state, it will transit to the initial state and become a live cell, and its timer will be initialized (see below). On the other hand, if the output index cell is already a live cell, its current state will not be affected, but its timer will be re-initialized. This is the third characteristic of the index cell: *posting an output message to the output index cells will activate these cells.*

The output index cells, once activated, may or may not accept the posted output message. The first output index cell that accepts the output message will remove this message from the output list of the current cell. (In case of a race, the outcome is nondeterministic.) If no output index cell accepts the posted output message, this message will stay indefinitely in the output list of the current cell. This is the fourth characteristic of the index cell: *an index cell does not always accept the input messages.*

After its computation, the index cell may remain active (live) or de-activate itself (dead). An index cell may also die, if no other index cells (including itself) posts messages to it. There is a built-in timer, and the cell will de-activate itself if the remaining time is used up before any message is received. This parameter – the time for the cell to remain live – is re-initialized each time it receives a new message and thus is once more activated. (Naturally, if this parameter is set to infinity, then the index cell becomes perennial and can remain live forever.) This is the fifth, and last, characteristic of the index cell: *a cell may die if it does not receive any message after a prespecified time.*

Although there can be many index cells, these cells may be all similar. For example, we may want to attach an index cell to an image, so that when a certain feature is detected, a message is sent to the index cell which will perform predetermined actions such as prefetching other images. If there are 10 such images, then there can be 10 such index cells, but they are all similar. These similar index cells can be specified by an *index cell type*, and the individual cells are the instances of the index cell type. We will use bold face letters to denote the index cell type, and ordinary letters to denote the index cell instance.

We have developed a tool called the *IC_Builder*, which helps the designer construct index cell types using a graphical user interface. An example is shown in Figure 12.1.

Figure 12.1(a) illustrates the state transition diagram of an index cell type under construction. This particular index cell type is **prefetch**. The **prefetch** index cell has two states: state 0 (which is also the initial state) and state -1 (which is the special dead state). The designer uses the graphical tool of the *IC_Builder* to draw this state transition diagram. For example, the designer can click on the fourth icon in the vertical icon menu (the one with a zigzag line) to draw a transition from state 0 to state 0. To avoid too many lines, only one transition line will be drawn, although the designer can specify multiple transitions from state 0 to state 0. In Figure 12.1(a), the designer creates two transitions from state 0 to state 0: transition1 and transition2. Now the details about the highlighted transition2 can be specified.

If the designer clicks on the "input_message" icon located to the right of the state transition specification dialog box in Figure 12.1(a), *IC_Builder* brings up the input message specification dialog box shown in Figure 12.1(b), so that the designer can specify the input messages. In this example, the designer specifies message 11 ("stenosis") as the only input message. A predicate can also be specified, and the input message is accepted only if this predicate is evaluated "true". In this example, there is no predicate, so the input message is always accepted.

Similarly, if the designer clicks on the "output_message" icon located to the right of the state transition specification dialog box in Figure 12.1(a), *IC_Builder* brings up the output message specification dialog box shown in Figure 12.1(c), so that the designer can specify the actions, output messages and output IC's. In this example, the designer specifies two actions: action 15 ("fetch_muga_image") and action 16 ("record_stenosis"). After the actions have been performed, the

(a)
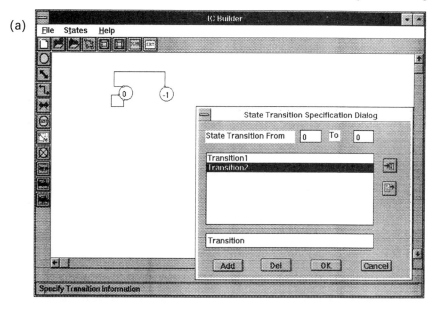

(b)
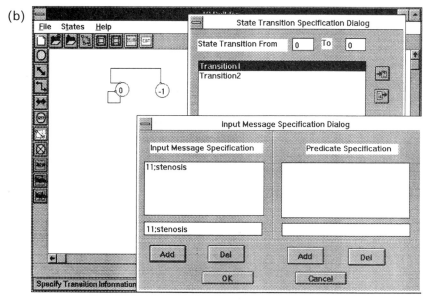

Figure 12.1 The *IC Builder*.

output message 11 ("stenosis") is posted to output index cell 3 whose type is **coronary_study**. This number 3 is the index cell's *identifier* or ic_id.

The formal definition of the active index is presented in Chang (1995). In the formal definition, one index cell can only post an output message to either a

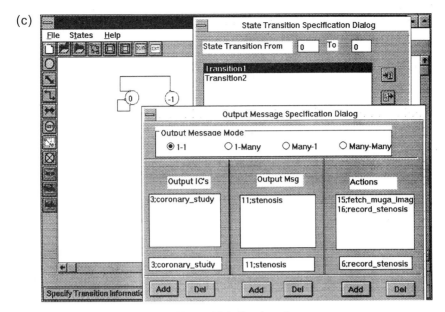

Figure 12.1 Continued.

single output index cell or a group of output index cells. However, based upon this basic index cell, we can compose more powerful index cells so that an index cell can post many output messages to one cell, or post many individual output messages to an equal number of output index cells respectively, so that message 1 goes to index cell 1, message 2 goes to index cell 2, etc. With this theoretical result, in practice the *IC_Builder* allows the designer to choose among the following four *output message modes*:

- 1–1: one message posted to one cell
- 1–many: one message posted to many cells
- many–1: many messages posted to one cell
- many–many: many messages posted individually to an equal number of cells.

In the example shown in Figure 12.1(c), the output message mode is 1–1, and the output message is posted to index cell 3. Notice the input message and the output message are the same. In other words, the current index cell will process the message 11 ("stenosis") and then pass this message to another index cell for further processing. This mechanism is useful in designing an index cell hierarchy. Of course, in general the input message and output message can be different.

Instead of posting an output message to a specific index cell whose identifier ic_id is known beforehand (in this example, index cell 3), the output message can also be posted to an index cell type. The notation is:

0;coronary_study

where the 0 indicates that the ic_id is unknown. In that case, an index cell of the type **coronary_study** is instantiated, and the output message is posted to the instantiated index cell. This mechanism is especially useful, when the active index is first constructed, and we do not yet know the ic_ids of the index cells.

Another useful mechanism is the ability to compute from the input message the resultant values of the actions, the output messages or the output ICs. For example, the input message may contain the image_id, which is copied to the output IC box so that the output message is posted to an IC with ic_id equal to image_id. The notation is:

$$copy(\ );coronary\_study$$

where copy( ) is a designer-supplied function included in the domain-specific part of the *IC_Manager* (see Section 12.3).

As illustrated by the above example, the *IC_Builder* can be used to construct the index cell types, which are stored in the index cell base ICB. Each index cell type can be instantiated as many times as that is needed. Therefore, although the ICB contains only a small number of index cell types, theoretically the ICB contains infinitely many instantiated index cells, if we regard every potentially instantiable index cell as a dead index cell in the ICB.

The designer's job is to design and construct a small number of index cell types pertaining to an application domain. In the applications to be described later, typically there are only a few index cell types.

12.3. THE ACTIVE INDEX SYSTEM

In the previous section, we have shown how the index cell type **prefetch** is specified. The first time when an image is examined by the end user, an index cell of the type **prefetch** will be activated to monitor this image. Thus the image examination event creates an external message, leading to the instantiation of the index cell type **prefetch**. Later, if a particular feature is detected in that image (either automatically or manually), a message will be posted to the index cell instance associated with that image. If the index cell accepts this message, it will perform the action to prefetch certain related images. In the previous example, the detected feature is "stenosis", and the "stenosis" message causes the index cell to prefetch the related Muga images of the same patient, and record this information.

The external environment therefore also posts messages to the active index system. In particular, the actions performed by an index cell may cause the external environment to post messages to some of the index cells, including the index cell that performs the said actions (i.e. circular messages). This can be handled by treating the external environment as an external index cell with ex_id $= -1$, which accepts messages posted to the external environment and in turn posts these messages to some of the index cells.

The engine of the active index system is the *IC_Manager*, which performs the functions of receiving incoming messages, activating index cells, performing actions, and handling outgoing messages. As illustrated in Figure 12.2, although in theory ic_1 can directly post message m_1 to ic_2, and ic_2 can directly post message m_2 to ic_3 residing in another machine, in practice every message must go through the local *IC_Manager*. Another implementation approach is to realize each cell as a separate process, but that will result in costly interprocess communication overheads. Since efficiency is a major concern, that approach was not adopted.

The core of the *IC_Manager* is described as follows:

```
IC_Manager(message)
begin
    if message contains ic_id
        begin /*the message is for a specific IC that should already exist*/
            locate ic_id in IX;
            add message to input_list; end;
    if message contains ic_type
        begin /*the message is for an IC to be created*/
            locate ic_type in ICB;
            create a new ic_id;
            add a new ic instance to IX;
            add message to input_list of this IC;
            add ic_id to the output_list of the output IC; end;
        while there is next ic_id in IX
            begin check whether message should be accepted;
            if message should be accepted
                if message has not been accepted by another IC
                    begin accept this message and remove it from output_list;
                        proces this IC; end
        end
end
```

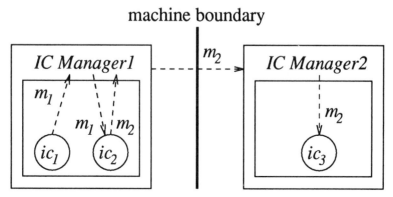

Figure 12.2 The *IC Manager*.

The *IC_Manager* has a domain-independent part and a domain-specific part. The domain-specific part contains the specific routines used by the ICs to perform predefined actions, and to compute output ICs and output messages from input messages. It also identifies and structures the external messages to be sent to the *IC_Manager*. The clear separation of domain-independent and domain-specific parts makes it easy to adapt the *IC_Manager* to a new application. The *IC_Manager* is written in standard C code and can easily be compiled together with the intended application system, on workstations as well as PCs, to produce a customized application system with built-in active index.

12.4. ACTIVE INDEX FOR SMART IMAGES

An active index is a dynamically changing net. As shown in Chang (1995), the active index can be used to realize Petri nets, generalized Petri nets (G-nets), etc., but its primary purpose is to serve as a dynamic index. In this section, we describe how to apply the active index to image information retrieval.

In current image information systems, images do not have the capabilities to respond automatically to situational changes occurred in their environments. With advances in software and hardware technologies, images can play a more active role in applications. For example, in the medical domain, after the examination of a patient's nuclear image, a physician may want to compare images of the same patient at different states (exercising, normal, excited, etc.), examine images in the time domain (past histories), and check images from other modalities. Instead of requiring the physician to retrieve these relevant images with explicit queries and to convert and highlight the images properly, a *smart image* can monitor the physician's actions and provide the necessary information in proper formats on time. Smart images should invoke actions by themselves. Depending upon applications, they can move themselves into proper local storage, preprocess themselves into the appropriate representations, and display themselves on the screen at the right time.

A smart image is an image with an associated knowledge structure, where knowledge includes attributes, routine procedures for how the image is used, and dynamic links to other objects for performing related actions. A smart image knows what actions to take based on the user's interaction with the image and on the environmental changes to the images (Chang *et al.*, 1992).

To realize a smart image, the image can be dynamically associated with a number of active index cells. It is important that the association is dynamic. In other words, if the image is never examined by the physician, there is no need to associate any index cell with it. Only when the image is examined will the index cell(s) be instantiated. If the image has not been accessed for some time, the associated cells will die. In what follows, we describe a three-level active index for the smart image, where each level has a small number of index cell types.

(a) *Level-1 Index*

The purpose of the level-1 index is to pre-perform certain operations and, more specifically, to prefetch images. To each image, exact one level-1 index cell can be associated. The **prefetch** index cell type is similar to the one described in section 12.2, but the transition diagram is more complicated because the cell has more states.

The first time the physician examines an image, the "examination" message is posted to an **MIC** type (medical IC) index cell, whose purpose is to record interaction *history*. An **MIC** index cell is instantiated. The same "examination" output message is also posted to the **prefetch** IC type. A **prefetch** index cell is instantiated, which from then on will be associated with that image.

When the physician detects certain feature in the image under examination, the physician clicks on a *hot spot* that triggers another message to be sent to the specific **prefetch** IC associated with that image. The **prefetch** IC accepts this message, and (1) activates the appropriate level-2 IC type, (2) posts an output message to the instantiated level-2 IC, and (3) changes state to the appropriate next state. It also performs the actions specified for this IC.

For example, if the IC is in the state s_{angio} and the input message indicates a stenosis condition, then the IC will activate the level-2 **coronary_study** IC by posting an output message "Image : Angio, Abnormality : Stenosis" to this IC type, and change state to s_{muga}. The action of this **prefetch** IC in state s_{angio} is to prefetch all Muga images of the patient.

In the above example, the states of this **prefetch** IC correspond to the different *image modalities*. When the physician is examining an image of a given modality, the index cell is in that state. Therefore, such states are observable and the interaction history can be recorded for the automatic construction of index cells (see below).

Figure 12.3 illustrates the relationships among images, hot spots and the level-1 active index. As illustrated in Figure 12.3, from the current state, depending upon the input message, the IC can prefetch all relevant images of a given modality.

(b) *Automatic Construction of Prefetch Index Cell*

When there are too many muga images to be prefetched, the computer's local memory may not be large enough and it becomes desirable to prefetch only a subset of these images that are the most relevant. To narrow down the range, we employ the following: (1) the filtering algorithm, (2) the image class hierarchy, and (3) the index cell construction algorithm.

Since an individual index cell is essentially a finite-state machine, if we can observe and record the input/output behavior pattern of a cell, there is an effective algorithm to construct the cell from the past *history* of user–system interactions. The algorithm enumerates permissible states and transitions to

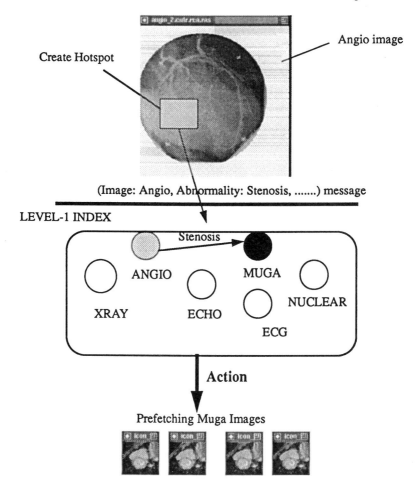

Figure 12.3 The Level-1 active index.

construct the *reachability tree* and derive the finite-state machine. In principle, we can record every click made by the user as well as every message issued by the user. In practice, we can use *filters* to extract the appropriate user messages and record them in the *history*. A moving window is kept, so that the recent history is analyzed by the algorithm to construct the finite-state machine for the index cell. For example, the filtering algorithm may only extract user's identification of abnormality and accessing of image data:

Abnormality = "Stenosis", Retrieve = "Muga image taken on date-x"

from the following history:

Doctor Name: Douglass A. Young
Patient Name: David Straker SSN: 152-83-2745 SEX: M Date of Birth: Sep
15 1953

Angio Image: Ang.Hrt.001, taken on Dec 19 1991, EID# = MHT-00010
CREATE_HS
Abnormality : Stenosis
RETRIEVE_IMAGE
Muga Image: Mug.Hrt.003, taken on Dec 10 1991, EID# = MHT-00030

In the smart image class hierarchy, images are divided into:

(a) recent images (within one month),
(b) fairly recent images (within one year),
(c) archival images (within ten years).

The simplest index cell described in Section 12.2 has only a single state besides the special dead state. The more sophisticated index cell has a next state corresponding to all images of a certain modality. The highly sophisticated index cell has next states corresponding to (a), (b) and (c) of each modality. The index cell construction algorithm tests whether date-x satisfies (a), (b) or (c) and then constructs the cell's next state(s). With this refined construction algorithm, only those images that are in (a), (b) or (c) will be prefetched.

(c) *Level-2 Index*

The purpose of the level-2 index is to perform hot-spot-triggered actions in a multimodality study. If the user makes known to the system what study is being conducted, such as coronary artery disease, ventricular function, and so on, the appropriate level-2 index cells will be activated based on the particular study.

Figure 12.4 illustrates the level-2 active index. In Figure 12.4, the active index is shown as a net of index cells. It should be emphasized that the connecting arcs in this net are dynamic. Output arcs are specified when a live cell accepts and processes the input. They can change dynamically. For example, the hot spot condition "LV_enlargement_abnormality", and heart volume quantitative data in nuclear image, and the hot spot condition "stenosis_abnormality", and low ejection fraction quantitative data in angio image, when triggered, will (1) cause the IC to activate the level-3 IC, (2) post the appropriate output message to the level-3 IC, and (3) de-activate the **left_ventricular_study** IC. Another hot spot may cause the IC to activate different output cells.

The spatial relations among objects can be tested before a level-2 index is activated. Some image processing functions may also require multiple images as input. The detection of LV enlargement and stenosis in two different images may require two separate image processing functions. The two functions may be disjoint, and no image fusion is required. We will just test the logical predicates in the above example. On the other hand, for some cases image fusion will be required, and we must register the images by performing nonlinear transformations to correlate images, etc. and to test their spatial relations.

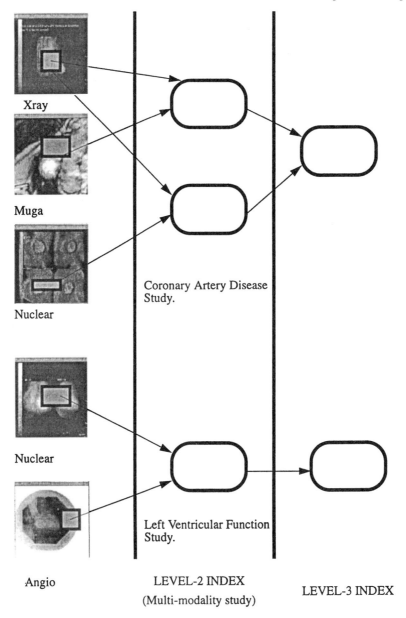

Figure 12.4 The Level-2 active index.

As explained in section 12.2, the same message may be posted to index cells at different levels. Thus the index cells form a hierarchy. By adding more specialized index cell types, the designer can add *private knowledge* to the smart image system, so that the smart image will behave differently for different users or user groups.

(d) *Level-3 Index*

The purpose of the level-3 index is to perform automatic linking and the retrieval of related information. When the user requests information, or an index cell decides the user may need certain information, a message is posted to the **retrieve** IC. The input to the **retrieve** IC is therefore the set of retrieval requests. An appropriate retrieval request will (1) cause the IC to activate another (possibly remote) IC, (2) post an output message to that IC, and (3) change to initial state. The action is to send information to the original requester. For example, the **tumor** IC, with the initial input message "tumor_ found", may initiate a retrieval request to retrieve all related information on that patient, and present it to the original requester (the physician).

12.5. REVERSIBLE INDEX FOR FEATURE-BASED INDEXING

The level-2 active index shown in Figure 12.4 can be used for feature-based indexing. When a feature is detected in an image, a hot spot is triggered to send a message to an index cell, which in turn may post output messages to other cells. Conversely, if we want to retrieve images having that feature, we need to *reverse* the flow in the index structure.

As an example, suppose we have two different projections for angio images: left anterior oblique (LAO) and right anterior oblique (RAO). Each projection may have many different image projection degrees, for example, LAO 45, LAO 30, and LAO 60.

Assume we have two images with LAO and RAO projections: Ang1.LAO45 and Ang2.RAO60. The physician may want to group those two images because both Ang1.LAO45 and Ang2.RAO60 give the physician indications of left circumflex (LCX) blood occlusion. Figure 12.5(a) shows how similar features between these two images can be found by using forward index and then reverse index.

With forward index, a hot spot with abnormality LCX in image Ang1.LAO45 triggers a new IC and causes the action to prefetch any MUGA image. Once this feature is detected, the index links can be reversed. Following the reverse index, Ang2.RAO60 is found.

The formal algorithm to construct reversible index is described in Chang (1995). Basically, the technique is to "tag" every input message, so that the input message also indicates where it comes from. When the active index is reversed, input messages become output messages and vice versa. However, the actions cannot be easily reversed, so the reverse actions are left to be specified by the designer. In information retrieval applications, the actions are replaced by a "record_node" action to record all the nodes (documents) visited in index reversal. A customized IC will then retrieve the recorded similar documents.

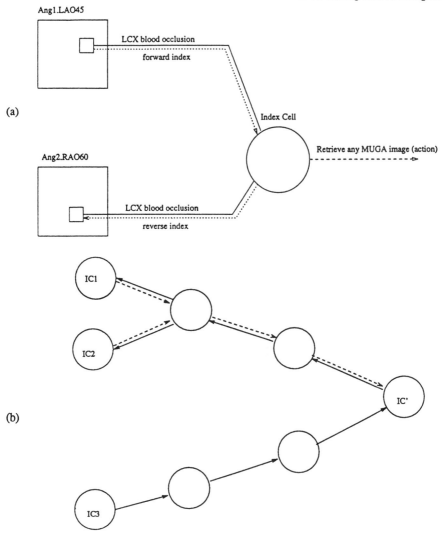

Figure 12.5 Reversible index.

The experimental active index system supports index reversal. The user must initiate the index reversal operation and specify the depth and width of the reversal, where the "reversal_depth" is the maximum number of cells to be reversed along a path, and the "reversal_width" is the maximum number of paths that will be searched after the "reversal_depth" has been reached. The reversal operation terminates when: (1) there is no more index node to reverse along a path, (2) the "reversal_depth" is reached along a path, or (3) the "reversal_width" is reached.

As shown in Figure 12.5(b), the user clicks on IC' to initiate the reversal operation. When IC1 is reached, the reversal operation terminates for that path. If the "reversal_depth" is 3, the reversal operation terminates at IC1, IC2 and IC3. If furthermore the "reversal_width" is 2, then depending upon the sequence of firing, the reversal operation may terminate at IC1 and IC2, or IC1 and IC3, or IC2 an IC3.

We can apply content-based feature extraction techniques (Faloutsous *et al.*, 1994) in forward indexing to detect a certain feature in an image, and then apply reverse indexing to find images having this feature, thus retrieving images that are similar – for example, similar in the sense of having the same feature and used by the physician in a similar medical protocol. Similarity may also be based upon satisfying the same spatial relations. We can first find the symbolic projections of the images, then apply the technique described in Section 6.3 (Lee and Hsu, 1992) to detect features (similar subpatterns), and finally apply reverse indexing to find images having similar spatial relations.

12.6. APPLYING ACTIVE INDEXING AND SPATIAL REASONING TO DISASTER MANAGEMENT

We illustrate the use of an active index with reference to a specific disaster event, the Pittsburgh oil spill (Comfort *et al.*, 1989). This scenario is based on the set of problems that emergency managers confronted during the oil spill in the Pittsburgh metropolitan region on 2 January 1988. The problems are essentially the same as those that placed the City of Pittsburgh and neighboring communities at substantial risk for two weeks in early January of 1988. The difference is that emergency managers, in this scenario, have access to an *interactive, intelligent, spatial information system for disaster management* (IISIS, Comfort and Chang, 1995). The IISIS prototype is designed to facilitate the search, synthesis, representation, analysis, and dissemination of changing information in this dynamic emergency response process involving multiple agencies and jurisdictions. In the initial stages of an emergency, information is vague and incomplete. The managers' task is to verify the conditions and degree of risk to the community, and to allocate resources and mobilize action effectively in order to reduce the risk and potential losses to the community. In any given instance, managers are both responding to existing damage and anticipating the likely occurrence of escalating risk if the existing problems are not solved.

Using the IISIS prototype, the scenario consists of the following sequence of events, revealing the alternatives to practicing emergency managers confronting a major release of a hazardous material in their community. The steps in the scenario where active indexing is applicable are in **bold face** type. Likewise, those steps where spatial reasoning is applicable are put in *italics*.

1. January 2, 1988: 5:02 p.m. Floreffe, PA.
   Coordinator's Menu: Select Field Status Board

2. Select Incident List
3. Select Incident Status
4. Report vague description: large fuel spill on Rte 837 near Floreffe, PA
5. Open second window: Allegheny County GIS: show map of southern Allegheny County, locate Floreffe, PA and Rte 837
6. Return to Coordinator's Menu with GIS map displayed: Open Notification Director
7. Follow sequence for notification of hazardous materials spill:
   a. Notify EPA Response Center;
   b. Notify local dispatch center of haz mat incident;
   c. Notify Allegheny County EMA of problem in Floreffe
8. In open GIS window: Locate source of spill at Ashland Tank Facility **Set hot spot at Ashland Tank Facility**
9. In GIS window, display Ashland facility, with site plan of tanks and fuel lines
10. In the first window, Coordinator accesses Field Status Board: Open Resources Window
11. **Active index prefetches information so that Coordinator can quickly identify available resources for hazardous materials response within a ten-mile radius of spill**: e.g. haz mat terms, fire departments, numbers of trucks, personnel with haz mat training
12. Coordinator updates Incident Status:
    a. Regional haz mat team arrives on scene;
    b. emergency coordination shifts to regional haz mat team; Floreffe coordinator becomes user, site coordinator
13. In second GIS window: identify collapsed tank and **set hot spot on tank** (see Figure 12.6)
14. **Active index uses CAMEO to identify substance in collapsed tank** (CAMEO is a computerized knowledge base of hazardous materials that is widely used by emergency managers to identify hazardous substances and their consequent effects upon unprotected people and communities)
15. **Active index uses CAMEO to identify substance in adjacent tanks**
16. In first window: regional coordinator updates information on Incident Status screen
    a. Hazardous Materials screen
    b. Incident characteristics screen
    c. Substance characteristics screen
    d. Risk Assessment: major spill
    e. Coordinator transmits updated risk assessment to Allegheny County Emergency Management Agency; ACEMA dispatches field coordinator to scene
17. In second GIS window: Display updates information on map of Ashland facility, showing collapsed tank and flooded containment area
18. In first window, regional coordinator updates Incident Status report;

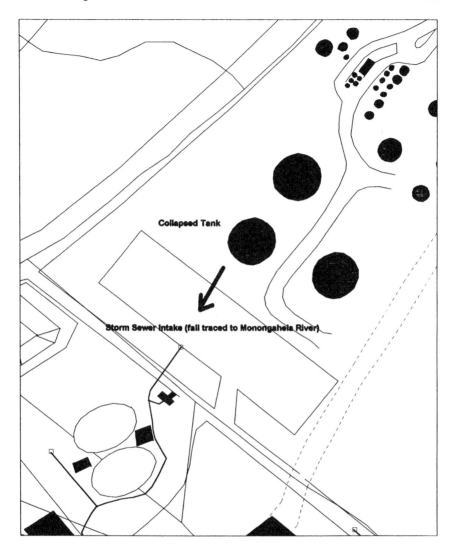

Figure 12.6 The emergency coordinator identifies collapsed tank to set hot spot.

ACEMA field coordinator arrives on scene; coordination shifts to County EMA; regional coordinator reverts to user status and site coordinator

19. ACEMA Coordinator updates Incident Status; following EPA guidelines, Ashland Oil Co. has notified USEPA of spill.
Nearest EPA response office is in Wheeling, WV
20. Diesel fuel sighted in river
21. In second GIS window: river shows small slick of fuel on water. **Active index notifies Coordinator**

22. ACEMA Coordinator notifies US Coast Guard, Marine Safety Office; USCG dispatches response team to scene
23. Coordinator's menu: bring up Jurisdictional Emergency Plans
24. Click on USEPA Response Plan; confirm exercise of authority, if spill is on land, EPA is coordinator; if spill is in water, USCG is coordinator
25. In second GIS window: show fuel spill on land at Ashland facility and in water in slick on Monongahela River
26. In second GIS window: bring up map of region, locate Wheeling, WV and calculate distance and routes to Floreffe
27. 9:00 p.m.: USCG, MSO assumes role of coordinator; ACEMA becomes user, site coordinator; EPA to assume responsibility upon arrival in morning
28. USCG follows EPA guidelines that make Ashland Company responsible for clean-up; meets separately with Ashland representatives who call private oil response company, O. H. Materials from Marion, Ohio
30. In second GIS window: open tri-state map and locate Marion, OH and O.H. Materials office. Using Oracle data base, show experience and qualifications of O.H. Materials Company: primary experience in clean-up of oil spills in Texas and on Gulf Coast, in still bays and lagoons
31. In second GIS window: show river with size of slick increasing: using database, bring up characteristics of river; fast-running current with locks and dams
32. ACEMA Coordinator uses *spatial reasoning program* to estimate the rate of flow of fuel into river, using the following characteristics: size of spill, rate of spread of slick on water, rate of river current, and timing of clean-up response
33. In second GIS window: ACEMA Site Coordinator at Floreffe queries knowledge base for Floreffe area to identify any additional hazardous materials that may interact with diesel fuel #2 within a five-mile radius of spill. Query identifies gasoline tanks at Ashland facility, Duquesne Light Plant next door, Hercules Plant 1/2 mile down road, and additional facilities in immediate area
34. ACEMA Site Coordinator checks facility plans for adjacent plants; shows facility plan for Duquesne Light; **sets hot spots of active index to detect possible paths to river**; plan reveals storm drain from containment area to river
35. ACEMA Site Coordinator updates Incident Status report, transmits report to USCG On-Scene Coordinator
36. In first window: ACEMA Coordinator updates Site report; 10.00 p.m. routine check of Ashland facility discovers gasoline leak from second damaged tank at Ashland facility; **active index sends alert messages to ACEMA Coordinator**
37. ACEMA Coordinator updates risk assessment: new threat of explosion from gasoline mixing with diesel fuel

38. In second GIS window: ACEMA Site Coordinator brings up plan of Ashland facility that shows damaged gasoline tank and piping under the huge pool of spilled fuel and points of possible leakage
39. Floreffe runs Likely Events *spatial reasoning program* to identify likely consequences of explosion on population of Floreffe; **hot spots of potential explosion area activates active index to highlight all affected areas**
40. ACEMA Coordinator notifies PEMA to threat to population of Floreffe; requests state declaration of emergency

The above scenario illustrates that active indexing supports the automatic prefetching of information. This ability, coupled with spatial reasoning, will allow the emergency managers to cope better with national emergencies.

12.7. ACTIVE INDEX FOR IMAGE INFORMATION SYSTEMS

The active index is a conceptual model. In actual implementation, the active index can be incorporated into almost any application system. For both the *smart image system* (SIS) and the *interactive, intelligent, spatial information system* (IISIS), for example, the hot spot lends itself to a natural coupling with the active index, in the sense that once a hot spot is triggered, a message is posted and the corresponding index cell is activated.

In implementing the experimental active index system, each cell is provided with an internal memory. Theoretically, the internal memory and the internal state together define the true state of the index cell. In practice, it is more convenient to have an internal working memory, so that the cells can cope with different situations flexibly. The internal memory is a C structure, so that the designer can include special routines in the domain-specific part of the *IC_Manager* to manipulate it. The ICB and IX are implemented as linked lists of C structures. Whenever there is a request to activate (or create) a new index cell, a new cell is obtained from an available list space. Conversely, a dead cell is returned to the available list space.

The major difference between an active index and a static index is that the active index is a dynamic collection of live index cells. The active index changes with time, as new index cells are activated and current index cells are de-activated.

The active index possesses the desirable characteristics discussed in Section 12.1:

- The active index can be used to initiate actions and is *active* rather than passive.
- Only the necessary index cells are activated as needed, so the index is *partial* rather than total.
- The index is *dynamic* and can evolve, grow and shrink.

- The index cell can post messages to the user in its action sequence and therefore the index can become *visible* to the user.
- Finally, the reversible index can be combined with content-based feature extraction techniques to process *imprecise* queries and perform similarity retrieval.

REFERENCES

Chang, S. K. (1995) Towards a theory of active index. *Journal of Visual Languages and Computing* **6**, 101–18.

Chang, S. K., Hou, T. Y. and Hsu, A. (1992) Smart image design for large image databases. *Journal of Visual Languages and Computing* **3**, 323–42.

Comfort, L. K. and Chang, S. K. (1995) An interactive, intelligent, spatial information system for disaster management: A national model. Paper presented at the National Emergency Management Association (NEMA) Information Technology and Communication Conference, Mt Weather, VA.

Comfort, L. K., Abrams, J., Camillus, J. and Ricci, E. (1989) From crisis to community: The Pittsburgh oil spill. *Industrial Crisis Quarterly* **3**, 17–39.

Faloutsous, C., Barber, R., Flickner, M. *et al.* (1994) Efficient and effective querying by image content. *Journal of Intelligent Information Systems* **3**, 231–62.

Grosky, W. I. (1986) Iconic indexing using generalized pattern matching techniques. *Computer Vision, Graphics, and Image Processing* **35**, 308–403.

Grosky, W. I. and Jiang, Z. (1992) A hierarchical approach to feature indexing. *SPIE/IS&T Conference on Image Storage and Retrieval Systems*, pp. 9–20.

Jagadish, H. V. (1991) A retrieval technique for similar shapes. *Proceedings of the 1991 ACM SIGMOD International Conference on Management of Data, SIGMOD Record* **20**, 208–17.

Klinger, A. and Pizano, A. (1989) Visual structure and data bases. In *Visual Database Systems*, pp. 3–25. North-Holland-Amsterdam.

Lee, S. Y. and Hsu, F. J. (1992) Spatial reasoning and similarity retrieval of images using 2D C-string knowledge representation. *Pattern Recognition* **25**, 305–18.

Mehrotra, R. and Grosky, W. I. (1989) Shape matching utilizing indexed hypothesis generation and testing. *IEEE Transactions on Robotics and Automation* **5**, 70–7.

Tagare, H. D., Gindi, G. R., Duncan, J. S. and Jaffe, C. C. (1991) A geometric indexing schema for an image library. *Computer Assisted Radiology '91*, pp. 513–18.

13

Extensions to Higher Dimensions

Several attempts to apply Symbolic Projection to higher dimensions have been made for various types of problems such as image sequences. A problem that has to be dealt with in relation to three dimensions is that the number of projection strings is extended to three pairs, in other words the information that need to be handled becomes three times as large as in the 2D case. This is easy to realize, since for the 3D case three planes must be considered with two coordinate axes each while in the 2D just one plane and two axes must be considered. Symbolic Projection is thus, when applied to 3D, merely an extension generalization of 2D, where each projection string pair of each plane can be handled separately. Image sequences are another and different problem on higher dimensions which differ from the 3D projection scheme in that the sequences encompass time as well.

The first two sections of this chapter are concerned with how to extend Symbolic Projection to three dimensions. The first approach, in section 13.1, was proposed by Costagliola *et al.* while the second, in section 13.2, is from work by Chang and Li. In section 13.3 Arndt and Chang show how Symbolic Projection can be applied to image sequences, thus encountering time in the subject. Finally, a logic-based approach in section 13.4, by del Bimbo and Vicario, shows how spatiotemporal reasoning can be facilitated and applied to sequences as well.

13.1. ICONIC INDEXING FOR 3D SCENES (COSTAGLIOLA *ET AL.*)

Costagliola *et al.* (1992) have proposed an extension to symbolic projection for the 3D case. Their work is in many ways a direct 3D extension of the original approach to Symbolic Projection by Chang *et al.* (see Chang, 1971; Chang and Li, 1995). Thus, this approach is primarily concerned with the problem of iconic indexing and much of the efforts are concerned with the ambiguity problems that may arise in conjunction with Symbolic Projection which was also pointed out by Chang *et al.* in the original work discussed in Chapter 3.

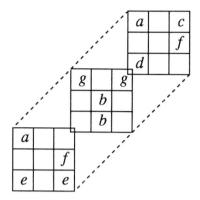

Figure 13.1 A symbolic scene in three dimensions with seven objects in three planes.

13.1.1. OPP3 and OPP2 Representations

The formal description of the method can be described as follows. Let V be a set of symbols, and $N = M_0 \times M_1 \times M_2$ where $M_i = \{1, 2, \ldots, n_i\}$ and $n_i \geq 0$ for $i = 0, 1, 2$. A symbolic scene over V and N can be defined as a mapping $s : N \to 2^V$, where 2^V is the power set of V.

Subsequently, it will be assumed that without loss of generality and for clarity, that $M_0 = M_1 = M_2 = M = \{1, 2, \ldots, n\}$, $N = M^3$.

The background here is a 3D scene with a set of objects in fixed positions. It is assumed that there is a given symbolic representation of the scene, in other words, there is a description of the scene in terms of Symbolic Projection. Figure

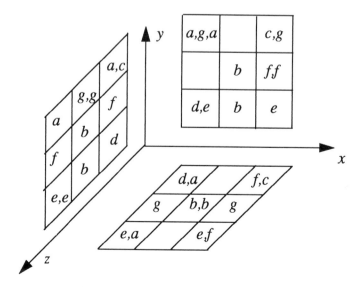

Figure 13.2 The orthogonal projections of the scene in Figure 13.1.

13.1 is an illustration of such a 3D symbolic scene including the object set $V = \{a,b,c,d,e,f,g\}$ and $N = \{1,2,3\}^3$.

In Figure 13.2, three symbolic planes s_{xy}, s_{yz} and s_{zx} represent the 2D projections of the scene in Figure 13.1. The planes are obtained by projecting each symbol in the scene onto the three cartesian planes x–y, y–z and z–x respectively. Costaglia *et al.* refer to this representation of the scene as OPP3, which stands for *orthogonal projection on planes for a 3D scene*. OPP3 is formally defined as follows.

Definition 13.1. Given a symbolic scene s over V, and $N = M^3$, its corresponding OPP3 representation is the triple (s_{xy}, s_{yz}, s_{zx}) where

$$s_{xy}(i,j) = \sum_{k \in M} s(i,j,k), \qquad \text{for } (i,j) \in M^2, \quad s(i,j,k) \in 2^V$$

$$s_{yz}(j,k) = \sum_{i \in M} s(i,j,k), \qquad \text{for } (j,k) \in M^2, \quad s(i,j,k) \in 2^V$$

$$s_{zx}(i,j(k,i)) = \sum_{j \in M} s(i,j,k), \qquad \text{for } (i,k) \in M^2, \quad s(i,j,k) \in 2^V$$

The sum $\sum$ when applied to symbols is assumed to denote the disjoint union operator, that is, if $S = \{a,b,c\}$ and $S' = \{a,b,d\}$ then $S + S' = \backslash\{a,b,c,a,b,d\backslash\}$.

Given a scene s, it can be shown that it is possible to construct the corresponding OPP3 representation. It can also be shown that, given an OPP3 expression, it is possible to reconstruct the symbolic scene. The number of slots in the OPP3 representation is reduced from n^3 to $3n^2$ but, on the other hand, an object redundancy will occur because each symbol will occur three times. One way of reducing this redundancy would be to reduce the number of projected planes and thus obtain an OPP2 representation. This is illustrated in Figure 13.3 where the number of slots is reduced to $2n^2$ and where each symbol is replicated twice. However, the representation in Figure 13.3 is not the only OPP2 that can be identified. Similar to what was pointed out in the work by Jungert and Chang (Chapter 14) three similar representations can be identified. Costagliola *et al.* call these OPP2, OPP2′ and OPP2″ for a scene s depending on which plane that is discarded. These three orthogonal projections are defined according to Definition 13.2.

Definition 13.2. Given a symbolic scene s over V and N, the OPP2 (respectively OPP2′, OPP2″) representation of s is the pair (s_{xy}, s_{zx}), (respectively (s_{xy}, s_{yz}), (s_{zx}, s_{yz})) where s_{xy}, s_{zx} and s_{yz} are defined in Definition 13.1.

For convenience, OPP2 is considered since there is no logical difference between the three orthogonal projections.

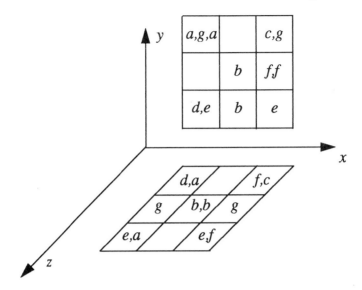

Figure 13.3 The OPP2 representation of the scene in Figure 13.1.

13.1.2. OPP1 Representations

A problem of concern is also that of projecting objects in a scene in just one direction along with the depth information of the objects with respect to their directions. However, here as well as in the OPP2 representation the choice of direction is meaningless. For this reason the depth information is always referred to as the z-direction. Illustrations to this type of projection can be seen in Figures 13.4(b) and (c) and is subsequently referred to as OPP1. The formal definition of OPP1 is as follows.

Definition 13.3. An OPP1 representation of a symbolic scene s over V and $N = M^3$ is given by the single mapping s'_{xy} where

$$s_{xy}(i, j) = \bigcup_{k \in M} c(i, j, k) \text{ with } (i, j) \in M^2$$

and

$$c(i, j, k) = \begin{cases} \varnothing & \text{if } s(i, j, k) = \varnothing \\ \{(s(i, j, k), k')\} & \text{otherwise} \end{cases}$$

where k' denotes the relative or absolute depth of the symbol $s(i, j, k)$ in the scene with respect to the z-axis. The concept of relative and absolute depth of a symbol of a scene can be seen in Figure 13.4 where in (b) the picture denotes the *absolute* OPP1 while in (c) the *reduced* OPP1 can be seen. Both are representations of the scene in (a).

The absolute OPP1 representation in Figure 13.4(b) takes into account the empty plane in the middle where $s_{z=2} = \varnothing$, independently of whether this plane

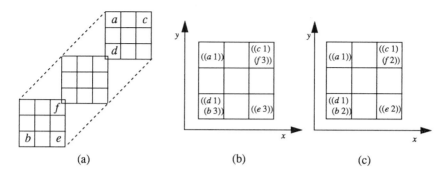

Figure 13.4 A symbolic scene (a) and its absolute OPP1 (b) and reduced OPP1 (c) representations.

in the reduced representation is stripped out or not. In the latter case, i.e. in (c), this changes the relative depth of the symbols b, e, and f from 3 to 2. A consequence of this is that when reconstructing the scene in Figure 13.4(c) just a reduced version can be obtained. Both the absolute and the reduced OPP1 representations are nonambiguous since each symbol is characterized by its depth k' corresponding to the z-direction but also by the pair (i, j) derived from the position of the symbol in the OPP1 representation. This information can thus be used to uniquely reconstruct either the original symbolic image of any scene or a description that is reduced compared to the original.

At this point, however, the OPP1 representation does not live up to the expectations of a suitable iconic index technique. It needs to be reduced further since the current representation contains too much data, i.e. the representation is not compact enough. Several possible solutions to this problem can be thought of. The approach taken is illustrated in Figure 13.5 and corresponds to a new linear representation subsequently called a generalized string or *Gen_string*. This string representation should not be confused with the general symbolic string type that was discussed in Chapter 7.

The Gen_string representation can be defined as follows. Let V be a set of symbols or the vocabulary; if ∂ is a positive integer then:

$$N' = M_1 \times M_2 \times \ldots \times M_{\partial-1} \tag{13.1}$$

where $M_i = \{1, \ldots, m_i\}$ and m_i is a positive integer and $M_i \cap V = \emptyset$ for

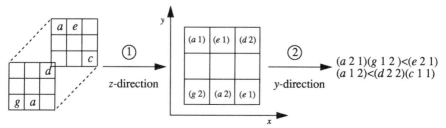

Figure 13.5 Steps in obtaining a linear representation of a symbolic scene.

$i = 1, \ldots, \partial - 1$; $A = \{\text{"}<\text{"}, \mathbf{b}\}$ where "$<$" is a special symbol not in V, and $\mathbf{b}$ denotes the empty string.

Definition 13.4. A generalized string (Gen_string) g over V, ∂, and N' is said to be $y_1 z_1 y_2 \ldots z_{n-1} y_n$ where n is an integer > 0; $y_i \in \{V^{pj} \times N', \mathbf{b}\}$ and $p_j > 0$ and $j = 1, \ldots, n$ and where $V^{pj} \times N' = \{x_1^j \ldots x_{pj}^j i_1^j \ldots i_{\partial-1}^j : (x_1^j \ldots x_{pj}^j) \in V^{pj}$ and $(i_1^j \ldots i_{\partial-1}^j) \in N'\} z_j \in A$, for $j = 1, \ldots, n-1$.

The total length of a Gen_string can be calculated from

$$(n - y^*)(\partial - 1) + (p_1 + p_2 + \ldots + p_n) + ((n-1) - z^*) \qquad (13.2)$$

where y^* and z^* represent those numbers of y_i and x_i that are equal to $\mathbf{b}$.

An illustration of the above is, for instance, the following Gen_string over $V = \{a, b, c, d\}$, $\partial = 3$ and $N' = \{1, 2, 3\}^2$:

$$g = cde13a22 < d32 < b33a21 < \qquad (13.3)$$

from which the following substrings and constants can be identified:

$$n = 6 \qquad (13.4)$$

$$p_1 = 3 \qquad (13.5)$$

$$y_1 = x_1^1 x_2^1 x_3^1 i_1^1 i_2^1 = cde13 \qquad (13.6)$$

$$z_1 = \mathbf{b} \qquad (13.7)$$

$$p_2 = 1 \qquad (13.8)$$

$$y_2 = x_1^2 i_1^2 i_2^2 = a22 \qquad (13.9)$$

$$z_2 = < \qquad (13.10)$$

$$p_3 = 1 \qquad (13.11)$$

$$y_3 = x_1^3 i_1^3 i_2^3 = d32 \qquad (13.12)$$

$$z_3 = < \qquad (13.13)$$

$$p_4 = 1 \qquad (13.14)$$

$$y_4 = x_1^4 i_1^4 i_2^4 = b33 \qquad (13.15)$$

$$z_4 = \mathbf{b} \qquad (13.16)$$

$$p_5 = 1 \qquad (13.17)$$

$$y_5 = x_1^5 i_1^5 i_2^5 = a21 \qquad (13.18)$$

$$z_5 = < \qquad (13.19)$$

$$p_6 = 0 \qquad (13.20)$$

$$y_6 = \mathbf{b} \qquad (13.21)$$

The length of g can now be determined by $(6-1)2+7+(5-2)=20$ characters.

Definition 13.5. Given a Gen_string g over V and ∂, then the rank of a symbol x_k^j, called $r(x_k^j)$ in g, is defined as one plus the number of occurrences of the symbol "$<$" preceding the x_k^j in g.

In the above example, x_1^2 has rank 1, while $r(x_1^5)=3$. Subsequently, a ∂-dimensional picture is referred to as an extension of the 3D symbolic scene that was defined earlier and where the dimension of N is rather ∂ than 3.

Each generalized string g over V, ∂, and N' corresponds to a ∂-dimensional symbolic picture over V and $N = M_0 \times N'$, where $M_0 = \{1, \ldots, m_0\}$ and m_0 is a positive integer. Given the Gen_string g, it is possible to construct the corresponding picture f by setting

$$m_0 = \text{number of "}<\text{"} + 1$$

$$m_1 = \max\{i_1^j \mid 1 \leqslant j \leqslant n\}$$

$$\vdots \tag{13.22}$$

$$m_{\partial-1} = \max\{i_{\partial-1}^j \mid 1 \leqslant j \leqslant n\}$$

and

$$f(k_0, k_1, \ldots, k_{\partial-1}) = \{x_1^j, \ldots, x_{p(j)}^j\} \tag{13.23}$$

$$\text{if } \exists j \text{ such that } k_0 = r(x_1^j) = \ldots = r(x_{p(j)}^j) \text{ and} \tag{13.24}$$

$$(k_1, \ldots, k_{\partial-1}) = (i_1^j, \ldots, i_{\partial-1}^j) \tag{13.25}$$

and

$$f(k_0, k_1, \ldots, k_{\partial-1}) = \varnothing \text{ otherwise.} \tag{13.26}$$

An illustration to this is, that from the Gen_string $g = cde13a22 < d32 < b33a21 <$ it is possible to produce a 3D picture corresponding to the one given in Figure 13.6 with the dimensions $m_0 \times m_1 \times m_2$ and

$$m_0 = 3 + 1 \tag{13.27}$$

$$m_1 = \max\{1, 2, 3, 3, 2\} = 3 \tag{13.28}$$

$$m_2 = \max\{3, 2, 2, 3, 1\} = 3 \tag{13.29}$$

Given a ∂-dimensional picture f over V and $N = m_0 \times \ldots \times m_{\partial-1}$, it is possible to uniquely construct the corresponding absolute Gen_string uniquely, except for the order of the x- and y-substrings, by setting $N' = M_1 \times \ldots \times M_{\partial-1}$ and observing that each $f(i_0, i_1, \ldots, i_{\partial-1}) = \{z_1, \ldots, z_p\}$ produces the substring $z_1 \ldots z_p i_1 \ldots i_{\partial-1}$ with $r(z_1) = r(z_2) = \ldots = r(z_p) = i_0$ and $p > 1$.

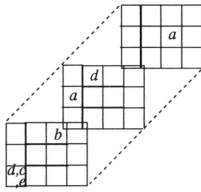

$$c\ d\ e\ 1\ 3\ a\ 2\ 2{<}d\ 3\ 2{<}b\ 3\ 3\ a\ 2\ 1{<}$$

Figure 13.6 A symbolic 3D scene and its corresponding Gen_string representation.

13.2. 3D STRINGS FOR BINARY PICTURES (CHANG AND LI)

13.2.1. Notations and Definitions

In this section another approach to 3D symbolic projection proposed by Chang and Li (1995) will be introduced. The following notations and definitions will be used. Let V be a set of symbols, $\mathbf{M} = \{1, 2, \ldots, n\}$, $A = \{:, =, <\}$. A volumetric function f is a mapping from M^3 to 2^V. A 3D string over V is defined as $\langle u, v, w \rangle$ where u, v and w are defined as:

$$u = x_1 a_1 x_2 a_2 \ldots a_{n-1} x_n \tag{13.30}$$

$$v = x_{p(1)} b_1 x_{p(2)} b_2 \ldots b_{n-1} x_{p(n)} \tag{13.31}$$

$$w = x_{q(1)} c_1 x_{q(2)} c_2 \ldots c_{n-1} x_{q(n)} \tag{13.32}$$

where $x_i \in V$, a_j, b_j, $c_j \in A$, $i = 1, 2, \ldots, n$, $j = 1, 2, \ldots, n-1$, p and q are permutations over $\mathbf{M}$.

Other notations are:

$f_x : \mathbf{M} \to 2^V$. For $i \in \mathbf{M}$,

$$f_x(i) = \bigcup_{i=1}^{n} \bigcup_{k=1}^{n} f(i, j, k)$$

$f_{xy} : \mathbf{M}^2 \to 2^V$. For $(i, j) \in \mathbf{M}^2$,

$$f_{xy}(i, j) = \bigcup_{k=1}^{n} f(i, j, k)$$

For $i \in \mathbf{M}$, $f_{x=i} : \mathbf{M}^2 \to 2^V$. For $(j, k) \in \mathbf{M}^2$,

$$f_{x=i}(j, k) = f(i, j, k)$$

For $S \in 2^V$, $f^S : \mathbf{M}^3 \to 2^V$. For $(i, j, k) \in \mathbf{M}^3$,

$$f^S(i, j, k) = f(i, j, k) \cap S.$$

Similarly, $f_y, f_z, f_{xz}, f_{yz}, f_{y=j}, f_{z=k}$ can be defined. If f is a binary picture, then in the definition, $\bigcup$ means "sum of".

The symbols are arranged in terms of 2D or 3D strings, while in terms of volumetric functions, symbols are arranged in sets. Here the strings and the sets are considered as two aspects of the same things, which means that a string can be regarded as a set or vice versa.

Definition 13.6. The 3D string of f is defined as $\langle u, v, w \rangle$, denoted by $s - 3d(f) = \langle u, v, w \rangle$, where $\langle u, v \rangle$ is the 2D string of f_{xz}, and $\langle v, w \rangle$ is the 2D string of f_{yz}.

Actually, $s - 3d(f)$ is $(f_x(1) < f_x(2) < \ldots < f_x(n), f_y(1) < f_y(2) < \ldots < f_y(n)$, $f_z(1) < f_z(2) < \ldots < f_z(n))$.

Suppose $u = u_1 < u_2 < \ldots < u_n$ such that there is no "<" in u_i, and where $i = 1, 2, \ldots, n$. Define $\text{subs}(i, u) = u_i$, which subsequently will be called the ith local substring of u, for $i = 1, 2, \ldots, n$. The reduced 3D string and the normal 3D string of picture f are defined in the same way as the reduced 2D string and the normal 2D string.

The rank of a symbol a in a string u, denoted by $r(a, u)$, is defined as the number of "<" preceding the symbol in $u + 1$. See also section 13.1.2, Definition 13.5.

13.2.2. Ambiguities and Switching Components of 3D Strings

Definition 13.7. A picture f is said to be ambiguous if there is a picture g such that $f \neq g$ and $s - 2d(f) = s - 2d(g)$ (or $s - 3d(f) = s - 3d(g)$).

Definition 13.8. f and g are said to be similar to each other if $f \neq g$ and $s - 2d(f) = s - 2d(g)$ (or $s - 3d(f) = s - 3d(g)$).

Theorem 13.1. A 2D symbolic picture f without subpattern of the form:

is ambiguous under the reduced 2D string if and only if its reduced 2D string contains the subsequence $(a < a, a < a)$, where a is some symbol in V.

The proof to Theorem 13.1 can be found in Section 3.5.

Corollary. If a picture f is ambiguous under a normal 2D string, then $s - 2d(f)$ must contain $(a < a, a < a)$ as a subsequence.

Corollary. If $s - 2d(f)$ contains $(S < S, S < S)$ as a subsequence, then f must be ambiguous under a normal 2D string, where S is a set in the local substring.

Theorem 13.2. A 3D symbolic picture f without the subpattern of the form:

is ambiguous under the reduced 3D string if and only if its reduced 3D string contains the subsequence $(a < a, a < a, a < a)$ or $(a < a, a < a, aa)$ or $(a < a, aa, a < a)$ or $(aa, a < a, a < a)$, where a is some symbol in V.

Proof. Suppose f is ambiguous, i.e. there is a g such that $f \neq g$ and $s - 3d(f) = s - 3d(g)$. If at least one of $f_{xy} \neq g_{xy}, f_{xz} \neq g_{xz}, f_{yz} \neq g_{yz}$ holds, then without loss of generality suppose $f_{yz} \neq g_{yz}$, then f_{yz} is ambiguous under the reduced 2D string, i.e. $s - 2d(f_{yz})$ has $(a < a, a < a)$ as a subsequence. Hence $s - 3d(f)$ either has $(aa, a < a, a < a)$ or $(a < a, a < a, a < a)$ as a subsequence. Otherwise, $f_{xy} = g_{xy}, f_{xz} = g_{xz}, f_{yz} = g_{yz}$. There is (ii, jj, kk) such that $f(ii, jj, kk) \neq g(ii, jj, kk)$. Then $f_{x=ii} \neq g_{x=ii}$, while $f_{x=ii\,y}(j) = f_{xy}(ii, j) = g_{xy}(ii, j) = g_{x=ii\,y}(j)$; $f_{x=ii\,z}(k) = f_{xz}(ii, k) = g_{xz}(ii, k) = g_{x=ii\,z}(k)$. This implies that $s - 2d(f_{x=ii})$ contains the subsequence $(a < a, a < a)$. Thus $s - 3d(f)$ has $(a < a, a < a, a < a)$ as subsequence.

The converse of the theorem can be proven in the same way as the 2D case. $\qquad\square$

Definition 13.9. A binary picture f is said to contain a type-i switching component if there are indices

$i_1, j_1, k_1, i_2, j_2, k_2, i_1 \neq i_2, j_1 \neq j_2, k_1 \neq k_2$ such that

(type-1) $f(i_1, j_1, k_1) = f(i_1, j_2, k_2) = 1 - f(i_1, j_1, k_2) = 1 - f(i_1, j_2, k_1)$
(*yz*-plane switching component)

(type-2) $f(i_1, j_1, k_1) = f(i_2, j_1, k_2) = 1 - f(i_1, j_1, k_2) = 1 - f(i_2, j_1, k_1)$
(*xz*-plane switching component)

(type-3) $f(i_1, j_1, k_1) = f(i_2, j_2, k_1) = 1 - f(i_1, j_2, k_1) = 1 - f(i_2, j_1, k_1)$
(*xy*-plane switching component)

(type-4) $f(i_1, j_1, k_1) = f(i_1, j_2, k_2) = 1 - f(i_2, j_1, k_1) = 1 - f(i_1, j_2, k_2)$
(*x*-direction switching component)

(type-5) $f(i_1, j_1, k_1) = f(i_2, j_2, k_2) = 1 - f(i_1, j_2, k_1) = 1 - f(i_2, j_1, k_2)$

(*y*-direction switching component)

(type-6) $f(i_1, j_1, k_1) = f(i_2, j_2, k_2) = 1 - f(i_1, j_1, k_2) = 1 - f(i_2, j_2, k_1)$

(*z*-direction switching component)

Theorem 13.3. If f and g are similar binary 3D pictures, at least one of the equations $f_{xy} = g_{xy}, f_{xz} = g_{xz}, f_{yz} = g_{yz}$ hold, then f has a switching component. If a binary 3D picture f has a switching component, it must be ambiguous.

Proof. Suppose f and g are similar. Then $f_x = g_x, f_y = g_y, f_z = g_z$, and there is a (ii, jj, kk) such that $f(ii, jj, kk) \neq g(ii, jj, kk)$. Assume that f has no switching components, thus a contradiction can be derived.

In Case 1, if just one of the three equations holds, e.g. $f_{xy} = g_{xy}$. Assume that f has no switching components and a contradiction can be derived. Then for the same reason as in Chang (1971) the following table can be identified:

$$
\begin{array}{ll}
f(i_1', j_1', k_1') = 1 & g(i_1', j_1', k_1') = 0 \\
f(i_2', j_2', k_1') = 0 & g(i_2', j_2', k_1') = 1 \\
f(i_2', j_2', k_2') = 1 & g(i_2', j_2', k_2') = 0 \\
f(i_3', j_3', k_2') = 0 & g(i_3', j_3', k_2') = 1 \\
f(i_3', j_3', k_3') = 1 & g(i_3', j_3', k_3') = 0 \\
\quad \vdots & \\
f(i_n', j_n', k_n') = 1 & g(i_n', j_n', k_n') = 0 \\
f(i_1', j_1', k_n') = 0 & g(i_1', j_1', k_n') = 1
\end{array}
$$

Since $f(i_1', j_1', k_n') = 0$, $f(i_n', j_n', k_n') = 1$ and $f(i_n', j_n', k_{n-1}') = 0$, then $f(i_1', j_1', k_{n-1}') = 0$ may exist when the same process is repeated, then $f(i_1', j_1', k_{n-2}') = \ldots = f(i_1', j_1', k_1') = 0$ will occur. This is contradictory to $f(i_1', j_1', k_1') = 1$. Thus f must have a switching component. In this case, it can be seen that the switching component is a directional switching component.

In Case 2, if at least two of the three equations hold, for instance, $f_{xy} = g_{xy}$, $f_{xz} = g_{xz}$, then consider the 2D binary picture $f_{x=ii}$, where $f_{x=ii} \neq g_{x=ii}$, while $f_{x=ii\,y}(j) = f_{xy}(ii, j) = g_{xy}(ii, j) = g_{x=ii\,y}(j)$, $f_{x=ii\,z}(k) = f_{xz}(ii, k) = g_{xz}(ii, k) = g_{x=ii\,z}(k)$. From Chang's reconstruction projections it is known that $f_{x=ii}$ has a 2D switching component. Thus f has a 3D switching component, which must be a planar switching component.

The second part of the theorem is obviously correct. ☐

Definition 13.10. f is a 3D picture. Define an *x-plane switching operation* $S[x: i_1, i_2](f)$ on f, for $i_1 \neq i_2$, to be a mapping from f to another 3D picture g such that for all $j, k, g(i_1, j, k) = f(i_2, j, k)$, $g(i_2, j, k) = f(i_1, j, k)$, and $g(i, j, k) = f(i, j, k)$ for $i \neq i_1, i_2$.

Similarly, $S[y: j_1, j_2](f)$ and $S[z: k_1, k_2](f)$ can be defined.

Lemma. f is a 3D binary picture, and g is the result of a series of plane switching operations on f. Then f has switching components if and only if g has switching components.

Definition 13.11. f is said to be equivalent to g under plane switching if it can be transformed to g by a series of plane switching operations.

Theorem 13.4. A 3D binary picture g has no switching components if and only if g is equivalent to a picture f which has the following properties:

$$f(i_1, j, k) \geqslant f(i_2, j, k) \quad \text{for all } (i_1, j, k) \text{ and } (i_2, j, k) \text{ with } i_1 < i_2$$
$$f(i, j_1, k) \geqslant f(i, j_2, k) \quad \text{for all } (i, j_1, k) \text{ and } (i, j_2, k) \text{ with } j_1 < j_2$$
$$f(i, j, k_1) \geqslant f(i, j, k_2) \quad \text{for all } (i, j, k_1) \text{ and } (i, j, k_2) \text{ with } k_1 < k_2$$

Proof. Suppose g has no switching components. A series of plane switching operations $S = S_1, S_2 \dots S_l$ on g can then be found to get f such that f_x, f_y, f_z decrease progressively.

If there are indices i_1, i_2, j, k, where $i_1 < i_2$ such that $f(i_1, j, k) < f(i_2, j, k)$, since $f_x(i_1) \geqslant f_x(i_2)$, then there must be (j_1, k_1) such that $f(i_1, j_1, k_1) > f(i_2, j_1, k_1)$. Thus (i_1, j, k), (i_2, j, k), (i_1, j_1, k_1), (i_2, j_1, k_1) form a switching component in f. Then g has a switching component, by the lemma above. This contradicts the premise. Hence, f has the properties stated by the inequalities.

The other part of the theorem is evidently true. $\square$

Definition 13.12. If a picture f has the property stated by the above inequalities, then f is said to be *straight convex*. ambiguous.

A pair of functions which are straight convex and similar to each other have been found. This indicates that there exist such functions which do not have any switching components but are ambiguous. This can be illustrated with some examples which can be seen in Figure 13.7.

$$f_x = g_x = (22, 20, 13, 10, 8, 6, 4, 2, 1) \tag{13.33}$$

$$f_y = g_y = (19, 16, 14, 12, 10, 8, 4, 2, 1) \tag{13.34}$$

$$f_z = g_z = (45, 28, 13) \tag{13.35}$$

From this example, it can be seen how tough the ambiguous characteristic problem is. Ambiguity has no relation to symmetry and is not directly related to the switching components defined earlier in this section. However, the example hints that the definition of switching components need to be expanded to include slantwise switchings as well.

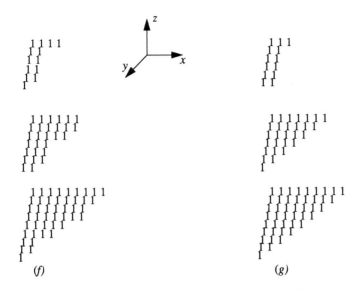

Figure 13.7 Examples of two ambiguous images without switching components.

13.3. IMAGE SEQUENCE COMPRESSION (ARNDT AND CHANG)

The processing of image sequences involving motion is of concern in many applications. Arndt and Chang (1989) have given an approach to this type of problem based on symbolic projection. The problem can be compared to indexing a film consisting of an ordered sequence of frames:

$$f_1, f_2, \ldots, f_n \tag{13.36}$$

If a 2D approach to Symbolic Projection, similar to the original by Chang *et al.*, is applied separately to each of the frames in order of appearance an index can be constructed for each of the frames. Thus the sequence of frames $f_1, f_2, \ldots, f_n$ may be divided into a number of series $s_1, s_2, \ldots, s_m$ where each series s_i denotes a *continuous* sequence of images that may or may not include operations such as zoom-in, and zoom-out. Since the flow of objects within each series is continuous, each frame need not be individually indexed. Instead the changes between the frames should be of more concern. Thus the beginning of each series s_1 should be marked with the frame f_{i1}. The 2D string method is then used to construct strings for the frame f_{i1}. The general form of the string pair $\langle u, v \rangle$ is:

$$u_{1,1}, u_{1,2}, \ldots, u_{1,k'} < u_{2,1}, u_{2,2}, \ldots, u_{2,k2} < \ldots < u_{m,1}, u_{m,2}, \ldots, u_{m,k^{(m)}} \tag{13.37}$$

$$v_{1,1}, v_{1,2}, \ldots, v_{1,k'} < v_{2,1}, v_{2,2}, \ldots, v_{2,k2} < \ldots < v_{n,1}, v_{n,2}, \ldots, v_{n,j^{(n)}} \tag{13.38}$$

From set theory:

$$\langle \{u_1\} \{u_2\} \ldots \{u_m\}, \{v_1\} \{v_2\} \ldots \{v_n\} \rangle \tag{13.39}$$

The beginning of the sequence labelled with the initial frame thus becomes:

$$f_{i1}\langle\{u_1\}\ \{u_2\}\ldots\{u_m\},\ \{v_1\}\ \{v_2\}\ldots\{v_n\}\rangle \tag{13.40}$$

The remainder of the index of each such sequence s_i consists of changes to the sets $1, 2, \ldots, m, m+1, m+2, \ldots, m+n$.

Definition 13.13. A *compressed index* is a string $A\Omega$ of the form: $f_{i1}\sigma_{i1}$, $f_{i2}\sigma_{i2}, \ldots, f_{ij}\sigma_{ij}, \ldots, f_{ik}\sigma_{ik}$ where each f is a marker of the frame number, and each $\sigma_j = \langle\{u_{j1}\}\ \{u_{j2}\}\ldots\{u_{jm}\},\ \{v_{j1}\}\ \{v_{j2}\}\ldots\{v_{jm}\}\rangle$ is a set of operations. The string $\Omega = f_{i1}\sigma_{i1}, f_{i2}\sigma_{i2}, \ldots, f_{ik}\sigma_{ik}$ represents the changes within that series.

Examples of changes that may occur between frames are insertion and deletion of objects to a series, merging and splitting of consecutive sets of frames, etc.

The sequences of sets σ can now be expressed in terms of the following syntax:

$$\sigma ::= \alpha\beta\gamma\delta \tag{13.41}$$

$$\alpha ::= \in\ |\ \langle+\rangle\ \text{OS OS\_list} \tag{13.42}$$

$$\beta ::= \in\ |\ \langle-\rangle\ \text{OS OS\_list} \tag{13.43}$$

$$\gamma ::= \in\ |\ \langle|\rangle\ \text{set pair\_list} \tag{13.44}$$

$$\delta ::= \in\ |\ \langle\rangle\ \text{set object\_list triple\_list} \tag{13.45}$$

$$\text{OS\_list} ::= \in\ |\ \text{OS OS\_list} \tag{13.46}$$

$$\text{OS} ::= \text{set}\ |\ \text{object\_list set} \tag{13.47}$$

$$\text{object\_list} ::= \text{object tail} \tag{13.48}$$

$$\text{tail} ::= \in\ |\ \text{object tail} \tag{13.49}$$

$$\text{pair\_list} ::= \in\ |\ \text{set pair\_list} \tag{13.50}$$

$$\text{triple\_list} ::= \text{set object\_list triple\_list} \tag{13.51}$$

where $\langle+\rangle$ denotes addition of a set of an object to a set and $\langle-\rangle$ denotes deletion of a set or an object of a set, $\langle|\rangle$ denotes the merging of two sets and finally $\langle\rangle$ denotes splitting of a set. "object" $\in V$ and each "set" is a number denoting a set.

Each object in a set $i < j$ is to the left of all the objects in j if i and j are part of the string u, or below all the objects in j if i and j are part of the string v.

The deletion of a set n implies the deletion of all objects $o \in n$.

A simple example of the method is present in Figure 13.8(a)–(d).

The 2D string of frames 1–99 is:

$$u = \text{man} < \text{door doorknob} \tag{13.52}$$

$$v = \text{man door doorknob} \tag{13.53}$$

for frames 100–199

$$u = \text{door doorknob man} \tag{13.54}$$

$$v = \text{man door doorknob} \tag{13.55}$$

for frames 200–299

$$u = \text{door doorknob} < \text{man} \tag{13.56}$$

$$v = \text{man door doorknob} \tag{13.57}$$

for frame 300:

$$u = \text{door doorknob out\_sign} \tag{13.58}$$

$$v = \text{door doorknob out\_sign} \tag{13.59}$$

Therefore

$$V = \text{man door doorknob out\_sign} \tag{13.60}$$

and

$$u_1 = \text{man} \tag{13.61}$$

$$u_2 = \text{door doorknob} \tag{13.62}$$

$$v_1 = \text{man door doorknob} \tag{13.63}$$

Consequently, the compressed index for the sequence is:

$$
\begin{aligned}
&1\,\{\text{man}\}\,\{\text{door doorknob}\},\,\{\text{man door doorknob}\} \\
&100\,\langle|\rangle\,1 \\
&200\,\langle\rangle\,\text{door doorknob} \\
&300\,\langle+\rangle\,\text{out-sign 1 out-sign 3}\,\langle-\rangle\,2\,\text{man 2}
\end{aligned}
\tag{13.64}
$$

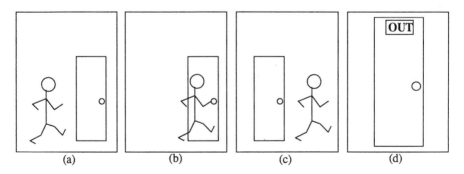

Figure 13.8 (a) frame 1, (b) frame 100, (c) frame 200, (d) frame 300.

13.4. SPATIOTEMPORAL INDEXING (DEL BIMBO AND VICARIO)

Del Bimbo and Vicario (1995) have described a frame work for spatiotemporal indexing which can be applied to image sequences as well. The approach is called *eXtended spatiotemporal logic* (XSTL) and is derived from the basic concepts of conventional temporal logic by Manna and Pnueli (1992). An earlier version of XSTL, called STL, was originally proposed by Del Bimbo *et al.* (1993). The purpose of both STL and XSTL was to develop a technique for identification and representation of spatio temporal relationships based on symbolic projection. The technique is intended for indexing of digital video sequences, as well as for a retrieval-by-contents system extended to permit metric expressivity. Thus XSTL, over time, progressively captures spatial relations between entities in a scene through metric qualifiers and assertions which are structured into two nested static and dynamic levels.

At static level, frame assertions capture ordering relations between entities within a single frame in a frame sequence. These relations are expressed through the use of spatial operators, e.g. *eventually in space* ($\diamond_e\pm$), *always in space* ($\Box_e\pm$) and *until in space* ($\text{unt}_e\pm$) which deal with the spatial coordinate axis e_i rather than with the time axis. An example of this is when referring to a 2D scene S, where r is a position and p an object. e_1 is a reference axis of the scene S. Then the assertion, which also is illustrated in Figure 13.9(a), can be expressed as

$$(r) \vDash_s \diamond_{e1} +p \tag{13.65}$$

The assertion expresses that when starting at position r and moving along the coordinate axis e_1 in a positive direction then there exists a position which is in object p.

At dynamic level, static frame assertions can be composed by means of the temporal operators defined by Manna and Pnueli, e.g. *eventually* ($\diamond_t$), *always* ($\Box_t$) and *until* (unt_t), to form sequence assertions capturing the evolution over time of the spatial relationships expressed as frame assertions. For instance, if σ is a sequential set of frames, j is a frame index and ϕ is a frame assertion, then

$$(j) \vDash_t \diamond_t \phi \tag{13.66}$$

means that, along the sequence σ, after the jth frame there is a frame in which the frame assertion ϕ holds.

Qualitative metric relations can be added both in frame and sequence assertions through the use of spatial and temporal freeze variables. The purpose of such freeze variables is to mark the positions of objects and to track their subsequent positions in the sequence. For instance, the frame assertion

$$(r) \vDash_s v_r \cdot [\phi] \tag{13.67}$$

means that $(r) \vDash_s \phi$ is satisfied if the position of r in the current frame is assigned to the freeze variable v_r. After the freeze operation, the variable v_r cannot be used in any future freeze operation but it can be used to refer to the position taken by r in subsequent frames of the current sequence.

Position values assigned to spatial freeze variables may be used to express metric values corresponding to distances between positions that can be tracked

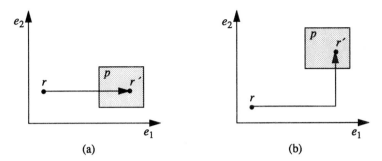

Figure 13.9 The meaning of the assertions in expression (13.65) (a) and in expression (13.68) (b).

in a sequence. The frame assertion

$$(r) \models_s v_r \cdot [\Diamond_{e_1} + \Diamond_{e_2} + v_{r'} \cdot [p]] \qquad (13.68)$$

which is illustrated in Figure 13.9(b), means starting in position r where the position is frozen in v_r. By a first movement along the coordinate axis e_1 in the positive direction followed by a second movement also in the positive direction but along e_2 then eventually a position is reached that is frozen in the variable $v_{r'}$ where the position in $v_{r'}$ is inside the object p. The assertion in (13.68) can be refined through what Del Bimbo and Vicario call *metric conditions* applied to the distance between the frozen positions of the variables v_r and $v_{r'}$. Thus

$$(r) \models_s v_r \cdot [\Diamond_{e_1} + \Diamond_{e_2} + v_{r'} \cdot [p]] \wedge \| v_r - v_{r'} \| = \text{NEAR} \qquad (13.69)$$

specifies that the positions frozen in v_r and $v_{r'}$ are close to each other. Obviously, the metric conditions are of qualitative nature and in this particular case the distance has a value lower than a certain, given threshold.

Temporal freeze variables are also used to retain the indices of the frames in which a certain condition holds and to express metric relations among these indices. This is illustrated in the example

$$(j) \models_t t_a \cdot [\Diamond_t t_b \cdot [\theta]] \wedge | t_a - t_b | = \text{SOON} \qquad (13.70)$$

which expresses that after the jth frame in the sequence a frame for which the assertion θ holds will be encountered *soon*. Hence, temporal freeze variables can be used to state qualitative temporal distances between frames in which different spatial conditions hold. By combining spatial and temporal metric relationships the relative velocities of the object can be determined.

XSTL captures spatiotemporal phenomena by taking spatial assertions as atomic propositions within temporal statements. Accordingly, XSTL assertions can be structures in two nested static and dynamic levels, which is demonstrated subsequently. Thus a sequence assertion Θ on a sequence σ is expressed as:

$$\Theta = (j) \models_t \theta \qquad (13.71)$$

In (13.71) j denotes the index of a frame in σ, and θ is a sequence formula. A sequence formula is recursively formed by composing frame assertions (Φ)

holding in the individual frames of σ and relational assertions on temporal freeze variables, through the freeze operator $t \cdot [\theta]$, the boolean connectives and the temporal operator:

$$\theta = \Phi \,|\, t \cdot [\theta] \,|\, \neg \theta \,|\, \theta_1 \wedge \theta_2 \,|\, \theta_1 \,\text{unt}_t\, \theta_2 \,|\, |t_a - t_b| = l \qquad (13.72)$$

where Φ is a frame assertion, t_a and t_b belong to a set T of temporal freeze variables and l is the qualitative distance between two instances belonging to an enumeration set which can be defined in any appropriate granularity to match the relevant phenomenon of the specific application context. Thus $l = \text{SOON} \,|\, \neg \text{SOON} \,|\, \ldots \,|\, \text{LATE} \,|\, \neg \text{LATE}$.

A frame assertion Φ on a generic frame of a sequence σ is expressed in the form:

$$\Phi = (r) \models_s \phi \qquad (13.73)$$

where r is a position in the scene represented by the frame and ϕ is a spatial formula. Furthermore, a spatial formula ϕ by inductively composing object identifiers and relational assertions on the values taken over time by a set V of spatial freeze variables through the freeze operator $v \cdot [\phi]$, the boolean connectives and the spatial "until" operators:

$$\phi = p \,|\, v \cdot [\phi] \,|\, \neg \phi \,|\, \phi_1 \wedge \phi_2 \,|\, \phi_1 \,\text{unt}_e \pm \phi_2 \,|\, \| v_1(t_a) - v_2(t_b) \|$$
$$= D \,|\, \| v_1(t_a) - v_2(t_a) \| \leqslant \| v_1(t_b) - v_2(t_b) \| \qquad (13.74)$$

where p is the identifier of an object, v is a spatial freeze variable belonging to the set V, $\text{unt}_e \pm$ is either a positive or negative "until" operator referring to the current direction along one of the coordinate axes, $v_1(t_a)$, $v_1(t_b)$, $v_2(t_a)$ and $v_2(t_b)$ are the values taken by the freeze variables v_1 and v_2 in the sequence frames whose indices are equal to the values of the temporal freeze variables t_a and t_b respectively, and D is the distance quantifier belonging to an enumeration set which can be defined in any granularity to match the actual characteristic of spatial phenomena: $D = \text{NEAR} \,|\, \neg \text{NEAR} \,|\, \ldots \,|\, \text{FAR} \,|\, \neg \text{FAR}$.

By using the XSTL language it becomes possible to deal with both temporal and spatial features of digital video sequences such that not only can temporal and spatial relations be identified but qualitative relations in space and time can be identified as well.

Descriptions of dynamic states are possible in XSTL since the temporal operators permit the description of relations between frames ordered in time and in which different spatial conditions hold. For instance, the sequence assertion

$$(t_a) \models_t ((q) \models_s \Diamond_{e_1} + \Diamond_{e_2} + p) \wedge \Diamond_t ((q) \models_s \Diamond_{e_1} - \Diamond_{e_2} + p) \qquad (13.75)$$

means that object p initially is right of object q and that, eventually, p will be left of q, which is illustrated in Figure 13.10.

Observe that this type of assertion permits any change in the spatial relations between the two objects, q and p, that may occur amidst the initial and final

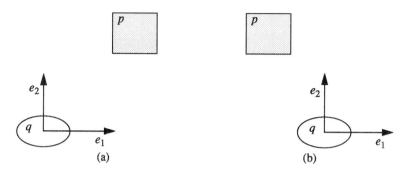

Figure 13.10 The initial (a) and final (b) states of an object q moving from the left to the right of an object p.

states of the movement of q. This is because, according to Del Bimbo and Vicario, the temporal eventually operator does not state a condition in the *next* state but rather in some state in the *future*. In general, the temporal eventually operator can be used to attain partial descriptions of a sequence in which only some relevant conditions are encountered. The temporal until operator, on the other hand, permits descriptions of sequences in which *all* subsequent conditions are defined.

Metric relations between time instances, in which different spatial relations hold, can be expressed by means of temporal freeze variables, e.g.

$$(t_a) \vDash_t (q) \vDash_s \Diamond_{e_1} +p) \wedge \Diamond_t t_b \cdot [((q) \vDash_s \Diamond_{e_1} -p)] \wedge |t_b - t_a| = \text{SOON} \quad (13.76)$$

means that object p initially is right of object q and that p will end up left of q within a period of time that is characterized as SOON.

Consequently, the joint use of spatial and temporal freeze variables makes it possible to represent relations and conditions where time and space are inherently tangled as, for instance, when advancing and approaching objects.

REFERENCES

Arndt, T. and Chang, S. K. (1989) Image sequence compression by iconic indexing. *Proceedings of the Workshop on Visual Languages, Rome*, pp. 177–82.

Chang, S. K. (1971) The reconstruction of binary patterns from their projections. *Communications of the ACM* **14**, 21–5.

Chang, S. K. and Li, Y. (1996) Representation of 3D symbolic and binary pictures using 3D strings. In *Intelligent Image Database Systems*, ed. S. K. Chang, E. Jungert and G. Tortora. World Scientific, Singapore.

Costagliola, G., Tortora, G. and Arndt, T. (1992) A unifying approach to iconic indexing for 2-D and 3-D scenes. *IEEE Transactions on Knowledge and Data Engineering* **4**, 205–22.

Del Bimbo, A. and Vicario, E. (1995) A logical framework for spatio-temporal indexing of image sequences. In *Intelligent Image Database Systems*, ed. S. K. Chang, E. Jungert and G. Tortora. World Scientific, Singapore, pp. 88–91.

Del Bimbo, A., Vicario, E. and Zingoni, D. (1993) Sequence retrieval by contents through spatio-temporal indexing. *Proceedings of the Workshop on Visual Languages*, pp. 88–91, Bergen.

Manna, Z. and Pnueli, A. (1992) *The Temporal Logic of Reactive and Concurrent Systems*. Springer-Verlag, New York.

14

The σ-Tree Spatial Data Model

The σ-tree is a symbolic hierarchical representation of a 3D space. The structure of the hierarchy is logically designed to support spatial reasoning and query techniques and serves therefore as a spatial data model. In what follows, three basic concepts of the σ-tree will be in focus, i.e. the universe, the cluster and the spatial model. Finally, two recursive query techniques are introduced of which one is developed to work directly on the σ-tree while the other was designed to work on a structure related to the σ-tree, called VTP.

This chapter discusses, in section 14.1, the basic principles of the σ-tree. The general view of the spatial data model is demonstrated in section 14.2 and the aspect of compacting σ-tree structures in section 14.3. Basic operations that can be applied to the σ-trees are presented in section 14.4. Section 14.5 shows how σR in various ways can be collapsed into two dimensions. Furthermore, the two related query methods are demonstrated in section 14.6. The first one is concerned with the σ-tree while the second, proposed by Lee, work on the VTP model. Finally, some concluding remarks concerning the query methods are given in section 14.7.

14.1. THE BASIC PRINCIPLES

The basic idea behind the σ-tree (Jungert and Chang, 1992) is simply to describe the universe in terms of a technique similar to that of using "Lego" blocks in conjunction with the symbolic algebra, see Chapter 7. Figure 14.1 partly illustrates the method. A universe containing objects or clusters is cut at all local extreme points towards the three planes y–z, z–x and y–x. This splits the universe in all three dimensions where those planes that are splitting the space are equivalent to the cutting lines described in Chapters 4 and 5. Once the cutting planes of the universe have been identified, a number of 3D building blocks, of which some include either objects or parts of the objects as well as clusters, are created from the space that may include empty space blocks as well. The empty space blocks are here denoted by **v** and are related to the **e**-objects which correspond to the empty space in the 2D representation. In the example

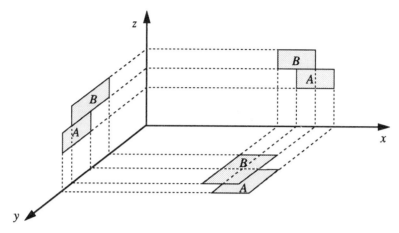

Figure 14.1 Projections of the two objects A and B onto the $y–x$, $z–x$ and $y–z$ planes.

given in Figure 14.1 a space containing two objects A and B is given. From this space the building blocks of the σ-tree can be identified as either A-blocks or B-blocks or empty space blocks. In Figure 14.1 the objects are projected on the $x–z$, $y–z$ and $x–y$ planes.

By means of the operator set $\{=,|,-,<\}$ it is possible to create a logical description of any universe. The σ-tree corresponding to the simple universe in Figure 14.1 can be described as:

$$\sigma R:\ \mathbf{v}=\mathbf{v}=\mathbf{v}=\mathbf{v}\,|\,\mathbf{v}=\mathbf{v}=\mathbf{v}=\mathbf{v}\,|\,\mathbf{v}=\mathbf{v}=\mathbf{v}=\mathbf{v}\,|\,\mathbf{v}=\mathbf{v}=\mathbf{v}=\mathbf{v}$$

$$-\,\mathbf{v}=\mathbf{v}=\mathbf{v}=\mathbf{v}\,|\,\mathbf{v}=\mathbf{v}=\mathbf{v}=\mathbf{v}\,|\,\mathbf{v}=\mathbf{v}=A=A\,|\,\mathbf{v}=\mathbf{v}=A=A$$

$$-\,\mathbf{v}=\mathbf{v}=\mathbf{v}=\mathbf{v}\,|\,\mathbf{v}=B=B=\mathbf{v}\,|\,\mathbf{v}=B=B=\mathbf{v}\,|\,\mathbf{v}=\mathbf{v}=\mathbf{v}=\mathbf{v} \quad (14.1)$$

The first row in expression (14.1) corresponds to the bottom "slice" in the $x–y$ plane, and the second row to the middle plane and so on. The sub-expressions $\mathbf{v}=\mathbf{v}=\mathbf{v}=\mathbf{v}$ correspond to projections of four empty space blocks to the x-axis where the empty space objects are projected along the y-axis. The "|"-operator is the concatenation of all such projected blocks, thus creating a "slice". "|" corresponds to the edge-to-edge operator but here the separating elements are planes not lines as in Chapter 5. Hence, "|" is a concatenation in the x-direction (using a $y–z$ cutting planes) while "−" is a concatenation along the z-axis (an $x–y$ cutting plane). The "−" operator is equivalent to "|" and is introduced to distinguish the projection in the x- and z-directions. To simplify the expressions in σR, here as well, the "=" operator can be left out. Hence, $\mathbf{v}=\mathbf{v}=\mathbf{v}=\mathbf{v}$ is equivalent to **vvvv**. The general set of operators, $\{=,|,<\}$, is thus extended to $\{=,|,-,<\}$. The "<" operator is included to allow stripping, i.e. elimination of all empty space objects. Finally, σR is called a "view" where the x-axis is the baseline and the direction of the view is along the y-axis.

14.2. THE GENERAL VIEW

In a 3D-space view the x-axis is the baseline and the direction goes along the y-axis, as mentioned in the previous section. However, this view is not the only one; in fact there are six such views, two for each plane. Generally speaking, the baseline and the view direction constitute the orientation of the slices in the space. Hence, a slice is described by means of $\{=, |\}$ which is a subset of the full set. Thus, any of the views of the universe can easily be determined. To differentiate between the different views a new projection operator is provided where a subscript indicates the baseline and the corresponding view direction, e.g.

$$\sigma_{xz} R: \text{vvv} \,|\, \text{vvv} \,|\, \text{vvv} \,|\, \text{vvv} - \text{vvv} \,|\, \text{vv}B \,|\, \text{vv}B \,|\, \text{vvv}$$

$$- \text{vvv} \,|\, \text{vv}B \,|\, \text{v}AB \,|\, \text{v}A\text{v} - \text{vvv} \,|\, \text{vvv} \,|\, \text{v}A\text{v} \,|\, \text{v}A\text{v} \qquad (14.2)$$

$\sigma_{xy} R$ is identical to σR, because this view is thought to be the most commonly used. Transformation between different views is fairly trivial and is merely a question of reordering rows and columns.

14.3. A COMPACT MODEL

A large number of densely arranged objects in a universe causes an even larger number of cutting lines to be drawn, and as a consequence the σ-tree grows in size. The obvious reason for this is that the space is split into many small subspaces. This was also indicated by Lee and Hsu who for that reason introduced the sparse cutting mechanism discussed in Chapter 5. Here *clusters* are introduced to overcome this problem. Thus, to avoid the problem of too many cuttings, cuttings inside the clusters are considered local and should not affect the global space. However, another quit different solution to the problem exist as well, which is called the *compact model*.

The idea behind the introduction of the compact model is simply to make the slices in the model independent of one another. This means that cuttings caused by an object in a neighbouring slice do not have any effect on the current slice. An independent slice then behaves like a 2D image. That is, the image algebra is not applicable to the complete string, since each slice is nonunified with respect to the other slices of the space. Transforming the model in (14.1) into a compact model with independent slices gives the result:

$$\sigma_c R: \text{v} - \text{v} \,|\, \text{v}A \,|\, \text{v}A - \text{v} \,|\, \text{v}B\text{v} \,|\, \text{v}B\text{v} \,|\, \text{v} \qquad (14.3)$$

The compact model obviously lacks in generality compared to the basic model since there is a loss of information in the former.

14.4. OPERATIONS

Two main types of operations are defined: *σ-tree operations*, which are applied to the tree directly, and *model operations* which are applied to the model, i.e. the strings.

14.4.1. σ-Tree Operations

Operations which directly can be applied to a σ-tree are:

- creation of new nodes
- deletion of existing nodes
- movement of existing nodes.

A node is either an object or a cluster. Certain constraints may apply; for instance, objects are not allowed to interfere with each other. For each cluster that is inserted a new part of the model must be created and for the opposite operation the corresponding part of the model is deleted. Thus, when a new cluster is created all nodes affected by that cluster also affect that corresponding part of the model, i.e. they are deleted from the string to which they belong and replaced by the new cluster.

Generation of clusters can be done in various ways. A new cluster is allowed to enclose any existing node. When a cluster is moved, on the other hand, the objects and the clusters inside that cluster will not be affected by the movement; they will just be included in it. This is a consequence of the fact that a cluster is not an explicit part of the universe; it is a user-defined part of the data model, i.e. σR. Objects and clusters may be inserted, deleted and moved arbitrarily, thus changing R and σR correspondingly. Finally, nodes may, if they are clusters, be merged, expanded and shrunk in an arbitrary way. When expanding a cluster, any type of object may be absorbed by that cluster. The opposite is applicable as well, i.e. when a cluster is shrunk its internal nodes may be swept out of the shrinking cluster.

14.4.2. Model Operations

A σ-tree operation is applied directly to the tree and affects, in most cases, the model of the universe, i.e. σR, indirectly. A model operation works in the opposite way, i.e. it is applied directly to the model and affects, if at all, the tree structure indirectly. Model operations are concerned with the detection, deletion and insertion of objects in σR, and movements of objects from one cluster to another. During these movements, optimization of the paths may be

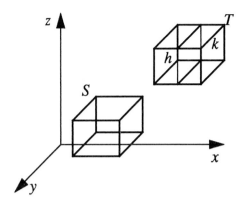

Figure 14.2 A universe with two clusters, *S* and *T*, where *S* is empty and *T* is filled up with the objects *h* and *k*.

necessary. Consider (14.4) from the example in Figure 14.2 which corresponds to a universe called *alpha* and where the view of the model is *xy*.

$$\sigma alpha: \quad \mathbf{vvv} \,|\, \mathbf{vvv} \,|\, \mathbf{vvv} - \mathbf{vvv} \,|\, \mathbf{v}C : S\mathbf{v} \,|\, \mathbf{vvv} - \mathbf{vvv} \,|\, \mathbf{vvv} \,|\, \mathbf{vvv} - \mathbf{vvv} \,|\, \mathbf{vvv} \,|\, C : T\mathbf{vv}$$

$$(14.4)$$

S and *T* are clusters whose descriptions correspond to (14.5) and (14.6) respectively:

$$\sigma S: \quad \mathbf{v} \tag{14.5}$$

$$\sigma T: \quad h \,|\, k \tag{14.6}$$

In this particular case *σS* is empty while *σT* includes the two objects, *h* and *k*, that fill up the cluster completely.

Consider a movement of object *k* from *T* into *S* such that *S* becomes:

$$\sigma S: \quad \mathbf{v}k \tag{14.7}$$

In this example, the translation of *k* is of no concern. Thus, after the movement of *k*, *σT* is changed into

$$\sigma T: \quad h \,|\, \mathbf{v} \tag{14.8}$$

The movement of *k* does not affect the top level of *σalpha*, only the contents of the two clusters *σT* and *σS* are affected.

14.5. PROJECTING σR ONTO 2D PLANES

Projecting *σR* onto a plane may correspond to one of the projection planes in Figure 14.1. Each one of these projections can be expressed in terms of symbolic 2D strings, that is either $\langle u, v \rangle$, $\langle u, w \rangle$ or $\langle v, w \rangle$. Projections can be made to any of the base planes of the coordinate system, e.g. for $z = 0$, and to any arbitrary plane parallel to a base plane.

When projecting a 3D model to a plane all objects perpendicular to that plane are projected according to the following rules, which support the collapse of a 3D image into 2D.

(i) $\mathbf{vv}\ldots\mathbf{v} \Rightarrow \mathbf{e}$
(ii) $\mathbf{v}\ldots a\ldots\mathbf{v} \Rightarrow a$
(iii) $\mathbf{v}\ldots aa\ldots\mathbf{v} \Rightarrow a$
(vi) $\mathbf{v}\ldots a\ldots b\ldots\mathbf{v} \Rightarrow \{ab\}$

Rule (i) corresponds to a sequence of empty volume objects projected down to an empty space object in two dimensions, i.e. $\mathbf{e}$. In rule (ii) the sequence involves an entity or a part of it, that is either a cluster or an object. The situation is the same in rule (iii), although it contains more than one part of an entity. Finally, rule (iv) includes two different entities which are overlapping.

Figure 14.3(a) and (b) illustrates the three alternative schemes that can be used to collapse a σ-tree into two dimensions. For $z = 0$ the direction is

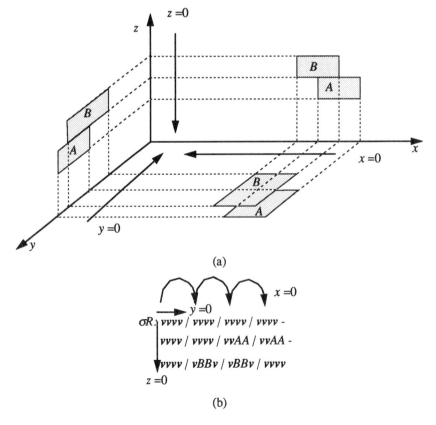

(a)

(b)

Figure 14.3 Directions of 3D space collapse into a 2D space (a) and the corresponding string collapse schemes (b).

"column-wise", while for $y = 0$ the collapse is performed unit by unit where a unit corresponds to all elements that starts and/or ends with an edge-to-edge operator. Finally, for $x = 0$ the collapse rules take elements "row-wise" at the same position in each unit. As an illustration, consider the $y = 0$ direction from the example in Figure 14.3(a) which results in

$$\mathbf{e}\,|\,\mathbf{e}\,|\,\mathbf{e}\,|\,\mathbf{e} - \mathbf{e}\,|\,\mathbf{e}\,|\,A\,|\,A - \mathbf{e}\,|\,B\,|\,B\,|\,\mathbf{e} \qquad (14.9)$$

The string in (14.9) is, however, not the final string since it can be simplified so as to better correspond to the traditional string type. First observe that the "$-$"-operator can be substituted by "$=$" since in (14.9) the string is made up by three rows on top of each other. This is obviously only true for the u-string. Law (xiv) from the image algebra in Chapter 7 can now be applied, thus changing the string in (14.9)

$$u_{xz}: \quad \mathbf{eee}\,|\,\mathbf{ee}B\,|\,\mathbf{e}\,AB\,|\,\mathbf{e}A\,\mathbf{e} \qquad (14.10)$$

Since the strings are unifyable the w_z-string can be directly created from u_{xy}. Thus after application of law (xiii) the resulting w_{xy}-string becomes:

$$w_{xz}: \quad \mathbf{eeee}\,|\,\mathbf{ee}AA\,|\,\mathbf{e}BB\mathbf{e} \qquad (14.11)$$

The projections to the two other planes, for $z = 0$ and $x = 0$, are generated in accordance with the same principles. The only difference is that for $z = 0$ an overlap element occurs in unit three, that is $\{AB\}$. Partial projection to a plane at a random level is also possible and can be done in an analogous way as was described here.

14.6. SPATIAL QUERY METHODS

Two query techniques which are both recursive will now be introduced of which the first is concerned with the σ-tree while the second was developed for a related structure called VPT; there are, however, many similarities between the two approaches.

14.6.1. Spatial Queries in the σ-Tree

Spatial reasoning is quite often concerned with the identification of spatial relationships between objects, such as "above", "left", "north of", etc. The problems that can be addressed in a σ-tree are similar, i.e. to define existing spatial relations in a universe by means of the discussed model. When identifying relations between objects their relative location in the universe must be considered, i.e. when identifying the relations the process must be concerned with relations where the objects may occur, for example in

- the same cluster
- parallel clusters
- on different levels of the same or different branches of the σ-tree.

Besides this, identification of arbitrary cluster-to-cluster relations will sometimes be of interest also. A simple query is generally described in terms of a predefined template, corresponding to a generalized symbolic projection string. The templates include, for example, information about the plane on which the 3D space is projected. For instance, if the problem is to verify an "above" relationship, then the projection string along the z-axis (the w-string) must be searched, i.e. the w_{x-z}-string. Applied to the example in Figure 14.1, it is easy to see that B is above A, which is evident from the string:

$$w_{xz}: \quad \textbf{eeee} - \textbf{ee}AA - \textbf{e}BB\textbf{e} \tag{14.12}$$

A template that can be used for identification of the "above" relation may have the following rule-like structure:

$$(\text{above } \$2 \ \$1) := \sigma_{xy}\Gamma : w_{xz} : \$1 < \$2 \quad \text{or} \quad \$1 - \$2 \tag{14.13}$$

where Γ corresponds to the model or to any submodel of the σ-tree. When matching the template against the string, the latter must first be stripped, i.e. the generalized empty space objects must be deleted, which gives:

$$w_{xz}: \quad AA - BB \tag{14.14}$$

After the stripping, equal subobjects can be merged according to the laws of reshaping in accordance with the symbolic image algebra. In this case, AA becomes A and BB becomes B. Hence, the string finally becomes:

$$w_{xz}: \quad A - B \tag{14.15}$$

which matches the template given above.

An alternative is to define a template which can be matched against σR. A problem is that when stripping is performed it becomes impossible to differentiate between "less than" in the z-direction and in the x-directions. This requires two different "less than" operators. The following convention is chosen: "$<$" means "less than" in the baseline direction, and "$-<$" means "less than" in the slice direction. This is a consequence of the stripping of the 3D space. Hence, the stripped string becomes:

$$\sigma R: \quad A - B \tag{14.16}$$

For instance, a query that matches the relation "neighbor of", i.e. two objects that touch each other in any direction, would look like:

$$
\text{(neighbour\_of \$1 \$2)} := \sigma_{xy}\Gamma: \quad
\begin{aligned}
(u_{xz}: & \quad \text{\$1 | \$2 or \$2 | \$1 and} \\
v_{yz}: & \quad \text{\$1 \$2 or \$2 \$1 and} \\
w_{xz}: & \quad \text{\$1 \$2 or \$2 \$1) or} \\
(w_{xz}: & \quad \text{\$1 − \$2 or \$2 − \$1 and} \\
u_{xz}: & \quad \text{\$1 \$2 or \$2 \$1 and} \\
v_{yz}: & \quad \text{\$1 \$2 or \$2 \$1) or} \\
(v_{yz}: & \quad \text{\$1 | \$2 or \$2 | \$1 and} \\
u_{yx}: & \quad \text{\$1 \$2 or \$2 \$1 and} \\
w_{yz}: & \quad \text{\$1 \$2 or \$2 \$1)}
\end{aligned}
\qquad (14.17)
$$

This particular rule looks at all six sides of \$1 to determine whether \$2 is a neighbour. To complete this template, three views are needed, namely the xy, the yz and the xz views.

A query in this formalism may look like:

$$\sigma R: \quad \text{(above 'B 'A)} \qquad (14.18)$$

which returns the assertion (above BA) which statement is linked to σR unless the answer of the query is false, in which case nil is returned. This query syntax is introduced to tell which part of the σ-tree that should be involved in the query. In other words, σR is internally bound to Γ.

Another type of query that is permitted is the following:

$$\sigma R: \quad \text{(above 'B ?x)} \qquad (14.19)$$

where $?x$ is a variable that will be bound to A as a result of the query, i.e. this query results in the same statement as in the first example. The query can, of course, be generalized even further, e.g.

$$\sigma R: \quad \text{(above ?y ?x)} \qquad (14.20)$$

In those cases when σR is left out, the default will be the full σ-tree including all its clusters at all levels.

$$\sigma R: \quad \text{(neighbor\_of } \mathbf{v} B) \qquad (14.21)$$

finally, is the answer to the question "is there a side of B which does not have a neighbor?". The generalized empty space object is used as parameter in the query. Internally $\mathbf{v}$ must be collapsed into an **e**-object since the query is defined to work just on planes. In the above, the resemblance with Prolog is obvious.

14.6.2. Spatial Queries in the VPT Model (Lee)

Lee (1992) has developed a query technique that can be applied to the virtual objects (VB), which were briefly discussed in Chapter 8. Generally speaking, a virtual object is a set of objects inside a common MBR where each single object can be represented by its own MBR. This structure is hierarchical and corresponds to the virtual picture tree model (VPT). However, the VPT model as such is of less concern here, instead the query technique will be in focus.

The string type used in the VPT model is the 2D C-string of Lee and Hsu, discussed in Chapter 5. This string type is not always unique contrary to the general string type in Chapter 4. The reason for this is that the sparse cutting mechanism does not necessarily lead to unique cuttings although they always will be minimal. For this reason, Lee has developed a method that generates a 2D C-string, not directly from the image, but from the general string type created from the image. The 2D C-string received in this way is called the *canonical* 2D string and it is always unique since the general string from which it is generated is always unique.

Lee points out in his work that when a query is applied to a 2D string of an image then subpicture query processing may not always be successful even if the substring of concern implicitly exists. The reason for this is that due to the cuttings multiple occurrences of some objects will appear in the strings. Furthermore, sometimes the image string cannot be successfully matched with the query because the image string contains objects, which are not participating in the match but are interfering with the process. As a consequence, Lee proposes an object dropping mechanism that allows dropping of redundant and unnecessary objects, i.e. noisy objects. The object dropping method involves a number of rules of which not all will be discussed here. The method can be described accordingly. Let $X o_1 (Y o_2 Z)$ be the type of expression that should be dealt with, where X, Y and Z are 2D strings and o_1 and o_2 are spatial operators. Then the following dropping rules can be defined.

- If X is dropped, Y and Z have the relationship $R_1 o_1 o_2$, where $R_1 o_1 o_2 = o_2$.
- If Y is dropped, X and Z have the relationship $R_2 o_1 o_2$ which can be determined in accordance with Table 14.1.

Lee has altogether identified four different tables for all occurring dropping rules.

In Table 14.1 a blank field means that there is no unique operator, i.e. when

Table 14.1 R_2

$o_1 \backslash o_2$	=	|	<	/	]	[	%
=	=	]	]	]	]	[	%
|	|	<	<	<	<	|	<
<	<	<	<	<	<	<	<
/	/						
]	]	]	]	]	]	%	%
[	[	%	%	%	%	[	%
%	%	%	%	%	%	%	%

dropping takes place, the 2D string may reveal ambiguous relations. That is, if X is dropped from the string $(((X \mid Y) = Z)[R$, then the string will be transformed into $((Z] Y)[R$ and an ambiguous format is received which is caused by the fact that the original 2D string was not in canonic form. In this particular case the general strings can, according to Lee, be one of $XZ \mid YZ \mid YZR$, $XZ \mid YZR$ or $XZ \mid XZR \mid YZR$ for which the 2D C-strings $Z(X \mid (Y] R))$, $Z(X \mid (YR))$ and $X((X] R) \mid YR$ can be generated respectively. If in any of these expressions X is dropped the resulting strings become $Z(Y] R)$, $Z](YR)$ and $Z(R] Y)$, respectively, which are not all ambiguous. For ambiguous strings the ambiguous parts must first be detected and then a new general 2D string generated. Then the canonical form is created once again and finally the dropping process can be applied again. However, the problem in the example above is due to the fact that the strings are not extracted from an image and for this reason a unique string does not necessarily have to be created. In the normal situation there is an image available from which a unique canonical string normally can be created.

Taking the canonical form and the dropping mechanism into account, then the means to apply spatial queries are available. Figure 14.4 is used to exemplify the query technique.

The canonical u-string from Figure 14.4 becomes:

$$u: D] A] E] C \mid (ACE)] B \mid B((C [E) < F) \mid F$$

$$\Rightarrow ((((D] (A] (E] C))) \mid ((A = (C = E))] B)) \mid (B = ((C [E) < F))) \mid F) \tag{14.22}$$

The transformed string in (14.22) is equivalent to the canonic form, and is created to better fit with the dropping mechanism since it naturally splits the

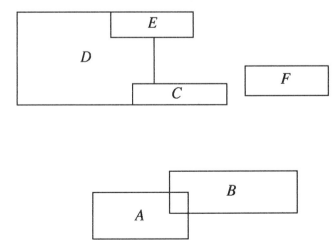

Figure 14.4 The image used for illustration of queries in the VPT model.

string into its basic cutting terms, i.e. the substring between two edge-to-edge operators. The object dropping in the substring will be illustrated below.

A query can now be applied to the image:

$$(?, \text{ where } \langle u, v \rangle = \langle D/E, E/D \rangle) \tag{14.23}$$

In this particular example just the u-string is dealt with. However, as can be seen, the u-string in the query, that is D/E, obviously does not match any substring in (14.22). As a consequence, all noisy objects, which in this particular case correspond to the set $\{A, B, C, F\}$ or alternatively all objects excluding those used in the query, must be eliminated or dropped. One step in this elimination process is illustrated in (14.24) where object C is dropped:

$$(D \,]\, (A \,]\, (E \,]\, C))) \Rightarrow (D \,]\, (A \,]\, E)) \tag{14.24}$$

By continuing drops of noisy objects eventually a compacted string, containing just the objects present in the query, will become available. Here the compact string created in this way is:

$$D \,]\, E \,|\, E \Rightarrow D/E \tag{14.25}$$

which corresponds to the compact canonical form that can be used to match the given query since it does not contain any noisy objects.

The query technique is, however, not just delimited to the type of query demonstrated so far. Other query types are applied as well. Examples of such types are "find all objects that overlap a given object". The other way around is also possible because here as well as in the σ-tree query technique recursive queries are permitted. Furthermore, queries of type "partial overlap", "adjacently contains" and "completely contained" are example of other possibilities.

14.7. CONCLUDING REMARKS

The two data models presented in this chapter, i.e. the σ-tree and the VPT models are similar in that they both demonstrate how queries can be formulated and answered in a *qualitative* way, which at this time is a unique characteristic that most query languages of today do not possess. Both data models are also similar in that they permit application of recursive queries; they differ in that only the σ-tree model is generalized into three dimensions.

REFERENCES

Jungert, E. and Chang, S. K. (1992) The σ-tree – a symbolic spatial data model. *Proceedings of the 11th IAPR International Conference on Pattern Recognition*, The Hague, pp. 461–465.

Lee, C.-M. (1992) The unification of the 2D string and spatial query processing. PhD thesis, Graduate College of the Knowledge Systems Institute, Skokie, Illinois.

15

A Survey of Image Information Systems and Future Directions

As presented in the previous chapters, the theory of Symbolic Projection provides a technique for iconic indexing as well as an approach for spatial reasoning. It also leads to image data structures for data representation and database accessing. Finally, it can be embedded in an object-oriented data model. In this chapter, we give a taxonomy of image indexing techniques, and present a survey of image information systems. We then discuss research issues in data models. Some thoughts on active image information systems are given to conclude this book.

15.1. INDEXING AND DATA STRUCTURES

The discussion of image indexing techniques could proceed in three directions: *index representation, index organization* and *index extraction*. The extraction of indexes could be manual, automatic or something in between (hybrid). For image information, index extraction is heavily dependent on the progress in image processing techniques. In this section, we concentrate on index representation and index organization.

In conventional database systems, keyword-based indexing techniques are sufficient to support users' needs. In image information systems, there are many applications which cannot be properly supported by keyword-based techniques. In addition to keywords, users often want to retrieve images by shape, texture, spatial relationships, etc. see for example Chang (1975), Grosky (1986), Grosky and Jiang (1992), Mehrotra and Grosky (1989), Tagare *et al.* (1991). That is, image features are used as indexes (called *image indexes*) and in many cases they cannot be represented as keywords. The representation of these image indexes possess some special characteristics:

- Image indexes are approximately represented.
- Image indexes do not have implicit order, in the sense that if *a*, *b* and *c* are

three index values and $a < b < c$, it does not mean that image b is more similar to image a than image c.

- Image index representations may have interrelated multiple attributes. That is, if $a1$ and $a2$ are two attributes of an index, result$(a1a2) \neq$ result$(a1) \cap$ result$(a2)$.

With these characteristics, it is difficult to use the conventional indexing structures such as B-tree, hashing, etc. cannot be used for the organization of image indexes. Image index structures must be explored (Klinger and Pizano, 1989). The image index structures should also support similarity retrieval (Jagadish, 1991).

The above considerations lead to the following three dimensions in classifying different approaches to image indexing:

- First dimension: How to structure image data? Image data structures include B-tree, K–D-tree, quad-tree, etc.
- Second dimension: How to select image indexes? Choices are keyword, shape descriptors, signatures, 2D strings, etc.
- Third dimension: How to acquire image index? We may distinguish automatic, hybrid or manual means of index construction.

The three dimensions are illustrated in Figure 15.1. An application area corresponds to a region in this three-dimensional space. They share some common method of data collection, indexing approach and data structuring. The combination of indexing and data structures provides the support for an image database system.

Based upon the theory of Symbolic Projection, data structures and efficient access methods for image databases were developed. The main idea is to encode

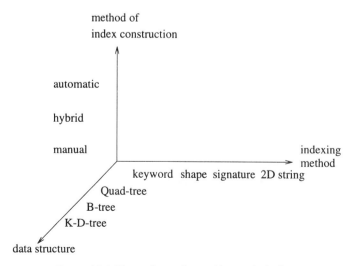

Figure 15.1 Three dimensions of image indexing.

a pairwise spatial relation as one bit in a bit string, so that the index structure consists of bit strings called signatures (Lee and Shan, 1990). For example, if image $f1$ satisfies spatial relations $r1$, $r3$ and $r4$, and image $f2$ satisfies spatial relations $r1$, $r2$ and $r4$, then their signatures (or record signatures) are 1011 and 1101, respectively. The images $f1$ and $f2$ can be grouped together and their block signature 1001 represents the spatial relations satisfied by both images. The block signatures and record signatures are organized into an index structure called the two-level signature file. Fast preselection algorithms were developed for such two-level signature files, so that unwanted images can be excluded from the detailed matching process, thus saving computational time.

Integrated index structures based upon the signatures for Type-0, Type-1, Type-2 and All-Type similarity retrievals are reported in Lee, Yang and Chen (1992). Tseng, Hwang and Yang (1994) then extended the technique to bit-sliced two-level signature files (BS2LSF) and developed fast image retrieval algorithms for the BS2LSF data structure. Combining Quad-Trees with 2D strings, Chou, Chang and Yang (1994) developed the 2D N-string, and proved that the storage requirement for N-strings are less than that for 2D H-strings in most cases. They also performed detailed experimental studies on the efficiency of retrieval algorithms for this index structure.

15.2. A SURVEY OF IMAGE INFORMATION SYSTEMS

A classification of some typical image information systems is given in Table 15.1. This table is not meant to be a comprehensive survey. Some of the systems are chosen for historical reasons, and the recent ones are chosen to illustrate the new approaches. Considered as a whole, Table 15.1 illustrates the general trends in current image database research.

IMAID, PICDMS, PSQL, GRIM_DBMS and IIDMS have been discussed earlier. The Visual Structure Database has some nice features, combining entity–relationship data model and the quad-tree (Klinger and Pizano, 1989). I-See is a software environment for image database research, emphasizing object-oriented approach, precompilation and query of image content (Fierens *et al.*, 1992). IDB is an object-oriented image database system emphasizing content-based archiving and retrieval with applications in PACS (Kofakis *et al.*, 1990). QBIC is IBM's system for querying by image and video content (Flickner *et al.*, 1995). FIBSSR performs feature-based similarity retrieval by matching shape boundaries (Mohrotra and Gary, 1995).

15.3. DATA MODELS

A data model is a collection of mathematically well-defined concepts to express both static and dynamic properties of data-intensive applications (Brodie 1984).

Table 15.1

System description	Query	Data model	Index method	Index extraction	Data structure	Application
IMAID: Integrate image and text data interfaced with image processing	QPE query by pictorial example by filling tables	relational				image processing
PSQL: Integrate image and text data but process them separately	command language PSQL SELECT STATE_NAME FROM STATE WHERE POPUL >50 000	relational	attributes	predefined or manual	R− tree or R+ tree	cartography
PICDMS: Picture DBMS using dynamic stacked image and gridded data	PICQUERY command language ADD (IMAGE FIDD FIX(8.0)) DIFF = BAND4-BAND5	stacked image	field name with current location	predefined or manual	flat file (3D matrix of stacked images)	image processing
IIDMS: Intelligent image database system using 2D strings as iconic index	iconic query by pictorial example by drawing pictorial query	relational	2D strings	automatic or manual (hybrid)	sigma-tree	image processing
Visual Structure Database	symbolic query Car in-front-of house	entity–relationship diagram	attributes	manual	quad-tree and entity–relationship records	cartography
I-See: Software environment emphasizes precompilation and query by image content	(a) Iconic query as guide to search (b) Symbolic query SHOW cities WEST-OF city name = "Pittsburgh"	object oriented	–	automatic using image analysis and AI technique		image processing
IDB: Image Archiving by Content	Match example image by similarity retrieval	object based	attributes	automatic using image analysis and AI technique		medical image database
QBIC: Query by Image Content	Example images, user-constructed sketches and drawings, color and texture patterns	object based	attributes (color, texture shape, camera and object motion	automatic or manual	R-frames (representative frames of objects)	digital library
FIBSSR: feature based similarity retrieval system	query by shape features	object based	multidimensional point access method	automatic	shape boundary records	defense (target shape) manufacturing (tool shape)

Static properties are objects, attributes, and relationships among objects. Dynamic properties are operations on objects, operation on properties and relationships amongst operations. Static properties are expressed using database schema (DDL), and dynamic properties are specifications for transactions (DML) and queries (QL). Integrity rules over objects (i.e. data states) and operations (state transitions) are sometimes also included in the data model.

From hierarchical network models to relational data models, traditional applications in the business environment with formatted data have enjoyed much success in information management. Even for nontraditional applications such as engineering information management, with extensions to the relational data model (e.g. long field, complex object, etc.) and the advances in object-oriented and semantic data models (Farmer *et al.*, 1985), some successes have also been claimed. With a growing list of new applications based on image information handling, limited success has been achieved in the direct management of image information. The standard approach for image data modeling is to model image data and text data separately, and image is usually stored in its entirety (Chang, 1990; Chock *et al.*, 1984; Tang, 1981). All the commercial image information systems treat images as black boxes (e.g. BLOBs or Binary Large OBjects). Special functions are provided to access the contents of images (e.g. beard(image_face)).

Traditionally, a database system could be considered having three basic levels: external models which support individual user views, a single conceptual model which defines an abstract representation of the database in its entirety, and an internal model which defines the database storage structures and cannot be seen by users. Generally speaking, a modeling technique provides a set of type constructs for users to model entities and relationships existing in application domains. It also provides mechanisms to construct external models from the conceptual model. The relational data model provides a single construct, relation. The (extended) entity–relationship model provides constructs for entity, aggregation, generalization and general user-named relationships.

Compared to alphanumeric data, image information carries some special characteristics:

1. The content of an image cannot be precisely described. The content of an image could be considered as a group of spatial objects (e.g. line, colour, point, etc.) with spatial relationships (e.g. adjacent, orientation, relative position, etc.) among them. In general, there are no precise ways to represent these objects and relationships. They can only be described by some approximate representations. For example, a closed contour could be represented as a set of vertexes of an interior polygon. On the other hand, alphanumeric data can always be precisely represented with basic data types as integer, string, etc. or even complex data types formed from basic data types. In either case, whenever a data type is assigned to an object or a relationship, it will not be changed. Therefore, a modeling technique with basic data types and the support of user-defined data

types is usually sufficient. But this is not enough to support the modeling of image data. Since a representation is approximate, that means it is not unique, and new and better representations may be adopted in the future. Furthermore, multiple representations are possible and the selection of the best representation cannot be fixed in the data model. For example, a closed contour could also be represented as a set of rectangles covering the inside of the contour.

The above discussion indicates that, an image modeling technique needs to hide the underlying data representations (data types) of spatial objects and relationships from users and provide mechanisms to map and select the proper representation dynamically. Many interesting issues need to be considered. Is the mapping one-to-many, many-to-one, or many-to-many? How should the consistency issue be considered? For multiple representations, should the equivalence issue be considered?

2. Spatial entities (objects) and relationships (we will call them image features) in images do not in themselves carry any semantic meanings. With alphanumeric information, entities and relationships carry semantic meanings through their given names. For example, the entity–relationship diagram

manager – manages → employee

clearly states that there are two entities, "manager" and "employee" with a relationship "manages" in between. Associating semantic meanings by naming will cause some problems with image information. First, the same image could be interpreted in different ways. Second, the same image could be used in different ways during different time periods. Third, since the image interpretation is an approximation in many cases, it may be changed due to better recognition techniques. In any of the above cases, directly associating semantic meanings to image entities and relationships will severely limit the usage of image information. A more feasible approach is to model image with spatial meanings, i.e. image features, and then associate semantic meanings, i.e. semantic features, to image features for different usages.

Without associating semantic meanings to spatial entities and relationships, different users may think of them very differently. For example, one person's definition of relative position may be very different from others. Therefore, in addition to traditional modeling constructs, spatial modeling constructs need to be defined for image modeling. Furthermore, since it is impossible to provide a complete set of image constructs for general applications, the modeling technique needs to provide mechanisms to support user defined image constructs and treat them as first class constructs. It should also support the building of complex image features from simple features.

The mapping between semantic features and image features raises further interesting research issues. Since computer vision technology has not matured (and may never mature), the above mapping cannot be automated. Therefore, the mapping becomes part of the image schema design. To facilitate database design and to avoid mistakes, mapping constructs need to be provided by an image modeling technique. Of course, it should also support the evolution of mapping constructs.

3. Image-based information could be queried by pictures. We often say that a picture is worth a thousand words. But, on the other hand, pictures may cause multiple or imprecise interpretations. Therefore, an image information system needs to provide some domain knowledge to help users to refine their intentions incrementally. A user's intention could be a single query, or it could be a complicated set of interrelated queries. One approach is to provide a knowledge-based module to guide users and give suggestions. The debate on whether semantics should be described in data models or by integrity constraints in traditional data modeling could be applied here. Many interesting research issues are raised here. First, what should be in data models and what should be in knowledge-based models? Second, how should this information be represented? Third, how do the two parts interact?

In recent years, researchers have begun to address some of the issues raised above. Researchers from the image processing community generally prefer generalized graph models as data models. For example, GRIM_DBMS uses attributed relational graphs as data model (Rabitti and Stanchev, 1989), and RDS promotes the relational data structure as the data model (Shapiro and Haralick, 1982). Researchers from the database camp advocate the use of extensible relational DBMS. Others begin to explore how to extend the database schema to define conceptual and external schemas for images and associated procedures (Mehrotra and Grosky, 1989). VIMSYS is probably the most extensive data model developed for image information management (Gupta *et al.*, 1991b). Based upon the object-oriented approach, this four-level model supports:

- image representation and relations
- image objects and relations
- domain objects and relations
- domain events and relations.

More recently, the object-oriented approach has become a favorite (Gupta *et al.*, 1991a; Gupta and Horowitz, 1991; Jagadish and Gorman, 1989; Liu Sheng and Wei, 1992) although adopting the object-oriented approach alone does not necessarily solve the data modeling problem. The σ-tree is a step in that direction, to provide a unified data model for active image information systems, capable of supporting multilevel image information processing and spatial reasoning.

15.4. CONCLUSION

To develop the next generation of active image information systems, we envision the confluence of active database, spatial reasoning, Petri nets, neural nets, image processing, artificial intelligence, data modeling, and object oriented systems. Many research issues need to be explored. The ongoing revolution in communications technology will make a profound impact on active image information system design. For example, the advances in wireless communications technology will bring multimedia data to mobile computers and even personal digital assistants. How to efficiently and effectively condense, compress, abstract and communicate image information will become important issues in the design of wireless communication systems. The theory of Symbolic Projection, its applications and systems incorporating advanced spatial reasoning and image information retrieval, may prove valuable when we consider the design of the next generation of active image information systems.

REFERENCES

Brodie, M. L. (1984) On the development of programming languages in *Conceptual Modelling: Perspective, Database, and Programming Languages* edited by Brodie, M. L., Mylopoulos, J. and Schmidt, W., Springer Verlag, New York, pp. 19–48.

Chang, S. K. (1990) *Principles of Pictorial Information Systems Design.* Prentice-Hall, Englewood Cliffs, NJ.

Chang, S. K. (1995) Towards a theory of active index. *Journal of Visual Languages and Computing* **6**, 101–18.

Chock, M., Cardena, A. F. and Klinger, A. (1984) Data base structure and manipulation capabilities of a picture data base management system (PICDMS). *IEEE Transactions on Pattern Analysis and Machine Language* **6**, 484–92.

Chou, Annie Y. H., Chang, C. C. and Yang, W. P. (1994) Symbolic indexing by N-Strings. Technical Report, Dept. of Computer and Information Science, National Chiao Tung University, Hsinchu, Taiwan, 1994.

Farmer, D. B., King, R. and Myers, D. A. (1985) The semantic database constructor. *IEEE Transactions on Software Engineering* **11**, 583–91.

Fierens, F., Van Cleynenbreugel, J., Sutens, P. and Oosterlinck, A. (1992) A software environment for image database research. *Journal of Visual Languages and Computing* **3**, 49–68.

Flickner, M., Sawhnoy, H., Niblack, W., *et al.* (1995) Query by image and video content: The QBIC system, *IEEE Computer*, September, pp. 23–32.

Grosky, W. I. (1986) Iconic indexing using generalized pattern matching techniques. *Computer Vision, Graphics, and Image Processing* **35**, 308–403.

Grosky, W. I. and Jiang, Z. (1992) A hierarchical approach to feature indexing. *SPIE/IS&T Conference on Image Storage and Retrieval Systems*, pp. 9–20.

Gupta, A., Weymouth, T. E. and Jain, R. (1991a) An extended object-oriented data model for large image data bases. *Proceedings of the 2nd Symposium on Large Spatial Databases (SSD'91)*, Zurich, pp. 45–61.

Gupta, A., Weymouth, T. E. and Jain, R. (1991b) Semantic queries with pictures: The VIMSYS model. *Proceedings of VLDB'91, Barcelona*, pp. 69–79.

Gupta, R. and Horowitz, E. (1991) *Object-Oriented Database with Applications to CASE, Networks and VLSI CAD.* Prentice-Hall, Englewood Cliffs, NJ.

Jagadish, H. V. (1991) A retrieval technique for similar shapes. *Proceedings of the 1991 ACM SIGMOD International Conference on Management of Data, SIGMOD Record* **20**, 208–17.

Jagadish, H. V. and O'Gorman, L. (1989) An object-model for image recognition. *Computer* **22**, 33–41.

Klinger, A. and Pizano, A. (1989) Visual structure and data bases. In *Visual Database Systems*, pp. 3–25. North-Holland, Amsterdam.

Kofakis, P., Karmirantzos, A., Kavaklis, Y. *et al.* (1990) Image archiving by content: An object-oriented approach. *SPIE Conference on Image Processing: Medical Imaging IV: PACS System Design and Evaluation, Newport Beach, CA*, Vol. 1234, pp. 275–86.

Lee, S. Y. and Shan, M. K. (1990) Access methods of image database. *International Journal of Pattern Recognition and Artificial Intelligence*, **4**, 27–44.

Lee, S. Y., Yang, M. C. and Chen, J. W. (1992) Signature file as spatial filter for iconic image database. *Journal of Visual Languages and Computing*, **3**, no. 4, 373–397.

Mehrotra, R. and Gary, J. E. (1995) Similar shape retrieval in shape data management, *IEEE Computer*, September, pp. 57–62.

Mehrotra, R. and Grosky, W. I. (1985) REMINDS: A relational model-based integrated image and text database management system. *Proceedings of Workshop on Computer Architecture for Pattern Analysis and Image Database Management, Miami Beach, FL*, pp. 348–54.

Mehrotra, R. and Grosky, W. I. (1989) Shape matching utilizing indexed hypothesis generation and testing. *IEEE Transactions on Robotics and Automation* **5**, 70–77.

Rabitti, F. and Stanchev, P. (1989) GRIM_DBMS: A GRaphical IMage DataBase Management System. *Visual Database Systems*, pp. 415–30. IFIP.

Shapiro, L. G. and Haralick, R. M. (1982) Organization of relationship models for scene analysis. *IEEE Transactions on Pattern Analysis and Machine Intelligence* **4**, 595–602.

Sheng, O. R. Liu and Wei, C. P. (1992) Object-oriented modelling and design of knowledge base/database systems. *Proceedings of the 8th International Conference on Data Engineering, Tempe, AZ*, pp. 87–105.

Tagare, H. D., Gindi, G. R., Duncan, J. S. and Jaffe, C. C. (1991) A geometric indexing schema for an image library. *Computer Assisted Radiology '91*, pp. 513–18.

Tang, G. Y. (1981) A management system for an integrated database of pictures and alphanumeric data. *Computer Graphics and Image Processing* **16**, 270–86.

Tseng, Judy C. R., Hwang, T. F. and Yang, W. P. (1994) Efficient image retrieval algorithms for large spatial databases. *International Journal of Pattern Recognition and Artificial Intelligence*, **8**, no. 4, 919–944, World Scientific, Singapore.

Bibliography

Allen, J. F. and Hayes, P. J. (1989) Moments and points in an interval-based temporal logic. *Comput. Intell.* 5, 225–238.

Arndt, T. and Chang, S. K. Image sequence compression by iconic indexing. *Proceedings of 1989 IEEE Workshop on Visual Languages*, 177–182, October 4–6, 1989, Rome, Italy.

Ballard, D. H. and Brown, C. M. (1982) Computer vision. Prentice-Hall, Englewood Cliffs, New Jersey.

Batini, C., Catarci, T., Costabile, M. F. and Levialdi, S. Visual query systems. *Technical Report N. 04.91, Dipartimento di Informatica e Sistemistica, Università di Roma "La Sapienza"*. Rome, Italy, 1991 (revised version 1993).

Bennett, B. (1994) Spatial reasoning with propositional logics. *Principles of Knowledge Representation and Reasoning*, pp. 51–62. Morgan Kaufmann, San Francisco, CA.

Benson, D. and Zick, G. Spatial and symbolic queries for 3D image data. *SPIE/IS&T Conference on Image Storage and Retrieval System*. February 1992.

Berry, J. K. (1995) *Spatial reasoning for effective GIS*. GIS World, Inc. Fort Collins, Colorado.

Del Bimbo, A., Vicario, E. and Zingoni, D. Sequence retrieval by contents through spatio temporal indexing. *Proc. of the 1993 IEEE workshop on Visual Languages*, pp. 88–91. August 1993, Bergen, Norway.

Del Bimbo, A., Campanai, M. and Nesi, P. (1993) *A three-dimensional iconic environment for image database querying, IEEE Transactions on Software Engineering*. **19**(10), 997–1011.

Del Bimbo, A., Vicario, E. and Zingoni, D. (1994) A spatial logic for symbolic description of image contents. *Journal of Visual Languages and Computing*. **5**(3), 267–286.

Del Bimbo, A. and Vicario, E. (1996) A logical framework for spatio-temporal indexing of image sequences. *Intelligent Image Database Systems*. Chang, S. K., Jungert, E. and Tortora, G. (eds) World Scientific Publishing Company.

BGFA91, Bizais, Y., Gibaud, B., Forte, A. M., Aubry, F., Paola, R. D. and Scarabin, J. M. A qualitative approach to medical image data bases. *SPIE Conf. on Image Processing, Medical Imaging V*, pp. 156–167. February 1991. San Jose, California.

Blattner, M. M. and Dannenberg, R. B. (eds.) (1992) *Multimedia Interface Design*. Addison-Wesley.

Borgefors, G. (1986) Distance transforms in digital images. *Computer Vision, Graphics and Image Processing* **34**: 344–371.

Brodie, M.L. (1984) On the Development of Data Models. On conceptual modeling: perspective, database, and programming languages. Brodie, M. L., Mylopoulos, J. and Schmidt, J. W. (eds.) Springer Verlag, New York, pp. 19–48, classification of data models.

Brolio, J., Draper, B. A. and Ross Beveridge, J. and Hanson, A. R. (1989) IRS: a database for symbolic processing in computer vision. *IEEE Computer* **22:** 22–30, object oriented DBMS.

Buehrer, D. J. and Chang, C. C. (1993) Application of a reciprocal confluence tree unit to similar-picture retrieval. *Advances in Spatial Databases, Lecture Notes in Computer Science*, pp. 437–442, Springer-Verlag.

Butterworth, P., Otis, A. and Stein, J. (1991) The gemstone object database management system. *Communications of the ACM* **34,** No. 10, 64–77.

Catarci, T., Chang, S. K. and Santucci, G. (1992) A unified framework providing multi-paradigm visual access to databases. Technical Report, University of Pittsburgh.

Catarci, T. and Costabile, M.F. (eds.) (1995) Special issue on visual query systems. *Journal of Visual Languages and Computing.* **6**(1).

Catarci, T., Chang, S. K., Costabile, M. F., Levialdi, S. and Santucci, G. (1996) A graph-based framework for multiparadigmatic visual access to databases. *IEEE Transactions on Knowledge and Data Engineering.*

Chang, C. C. (1991) Spatial match retrieval of symbolic pictures. *Journal of Information Science and Engineering.* **7,** 405–422, iconic indexing.

Chang, C. C. and Lee, S. Y. (1991) Retrieval of similar pictures on pictorial databases. *Pattern Recognition* **24**(7), 657–680.

Chang, C. C. (1992) A fast algorithm to retrieve symbolic pictures. *International Journal of Computer Mathematics* **43**(1), 133–138.

Chang, C. C. and Wu, T. C. (1992) Retrieving the most similar symbolic pictures from pictorial databases. *Information Processing and Management* **28** (5), 581–588.

Chang, C. C. and Buehrer, D. J. (1992) A survey of some spatial match query algorithms. *Advanced Database Research and Development Series database systems for next-generation applications – practice.* World Scientific Publishing Company, Singapore, pp. 218–223.

Chang, C. C. and Buehrer, D. J. (1992) De-clustering image databases. *Future databases '92, Advanced Database Res. and Devel. Series* **3,** 142–145, World Scientific Publishing Company, Singapore.

Chang, C. C. and Lin, D. C. (1994) Utilizing the concept of longest common subsequence to retrieve similar chinese characters. *Computer Processing of Chinese and Oriental Languages* **8**(2), 177–191.

Chang, H., Hou, T., Hsu, A. and Chang, S. K. (1995) Management and applications of tele-action objects. *ACM Multimedia Systems Journal.* **3**(5–6), 204–216, Springer-Verlag.

Chang, H., Hou, T., Hsu, A. and Chang, S. K. Tele-action objects for an active multimedia system. *Proceedings of Second Int'l IEEE Conf. on Multimedia Computing and Systems*, May 15–18, 1995, pp. 106–113, Washington, D.C.

Chang, N. S. and Fu, K. S. (1980) Query-by-pictorial example. *IEEE Trans. Software Eng.* SE-6, pp. 519–524, spatial relational data model.

Chang, S. K. and Shelton, G. L. (1971) *IEEE Trans. on Systems, Man, and Cybernetics* **17**(1), 90–94, two algorithms for multiple-view binary pattern reconstruction.

Chang, S. K. (1971) The reconstruction of binary patterns from their projections. *Communications of the ACM* **14**(1), 21–25.

Chang, S. K. and Ke, J. S. Translation of fuzzy queries for relational database system. *IEEE Transactions on Pattern Analysis and Machine Intelligence* PAMI-1 (3), pp. 281–294, July 1979.

Chang, S. K. and Kunii, T. (1991) Pictorial database systems. *Computer – Special Issue on Pictorial Information Systems.* (ed.) Chang, S. K. 13–21.

Chang, S. K. and Liu, S. H. (1984) Indexing and abstraction techniques for pictorial databases. *IEEE Transactions on Pattern Analysis and Machine Intelligence* PAMI-6, 475–484.

Chang, S. K. and Clarisse, O. (1984) Interpretation and construction of icons for man-machine interface in an image information system. *IEEE Proceedings of Languages for Automation*, 38–45.

Chang, S. K. and Jungert, E. A spatial knowledge structure for image information systems using symbolic projections. *Proceedings of the Fall Joint Computer Conference (FJCC)*, 79–86. November 2–6, 1986, Dallas, Texas.

Chang, S. K. (1987) Icon semantics – a formal approach to icon system design. *International Journal of Pattern Recognition and Artificial Intelligence* 1(1), 103–120.

Chang, S. K., Shi, Q. Y. and Yan, C. W. Iconic indexing by 2D strings. *IEEE Transactions on Pattern Analysis and Machine Intelligence* PAMI-9 (3), 413–428, May 1987.

Chang, S. K. and Li, Y. (1988) Representation of multi-resolution symbolic and binary pictures using 2DH strings. *Proceedings of IEEE Workshop on Language for Automation*. 190–195.

Chang, S. K., Yan, C. W., Arndt, T. and Dimitroff, D. An intelligent image database system. *IEEE Trans. on Software Engineering, Special Issue on Image Database* 681–688, May 1988.

Chang, S. K., Jungert, E. and Li, Y. Representation and retrieval of symbolic pictures using generalized 2D strings. *Proc. of SPIE Visual Communications and Image Processing Conference* November 5–10 1989, 1360–1372.

Chang, S. K. (1990) *Principles of Pictorial Information Systems Design*, Prentice-Hall.

Chang, S. K. (1990) Visual reasoning for information retrieval from very large databases. *Journal of Visual Languages and Computing* 1(1), 41–58.

Chang, S. K. and Jungert, E. (1990) A spatial knowledge structure for visual information systems, *Visual Languages and Applications* (eds.) Ichikawa, T., Korfhage, R. and Jungert, E., pp. 277–304, Plenum Publishing Company.

Chang, S. K., Jungert, E. and Li, Y. (1990) The design of pictorial databases based upon the theory of symbolic projections. *Design and Implementation of Large Spatial Databases* (eds.) Buchmann, A., Gunther, O. and Smith, T. R., pp. 303–323, Springer-Verlag, Berlin.

Chang, S. K. and Jungert, E. Pictorial data management based upon the theory of symbolic projections. *Journal of Visual Languages and Computing* 2(3), pp. 195–215, September 1991.

Chang, S. K., Hou, T. Y. and Hsu, A. Smart image design for large image databases. *Journal of Visual Languages and Computing* 3(4), 323–342, December 1992.

Chang, S. K., Lee, C. M., Dow, C. R. A 2D-string matching algorithm for conceptual pictorial queries. *Proc. of SPIE/IS&T 1992 Symposium on Electronic Image: Science and Technology, Conf. on Image Storage and Retrieval Systems* pp. 47–58, Feb 10–14 1992, San Jose, California.

Chang, S. K., Costabile, M. F., Levialdi, S. A framework for intelligent visual interface design for database systems. *Proceedings of Int'l Workshop on Interfaces to Database Systems (IDS92)* pp. 377–391, University of Glasgow, Scotland, July 1–3 1992.

Chang, S. K. and Hsu, A. Image information systems: where do we go from here? *IEEE Trans. Knowledge and Data Engineering*, 431–442, October 1992.

Chang, S. K., Orefice, S., Tucci, M. and Polese, G. (1994) A methodology and interactive environment for iconic language design. *International Journal of Human-Computer Studies* 41, 683–716.

Chang, S. K., Costabile, M. F. and Levialdi, S. Reality bites – progressive querying and result visualization in logical and vr spaces. *Proc. of IEEE Symposium on Visual Languages*, 100–109, October 1994, St. Louis.

Chang, S. K. (1995) Towards a theory of active index. *Journal of Visual Languages and Computing* 6(1), 101–118.

Chang, S. K., Costagliola, G., Pacini, G., Tucci, M., Tortora, G., Yu, B. and Yu, J. S. Visual language system for user interfaces. *IEEE Software* 33–44, March 1995.

Chang, S. K. and Costabile, M. F. (1996) Visual interface to multimedia databases. *Handbook of Multimedia Information Management* (eds.) Grosky, W. I., Jain, R. and Mehrotra, R. Prentice Hall.

Chang, S. K. and Li, Y. (1996) Representation of 3D symbolic and binary images using 3D strings. *Intelligent Image Database Systems* (eds) Chang, S. K., Jungert, E. and Tortora, G. World Scientific Publishing Company.

Chen, C. Y. and Chang, C. C. An object-oriented similarity retrieval algorithm for iconic image databases. *Pattern Recognition Letters* **14**, 465–470, June 1993.

Chock, M., Cardenas, A. F. and Klinger, A. (1981) Manipulating data structures in pictorial information systems. *IEEE Computer* **14**(11), 43–50, pictorial information systems.

Chock, M., Cardenas, A. F. and Klinger, A. Data base structure and manipulation capabilities of a picture data base management system (PICDMS). *IEEE Transactions on Pattern Analysis and Machine Language* PAMI-6 (4), 484–492, July 1984, pictorial information systems.

Chou, A. Y. H., Chang, C. C. and Yang, W. P. (1994) Symbolic indexing by N-strings. *Technical Report, Dept. of Computer and Information Science*, National Chiao Tung University, Hsinchu, Taiwan.

Choudhary, A. and Ranka, S. (1992) Mesh and pyramid algorithms for iconic indexing. *Pattern Recognition* **25**(9), 1061–1067.

Clarke, B. L. (1981) A calculus of individual based on connection. *Notre Dame Journal of Formal Logic* **2**(3).

Cohen, P. R., Greenberg, M. L., Hart, D. M. and Howe, A. E. Trial by fire: understanding the design requirements for agents in complex environments. *AI Magazine*, 32–48, Fall 1989.

Comfort, L. K., Abrams, J., Camillus, J. and Ricci, E. (1989) From crisis to community: the Pittsburgh oil spill. *Industrial Crisis Quarterly* **3**(1), 17–39.

Comfort, L. K. and Chang, S. K. An interactive, intelligent, spatial information system for disaster management: a national model. *Paper presented at the National Emergency Management Association (NEMA) Information Technology and Communication Conference*, August 22–24 1995. Mt. Weather, Virginia.

Costagliola, G., Tucci, M. and Chang, S. K. (1992) Representing and retrieving symbolic pictures by spatial relations. *Visual Database Systems II* (eds.) Knuth, E. and Wegner, L. M., pp. 49–59, North-Holland.

Costagliola, G., Tortora, G. and Arndt, T. (1992) A unifying approach to iconic indexing for 2D and 3D scenes. *IEEE Transactions on Knowledge and Data Engineering* **4**, 205–222.

D'Atri, A. and Tarantino, L. From browsing to querying. *Data Engineering* **12**(2), 46–53, June 1989.

Dorf, M. L., Mahler, A. F. and Lehmann, P. F. (1994) Incorporating Semantics into 2D Strings. *Proceedings of 22nd Annual ACM Computer Science Conference* ed. Cizmar, D. 110–115, Phoenix, Arizona.

Drakopoulos, J. and Constantopoulos, P. An exact algorithm for 2D string matching. *Technical Report 21, Institute of Computer Science, Foundation for Research and Technology*, November 1989, Hellas, Heraklion, Greece.

Dutta, S. Qualitative spatial reasoning: a semi-quantitative approach using fuzzy logic. *Conference Proceedings on Very Large Spatial Databases*, July 17–19, 1989, 345–364, Santa Barbara.

Eco, U. (1975) *A theory of semiotics*, Indiana University Press.

Egenhofer, M. J. (1994) Deriving the combination of binary topological relations. *Journal of Visual Languages and Computing* **5**(2), 133–149.

Egenhofer, M. J. and Herring, J. R. (eds.) (1995) *Advances in spatial databases (Proceedings of 4th international symposium, SSD'95, Portland, ME., August 1995)* Springer-Heidelberg.

Faloutsous, C., Barber, R., Flickner, M., Hafner, J., Niblack, W., Petkovic, D. and Equitz, W. (1994) Efficient and effective querying by image content. *Journal of Intelligent Information Systems* 3, 231–262.

Farmer, D. B., King, R. and Myers, D. A. (1985) The Semantic Database Constructor. *IEEE Trans. on Software Eng.* SE-11 (7) 583–591, semantic data model.

Feri, R., Foresti, G. L., Murino, V., Regazzoni, C. S. and Vernazza, G. (1990) Spatial reasoning by knowledge-based integration of visual and fuzzy cues. *Signal Processing V: Theories and Applications* (eds.) Torres, R. L., Masgrau, E. and Lagunas, M. A. Elsevier Science Publishers.

Fierens, F., Van Cleynenbreugel, J., Suetens, P. and Oosterlinck, A. (1992) A software environment for image database research. *Journal of Visual Languages and Computing* 3, 49–68, pictorial database, object oriented DBMS.

Fishman, D. H., Beech, D., Cate, H. P., Chow, E. C., Connors, T., Davis, J. W., Derrett, N., Hoch, C. G., Kent, W., Lyngbnek, P., Mahbod, B., Neimat, M. A., Ryan, T. A. and Shan, M. C. IRIS: an object-oriented database management system. *ACM Trans. on Office Information Systems ACM* 5(1), 48–69, January 1987, object oriented DBMS.

Flickner, M., Sawhney, H., Niblack, W., *et al.* Query by image and video content: the QBIC system. *IEEE Computer*, 23–32, September 1995, FBLB92.

Fox, E. A. Advances in interactive digital multimedia systems. *IEEE Computer* 24(10), 9–21, October 1991.

Frank, A. U. (1992) Qualitative spatial reasoning about distances and directions in geographical information systems. *Journal of Visual Languages and Computing* 3(4), 343–371.

Frank, A. U. and Kuhn, W. (eds.) (1994) *Spatial information theory, a theoretical basis for GIS (Proceedings of the International Conference COSIT'95, Semmering, Austria, September 1995)* Springer-Verlag, Heidelberg.

Freksa, C. and Zimmerman, K. Enhancing spatial reasoning by the concept of motion. *Visual Languages and Computing* 5, 13–149.

Freksa, C. (1991) Conceptual Neighbourhood and its role in temporal and spatial reasoning. *Decision Support Systems and Qualitative reasoning* (eds.) Singh, M. G. and Travé-Massmyès, L., pp. 181–187, Elsevier Science Publishing Company.

Freksa, C. (1992) Using orientation information for qualitative spatial reasoning. *Theories and Methods of Spatio-Temporal Reasoning in Geographic Space*, 162–178, Springer-Verlag, Heidelberg

Freksa, C. and Rohrig, R. (1993) Dimensions of qualitative reasoning. *Proceedings of the Workshop on Qualitative Reasoning and Decision Technologies (QUARDET'93)*, ed. Carreté, N. and Singh, M. G. CIMNE, Barcelona, 483–492.

Fu, K. S. (1982) *Syntactic pattern recognition and applications.* Prentice-Hall, Englewood Cliffs, New Jersey.

Fu, K. S. (1982) *Syntactic methods in pattern recognition.* Academic Press, New York.

Gargantini, I. (1982) An efficient way to represent quadtrees. *Commun. ACM* 12, 905–910. 25 image storage structures.

Gary, J. E. and Mehrotra, R. (1993) Similar shape retrieval using a structural feature index. *Information Systems* 18(7), 525–537.

Gould, L. and Finzer, W. Programming by rehearsal. *Byte*, 187–210, June 1984.

Grosky, W. I. (1984) Toward a data model for integrated databases. *Computer Vision, Graphics and Image Processing* 25, 371–382.

Grosky, W. I. (1986) Iconic indexing using generalized pattern matching techniques. *Computer Vision, Graphics, and Image Processing* 35, 308–403.

Grosky, W. I. and Mehrotra, R. Image database management. *IEEE Computer Magazine* **22**(12), 7–8, December 1989, image data model.

Grosky, W. I. and Mehrotra, R. (1990) Index-based object recognition in pictorial data management. *Computer Vision, Graphics, and Image Processing* **52**, 416–436.

Grosky, W. I. and Jiang, Z. A hierarchical approach to feature indexing. *SPIE/IS&T Conference on Image Storage and Retrieval Systems*, 9–20, February 1992.

Grosky, W. I., Neo, P. and Mehrotra, R. A pictorial index mechanism for model-based matching. IEEE Transactions on Data and Knowledge Engineering **5**, 3–14.

Gudivada, V. N. and Raghavan, V. V. (eds.) Special issue of IEEE computer magazine on content-based image retrieval systems. *IEEE Computer Society*, September 1995.

Gupta, A., Weymouth, T. E. and Jain, R. An extended object-oriented data model for large image data bases. *Proc. of Second Symposium on Large Spatial Database (SSD'91)*, 45–61. Zurich, Switzerland. 28–30 August, 1991, R++tree, Object-oriented, remote sensing images.

Gupta, A., Weymouth, T. E. and Jain, R. Semantic queries with pictures: the VIMSYS model. *Proceedings of VLDB'91*, 69–79. September 1991, Barcelona, Spain.

Gupta, R. and Horowitz, E. (1991) Object-oriented database with applications to CASE, networks and VLSI CAD. Prentice-Hall, object oriented DBMS.

Guttman, A. R-Trees: a dynamic index structure for spatial searching. *Proceedings ACM-SIGMOD 1984 International Conference on Management of Data* 47–57, 18–21 June 1984.

Halbert, D. C. Programming by example, *Xerox Office Systems Division, TR OSD-T8402*, December 1984.

Hanne, K. and Bullinger, H. (1992) Multimodal communication: integrating text and gestures. *Multimedia Interface Design* (eds.) Blattner, M. M. and Dannenberg, R. B., 127–138, Addison-Wesley.

Helm, R., Marriott, K. and Odersky, M. (1991) Constraint-based query optimization for spatial databases. *Proc. ACM Symposium on Principles of Database Systems*, 181–191, Denver, Colorado, query optimization.

Hernández, D., Clementini, E. and Di Felice, P. (1994) Qualitative distances. *Spatial Information Theory, a Theoretical Basis for GIS (Proceedings of the International Conference COSIT'95, Semmering, Austria, September 1995)* (eds.) Frank, A. U. and Kuhn, W., 45–58, Springer-Verlag, Heidelberg.

Hildebrandt, J. W. and Tang, K. Symbolic two and three dimensional picture retrieval. *Workshop on Two and Three Dimensional Spatial Data: Representation and Standards*, 7–8 December 1992, Perth, Western Australia. Australian Pattern Recognition Society.

Hill, W., Wroblewski, D., McCandiess, T. and Cohen, R. (1992) Architectural qualities and principles for multimodal and multimedia interfaces. *Multimedia Interface Design* (eds) Blattner, M. M. and Dannenberg, R. D., 311–318, Addison-Wesley.

Hirakawa, M. and Jungert, E. An image database system facilitating icon-driven spatial information definition and retrieval. *Proceedings of the 1991 Workshop on Visual Languages*, 192–198, 8–11 October 1991, Kobe, Japan.

Hirschberg, D. S. (1977) Algorithms for the longest common subsequence problem. *Journal of the ACM* **24**, 350–353.

Holmes, P. D. and Jungert, E. Shortest paths in a digitized map using a tile-based data structure. *Proceedings of Third Inter. Conf. on Engineering Graphics and Descriptive Geometry*, 238–245, July 1988. Vienna, Austria.

Holmes, P. D. and Jungert, E. Symbolic and Geometric Connectivity Graph Methods for Route Planning in Digitized Maps. *IEEE Transaction on Pattern Analysis and Machine Intelligence (PAMI)*, **14**(5), 549–565, May 1992.

Hou, T. Y., Hsu, A., Liu, P., Chiu, M. Y. A content-based indexing technique using relative geometry features. *SPIE/IS&T Symposium on Electronic Imaging Science and*

Technology, Conference on Image Storage and Retrieval Systems, 59–68, February 1992, San Jose, California.

Huang, K. T. (1990) Visual interface design systems. *Principles of Visual Programming Systems* (ed.) Chang, S. K. Prentice-Hall.

Hunt, J. W. and Szymanski, T. G. (1977) A fast algorithm for computing longest common subsequences. *Communications of the ACM* **20**, 350–353.

Hunter, G. M. and Steiglitz, K. (1979) Linear transformation of pictures represented by quadtrees. *Computer Graphics and Image Processing* **10**, 289–296.

Ignatius, E., Senay, H. and Favre, J. An intelligent system for task-specific visualization assistance. *Journal of Visual Languages and Computing* **5**(4) 321–338, December 1994.

Jackson, C. L. and Tanimoto, S. L. (1980) Oct-trees and their use in representing three-dimensional objects. *Computer Graphics and Image Processing* **14**, 249–270.

Jagadish, H. V. and O'Gorman, L. (1989) An object-model for image recognition. *Computer* **22**, 33–41, object oriented DBMS.

Jagadish, H. V. A retrieval technique for similar shapes. *Proceedings of the 1991 ACM SIGMOD International Conference on Management of Data, SIGMOD Record* **20**(2), 208–217, June 1991.

Joseph, T. and Cardenas, A. F. Picquery: a high level query language for pictorial database management. *IEEE Transactions Software Engineering* **14**, 630–638, May 1988, query language pictorial information systems.

Jungert, E., Borgefors, G., Fransson, J., Lindgren, T., Olsson, L., Roldan-Prado, R. and Toller, E. Vega – a geographical information system. *Proceedings of the Scandinavian Research Conference on Geographical Information Systems*, 118–133, 13–14 June 1985. Linkoping, Sweden.

Jungert, E. Run length code as an object-oriented spatial data structure. *Proceedings of IEEE Workshop on Languages for Automation*, 66–70, 27–29 August 1986, Singapore.

Jungert, E. and Holmes, P. D. (1988) *IEEE Workshop on Visual Languages*, 248–255.

Jungert, E. and Kittler, J. (1988) Pattern Recognition, 343–351, Extended symbolic projections as a knowledge structure for spatial reasoning and planning. Springer-Verlag, Berlin and Heidelberg.

Jungert, E., Chang, S. K. and Kunii, T. L. (1989) An algebra for symbolic image manipulation and transformation. *Visual Database Systems*, 301–317, North-Holland.

Jungert, E. Symbolic expressions within a spatial algebra: unification and impact upon spatial reasoning. *Proceedings of the 1989 IEEE Workshop on Visual Languages*, 157–162, 4–6 October 1989, Rome, Italy.

Jungert, E. and Hampus, P. A database structure for an object-oriented raster-based geographical information system. *Proc. of the 1st European Conference on Geographical Information Systems*, 526–533, 1–13 April 1990, Amsterdam, The Netherlands.

Jungert, E. and Chang, S. K. The sigma-tree – a symbolic spatial data structure. *Proceedings of the 11th IAPR International Conference on Pattern Recognition*, 461–465, 29–31 August 1992, Hague.

Jungert, E. The observer's point of view: an extension of symbolic projections. *Proc. of GIS, From Space to Territory, Theories and Methods of Spatio-Temporal Reasoning, International Conference*, 179–195, 21–23 September 1992, Springer-Verlag, Pisa, Italy.

Jungert, E. (1993) Symbolic reasoning on object shapes for qualitative matching. *Spatial Information Theory – A Theoretical Basis for GIS* (eds.) Frank, A. D. and Campari, I. 444–462, Springer-Verlag, Heidelberg.

Jungert, E. Qualitative spatial reasoning for determination of object relations using symbolic interval projections. *Proc. of the IEEE Conference on Visual Languages*, 83–87, 24–26 August 1993, Bergen, Norway.

Jungert, E. Graqula – a visual information-flow query language for a geographical information system. *Journal of Visual languages and Computing*, **4**(4), 383–401, December 1993.

Jungert, E. Rotation invariance in symbolic projections as a means for determination of binary object relations. *Proceedings of the Workshop on Qualitative Reasoning and Decision Technologies (QUARDET'93)*, 503–512, 16–18 June 1993, Barcelona, Spain.

Jungert, E. Spatial reasoning about landmarks with uncertain positions. *Proceedings of the ACM workshop on Advances in Geographic Information Systems*, 83–87, November 5 1993, Arlington, Virginia.

Jungert, E. and Chang, S. K. An image algebra for pictorial data manipulation. *Journal of Computer Vision, Graphics and Image Processing (CVGIP): Image Understanding*, **58**(2), 147–160, September 1993.

Jungert, E. Using symbolic projection for spatio-temporal reasoning on tracing objects with uncertain positions. *Proceedings of AAAI workshop on spatial and temporal reasoning*, 101–108, 31 July 1994, Seattle, Washington.

Jungert, E. Qualitative spatial reasoning from the observer's point of view – towards a generalization of symbolic projection. *Journal of Pattern Recognition*, **27**(6), 801–813, June 1994, Pergamon Press.

Jungert, E. Rotation invariance in symbolic slope projection as a means to support spatial reasoning. *Proc. of the third Int. Conf. on Automation, Robotics and computer vision (ICARCV'95)*, 364–368, 9–11 November 1994, Singapore.

Jungert, E. (1996) Determination of the views of a moving agent by means of symbolic projection. *Intelligent Image Database Systems* (eds.) Chang, S. K., Jungert, E. and Tortora, G. World Scientific Publishing Company, Singapore.

Kim, Y. and Haynor, D. R. (1991) Requirements for PACS workstations. *Proc. of 2nd Int. Conf. on Image Management and Communication in Patient Care*, 36–41, IEEE Computer Society Press.

Klinger, A., Rhode, M. L. and To, V. T. Accessing image data. *Int. J. Policy Analysis Information Syst.* **1**(2), 171–189, January 1978.

Klinger, A., Pizano, A. and Kunii, T. L. (1989) Visual Structure and Data Bases. *Visual Database Systems*, 3–25, North-Holland, pictorial database system.

KKKP90. Kofakis, P., Karmirantzos, A., Kavaklis, Y., Petrakis, E. G. M. and Orphanoudakis, S. Image archiving by content: an object-oriented approach, *SPIE Conf. on Image Processing, Medical Imaging IV: PACS System Design and Evaluation*, 275–286, **1234**, February 1990, Newport Beach, California.

Kostomanolakis, S., Lourakis, M., Chronaki, C., Kavaklis, Y. and Orphanoudakis, S. C. (1993) Indexing and retrieval by pictorial content in the I2C image database system. *Technical Report, Institute of Computer Science, Foundation for Research and Technology-HELLAS*, Heraklion, Crete, Greece.

Kundu, S. (1988) The equivalence of the subregion representation and the wall representation for a certain class of rectangular dissections – *Communications of the ACM*, **31**, 752–763.

Laurini, R. and Thomson, D. (1992) Fundamentals of spatial information systems. Academic Press, London.

Lee, C. M. (1992) The unification of the 2D string and spatial query processing. *PhD Thesis, Graduate School of the Knowledge Systems Institute*, Skokie, Illinois.

Lee, D. T. and Preparata, F. P. Euclidean shortest paths in the presence of rectilinear barriers. *Networks* **14**, 393–410.

Lee, S. Y., Shan, M. K. and Yang, W. P. (1989) Similarity retrieval of iconic image database. *Pattern Recognition* **22**, 675–682.

Lee, S. Y. and Shan, M. K. (1990) Access methods of image database. *International Journal of Pattern Recognition and Artificial Intelligence*, **4**, 27–44.

Lee, S. Y. and Hsu, F. J. (1990) 2D C-string: a new spatial knowledge representation for image database systems. *Pattern Recognition, 23*(10), 1077–1087.

Lee, S. Y. and Hsu, F. J. Picture algebra for spatial reasoning of iconic image represented in 2D C-string. *Pattern Recognition Letters* **12,** 425–435.

Lee, S. Y., Yang, M. C. and Chen, J. W. (1991) 2D B-string: a new spatial knowledge representation for image database systems. *Technical Report, Department of Computer and Information Science,* National Chiao Tung University, Hsinchu, Taiwan.

Lee, S. Y., Yang, M. C. and Chen, J. W. (1992) 2D B-string knowledge representation and picture retrieval for image database. Second International Computer Science Conference – Data and Knowledge Engineering. Theory and Applications, Hong Kong, December 13–16, 609–615.

Lee, S. Y., Yang, M. C. and Chen, J. W. (1992) Signature file as spatial filter for iconic image database. *Journal of Visual Languages and Computing* **3**(4), 373–397.

Lee, S. Y. and Hsu, F. J. (1992) Spatial reasoning and similarity retrieval of images using 2D C-string knowledge representation. *Pattern Recognition* **25** (3), 305–318.

Levin, K. and Fielding, R. Methods to prefetch comparison images in image management and communication systems (IMACS). *SPIE conf. on Image Processing, Medical Imaging IV,* 270–274, February 1990, Newport Beach, California, USA.

Li, X. R. and Chang, S. K. Visualization of image from 2D strings by visual reasoning. *Proceedings of SPIE visual communications and image processing conference,* April 1991, Orlando, Florida, USA.

Liang, J. and Chang, C. C. (1993) Similarity retrieval on pictorial databases based upon modulo operations. *Advanced Database Research and Development Series – Vol. 4 Database Systems for Advanced Applications '93,* **4,** 19–26, World Scientific Publishing Company.

Lodder, H., Poppel, B. M. and Bakker, A. R. Prefetching: PACS image management optimization using HIS/RIS information. *SPIE conf. on Image Processing, Medical Imaging V,* 227–233, February 1991, San Jose, California, USA.

Lohse, G. L., Biolsi, K. A., Walker, N., Rueter, H. H. (1994) A classification of visual representations. *Communications of the ACM* **37** (12), 36–49.

Lozano-Perez, T. Automatic planning of manipulator transfer movements. *IEEE Trans. on Systems, Man and Cybernetics,* 681–698, October 1981, **SMC-11**(10).

Lozano-Perez, T., Spatial planning: a configuration space approach. *IEEE Transactions on Computers,* 108–120, **C-32**(2), February 1983.

Madhyastha, T. M., Reed, D. A. Data sonification: do you see what I hear? *IEEE Software,* **12**(2), 45–56, March 1995.

Manna, Z. and Pnueli, A. (1992) *The Temporal Logic of Reactive and Concurrent Systems.* Springer-Verlag, New York, USA.

McKeown, D. M. The role of artificial intelligence in the integration of remotely sensed data with geographic information systems. *IEEE Transactions on Geoscience and Remote Sensing* **GE-25,** 330–348, May 1987.

Mehrotra, R. and Grosky, W. I. REMINDS: A relational model-based integrated image and text database management system. *Proceedings of Workshop on Computer Architecture for Pattern Analysis and Image Database Management,* 348–354, November 1985, Miami Beach, Florida, USA.

Mehrotra, R. and Grosky, W.I. Shape matching utilizing indexed hypotheses generation and testing. *IEEE Trans. on Robotics and Automation,* **5**(1), 70–77, February 1989.

Mehrotra, R. and Gary, J. E. (1995) Feature-index based similar shape retrieval. *Visual database systems III* (eds.) Spaccapietra, S. and Jain, R. 46–65, Chapman & Hall.

Mehrotra, R. and Gary, J. E. Similar shape retrieval in shape data management. *IEEE Computer,* 57–62, September 1995.

Meyer-Wegener, K., Lum, V. Y. and Wu, C. T. (1989) Image management in a multimedia database system. *Visual Database Systems* (ed.) Kunii, T. L. 497–523. Multimedia database system.

Motro, A. BAROQUE: an exploratory interface to relational databases. *ACM Trans. on Office Information Systems*, **4**(2), 164–181, April 1986.

Motro, A. VAGUE: a user interface to relational database that permits vague queries. *ACM Trans. on Office Information Systems* **6**(3), 187–214, July 1988. Imprecise DBMS query processing.

Mukerjee, A. and Joe, G. (1986) A qualitative model for space. *Proc of AAAI-90*, pp. 721–727. July 29 to August 3 1990, Boston, Massachusetts, USA.

Mun, S. K. (1991) Image management and communication (IMAC) in 1991. *Proc. of 2nd Int. Conf. on Image Management and Communication in Patient Care*, 6–15, IEEE Computer Society Press.

Myers, B. A. (1988) *Creating User Interfaces by Demonstration*. Academic Press, Boston, USA.

Myers, B. A. Visual programming, programming by example, and program visualization: a taxonomy. *Proceedings of SIGCHI'86*, 59–66, 13–17 April 1986, Boston, Massachusetts, USA.

Nahouti, E. and Petry, F. (eds.) (1991) *Object-Oriented Databases. IEEE Computer Society Press, Los Alamitos, California, USA.*

Nielsen, J. The art of navigation through hypertext. *Communications of ACM*, 296–310, March 1990. Image data model.

Nilsson, N. J. (1979) *Principles of Artificial Intelligence*. Tioga Publishing, Palo Alto, California, USA.

Olson, J. R. and Olson, G. M. (1990) The growth of cognitive modeling in human–computer interaction since GOMS. *Human Computer Interaction*, **5**(2, 3), 221–265.

Orenstein, J. (1986) Spatial query processing in an object-oriented database system. *Proceedings ACM-SIGMOD 1986 International Conference on Management of Data*, 326–336, 28–30 May 1986.

Papadias, D. and Sellis, T. (1995) Pictorial query-by-example. *Journal of Visual Languages and Computing*, **6**(1), 53–72.

Persson, J. and Jungert, E. (1992) Generation of multi-resolution maps from run-length encoded data. *International Journal of Geographical Information Systems*, **6,** 497–510.

Petraglia, G., Sebillo, G., Tucci, M. and Tortora, G. (1996) A normalized index for image databases. *Intelligent Image Database Systems*, (eds.) Chang, S. K., Jungert, E. and Tortora, G. World Scientific Publishing Company, Singapore.

Petrakis, E. G. M. (1993) *Image Representation, Indexing and Retrieval Based on Spatial Relationships and Properties of Objects*. March 1993. Ph.D. Thesis, Technical Report FORTH-ICS/TR-075, Institute of Computer Science, Foundation of Research and Technology (FORTH).

Petrakis, E. G. M. and Orphanoudakis, S. C. (1993) Methodology for the representation, indexing, and retrieval of images by content. *Image and Vision Computing*, **8**(11), 504–521.

Petrakis, E. G. M. and Orphanoudakis, S. C. (1996) A Generalized Approach for Image Indexing and Retrieval based on 2D Strings. *Intelligent Image Database Systems*, (eds.) Chang, S. K., Jungert, E. and Tortora, G. World Scientific Publishing Company, Singapore.

Peuquet, D. J. and Zhang, C. X. (1987) *An Algorithm to Determine the Directional Relationship between Arbitrarily-Shaped Polygons in the Plane*, **20**(1), 65–74.

Pizano, A., Klinger, A. and Cardenas, A. F. (1989) Specification of spatial integrity constraints in pictorial databases. *IEEE Computer*, 59–71, December.

Rabitti, F. and Stanchev, P. (1989) GRIM–DBMS: A Graphical Image Database

Management System. *Visual Database Systems*, ed. Kunii, T. L., 415–430, North-Holland, Amsterdam.

Rabitti, F. and Savino, P. (1991) *Image Analysis for Semantic Image Databases*, IEI-CNR Technical Report B4-35, October 1991, Pisa, Italy.

Randell, D. A., Cui, Z. and Cohn, A. G. (1992) A spatial logic based on regions and connection. *Third International Conference on Knowledge Representation and Reasoning*, Morgan Kaufmann, San Francisco, California, USA.

Robertson, G. G., Card, S. K. and Mackinlay, J. D. (1993) Nonimmersive virtual reality. *IEEE Computer*, **26** (2), 81–83.

Robertson, G. G., Card, S. K. and Mackinlay, J. D. (1993) Information visualization using 3D interactive animation. *Communications of the ACM*, **36**(4), 57–71.

Roussopoulos, N. and Leifker, D. (1985) Direct spatial search on pictorial databases using packed R-trees. *Proceedings ACM-SIGMOD 1985 International Conference on Management of Data*, 28–31 May 1985, 17–31.

Roussopoulos, N., Faloutsos, C. and Sellis, T. (1988) An efficient pictorial database system for PSQL. *IEEE Transactions Software Engineering*, **14,** 639–650.

Sabharwal, C. L. and Bhatia, S. K. (1995) Perfect hash table algorithm for image databases using negative associated values. *Pattern Recognition*, **28**(7), 1091–1101.

Samet, H. (1984) The quadtree and related hierarchical data structures. *ACM Computing Surveys*, **16**(2), 187–260.

Samet, H. (1990) *The Design and Analysis of Spatial Data Structures*, Addison-Wesley, Reading, Massachusetts, USA.

Samet, H. (1990) *Applications of Spatial Data Structures: Computer Graphics, Image Processing and GIS*, Addison-Wesley, Reading, Massachusetts, USA.

Samet, H. (1995) Spatial data structures. *Modern Database Systems: The Object Model, Interoperability and Beyond*. (ed.) Kim, W., pp. 361–385.

Schlieder, C. (1996) Ordering information and symbolic projection, *Intelligent Image Database Systems*, (eds) Chang, S. K., Jungert, E. and Tortora, G., World Scientific Publishing Company, Singapore.

Sethi, I. K. and Han, K. (1995) Use of local structural association for retrieval and recognition of signature images. *SPIE Proceedings of Storage and Retrieval for Image and Video Databases III*. ed. Niblack, W. and Jain, R. C., **2420**, 125–134.

Shapiro, L. G. and Haralick, R. M. (1979) A spatial data structure. *Technical Report CS 79005-R*. Department Computing Science, Virginia Polytechnic Institute and State University, USA.

Shapiro, L. G. and Haralick, R. M. (1982) Organization of relational models for scene analysis. *IEEE Trans. on Pattern Analysis and Machine Intelligence*, **4**(6), 595–602.

Sheng, O. R. L., Wang, H. and Garcia, H. M. C. (1990) IRES – image retrieval expert system. *SPIE conf. on Image Processing, Medical Imaging IV*, 832–841, February, Newport Beach, California, USA.

Liu Sheng, O. R. and Wei, C. P. (1992) Object-oriented modelling and design of knowledge-base/database systems. *Proceedings of the Eights International Conference on Data Engineering*, 87–105, 3–7 February 1992, Tempe, Arizona, USA.

Shneiderman, B. (1992) *Designing the User Interface*, Addison Wesley Publishing Company.

Sirovich, L. and Smith, D. C. (1977) *Pygmalion: A Computer Program to Model and Stimulate Creative Thought*, Birkhauser, Stuttgart, Germany.

Smith, T. R. and Park, K. (1992) Algebraic approach to spatial reasoning. *International Journal for Geographic Information Systems*, **6**(3), 177–192.

Spaccapietra, S. and Jain, R. (eds.) (1995) *Visual Database Systems III – Visual Information Management (Proceedings of the third IFIP 2.6 Working Conference on Visual Database Systems)*, Chapman & Hall.

Sties, M., Sanyal, B. and Leist, K. (1976) Organization of object data for an image information system. *Proceedings 3rd Int. Joint Conference on Pattern Recognition*, 863–869.

Stoakley, R., Conway, M. J. and Pausch, R. (1995) *Virtual reality on a WIM: interactive worlds in miniature. Proc. of CHI-95*, 265–272, 7–11 May 1995, Denver, Colorado, USA.

Sun, Y. Guo. (1993) *Description of Topological Spatial Relations and Representation of Spatial Relations using 2D T-String*, Ph.D. Thesis, Wuhan Surveying Technology University. Wuhan, China. Mr. Y. Guo Sun, China Siwei Surveying & Mapping Technology Corporation, 1 Baishengcun, Zizhuyuan, Beijing 100044, P.R. China.

Tabuchi, M., Yagawa, Y., Fujisawa, M., Nigishi, A. and Muraoka, Y. (1991) Hyperbook: a multimedia information system that permits incomplete queries. *Proc. of International Conference on Multimedia Information Systems*, 3–16, Singapore.

Tagare, H. D., Gindi, G. R., Duncan, J. S. and Jaffe, C. C. (1991) A geometric indexing schema for an image library. *Computer Assisted Radiology '91*, 513–518.

Tagare, H. D., Jaffe, C. C. and Duncan, J. (1992) *Requirements for medical image databases*, Technical Report, School of Medicine, Yale University, USA.

Tang, G. Y. (1981) A management system for an integrated database of pictures and alphanumeric data. *Computer Graphics and Image Processing*, **16**, 270–286.

Tang, M. and Ma, S. D. (1992) A new method for spatial reasoning in image databases. *Visual Database Systems II*, (eds.) Knuth, E. and Wegner, L. M., 37–48, North-Holland.

Tanimoto, S. L. (1976) An iconic/symbolic data structuring scheme. *Pattern Recognition and Artificial Intelligence*, pp. 452–471, Academic Press, New York, USA.

Tseng, J. C. R., Hwang, T. F. and Yang, W. P. (1994) Efficient image retrieval algorithms for large spatial databases. *International Journal of Pattern Recognition and Artificial Intelligence*, **8** (4), 919–944, World Scientific Publishing Company, Singapore.

Ullman, J. D. (1988) *Principles of Database and Knowledge-based Systems*, Computer Science Press, Rockville, Maryland, USA.

View, L. (1992) A logical framework for reasoning about space, *Proceedings of the Conference on Spatial Information Theories*, 25–53, Elba, Italy.

Wagner, R. A. and Fischer, M. J. (1974) The string-to-string correction problem. *J. Assoc. Comput. Mach. (JACM)* **21**(1), 168–173.

Wald, J. A. and Sorenson, P. G. (1984) Resolving the query inference problem using Steiner trees. *ACM Trans. on Database Systems*, **9**(3), 348–368.

Williams, M. D. (1984) What makes RUBBIT run? *International Journal on Man-Machine Studies*, **21**(4), 333–352.

Winston, P. H. (1984) *Artificial Intelligence*, Addison-Wesley, Reading, Massachusetts, USA.

Wirth, N. (1977) *Algorithms + Data Structures = Programs*, Prentice-Hall, Englewood Cliffs, New Jersey, USA.

Wong, E. K. and Fu, K. S. (1986) A hierarchical orthogonal space approach to three-dimensional path planning. *IEEE Transactions on Robotics and Automation*, **RA-2**(1), 43–53.

Wu, T. C. and Chang, C. C. (1994) Application of geometric hashing to iconic database retrieval. *Pattern Recognition Letters*, **15**, 871–876.

Zink, S. and Jaffe, C. C. (1993) Medical imaging databases – a national institutes of health workshop. *Investigative Radiology*, **28**(4), 366–372, J. B. Lippincott Company.

Zloof, M. M. (1977) Query by example. *IBM Systems Journal*, **16**(4), 324–343.

Index

Algorithms, in the index, are given in bold. Some special symbols are either given in bold or in italic, depending on their given context.